# Covalent Catalysis by Enzymes

Leonard B. Spector

# Covalent Catalysis by Enzymes

With 61 Figures

Springer-Verlag
New York   Heidelberg   Berlin

Dr. Leonard B. Spector
The Rockefeller University
York Avenue
New York, New York 10021, USA

Library of Congress Cataloging in Publication Data

Spector, Leonard B.
  Covalent catalysis by enzymes.
  Bibliography: p.
  Includes index.
  1. Enzymes.  I. Title.
QP601.S5623     574.19′25     81-23251
                               AACR2

Printed in the United States of America.

9 8 7 6 5 4 3 2 1

ISBN 0-387-**90616-9**  Springer-Verlag  New York  Heidelberg  Berlin
ISBN 3-540-**90616-9**  Springer-Verlag  Berlin  Heidelberg  New York

*To F. Lipmann and H. Bunche*

# Preface

Some years ago one of my students and I reported that the acetate kinase reaction is mediated by a phosphorylated form of the enzyme [R. S. Anthony and L. B. Spector, *JBC* 245, 6739 (1970)]. The reversible reaction between ATP and acetate to give acetyl phosphate and ADP had hitherto been thought to proceed by direct transfer of a phosphoryl group from ATP to acetate in a single-displacement reaction. But now it became clear that acetate kinase was one of that substantial number of enzymes whose mechanism is that of the double displacement. For some reason, I began to wonder about the possibility that *all* enzymes, like acetate kinase, are double-displacement enzymes, and do their work by covalent catalysis. For one thing, I could not think of a single instance of an enzyme for which single-displacement catalysis had been proved, and inquiries on this point among knowledgeable friends elicited the same negative response. Moreover, it was long known that the two other kinds of chemical catalysis—homogeneous and heterogeneous—occur through the intermediary formation of a covalent bond between catalyst and reactant. I began to feel confident that chemical catalysis by enzymes must happen the same way. But how could one be sure? There seemed to me to be only one way: to search the literature for authentic cases of covalent catalysis by enzymes and to see if their number and chemical diversity are sufficiently large to warrant my confidence. The results of that literature search form the substance of this book.

In Chapter 1, I set out the main reasons for believing in covalent catalysis by all enzymes. The next six chapters give the evidence for this belief, with examples of individual enzymes which seem to me to be pertinent or interesting or both, and which illustrate the diverse forms which covalent catalysis takes among enzymes. The treatment given each enzyme is short and terse.

Inevitably, there is speculation on controversial subjects; but speculation and controversy are, after all, inseparable from an advancing science, and are often an impulse to new advances. Chapter 8 sums it all up. The reader who is in a hurry can acquaint himself with the main arguments by simply reading Chapters 1 and 8.

To Drs. O. W. Griffith and E. B. Keller I am grateful for their many helpful comments.


New York City                                                                      L. B. S.
February 1982

# Contents

# Chapter 1

# The Thesis

This book is built upon a simple theme; namely, that all enzyme reactions proceed through at least one intermediate in which the enzyme is covalently joined to its substrate or a fragment thereof. There are two main reasons for believing this. One is that the two other kinds of chemical catalysis—nonenzymic catalysis in solution and heterogeneous catalysis on a solid surface—proceed in just this way. The other reason rests on a substantial body of evidence showing that this same mode of catalysis is used by more than 400 of the 2200 enzymes currently listed by the Enzyme Commission of the International Union of Biochemistry (1). By contrast, there is not a single authenticated instance of an enzyme that uses the single-displacement mode of action, in which a fragment of substrate is passed directly between donor and acceptor (2). A corollary theme of the book therefore is that the single-displacement mechanism figures little, if at all, in enzymic catalysis (3).

The six chapters following the present one are devoted each to one of the six main classes of enzymes as they are designated by the Enzyme Commission. These classes are, in order: oxidoreductases, transferases, hydrolases, lyases, isomerases, and ligases. Each of the six chapters includes at the end

(1) Enzyme Nomenclature. Recommendations (1978) of the nomenclature committee of The International Union of Biochemistry. Academic Press, New York, 1979.

(2) For documentation of this statement I can do no more than refer the reader to the whole corpus of literature on the chemical mechanism of enzyme action.

(3) In this book we confine our discussion of catalysis to *mass-law* catalysis, in which the concentration of the catalyst appears in the rate equation for the reaction. In such catalysis it is thought that the chemical route taken by the reaction is different from the one taken by the uncatalyzed reaction (4). In this respect mass-law catalysis may differ from the rate accelerations associated with solvent and salt (that is, medium) effects.

(4) C. N. Hinshelwood, *The Structure of Physical Chemistry*, Oxford University Press, Oxford, 1951, p. 398.

an appropriate table listing those enzymes for which some positive evidence exists pointing to a covalent enzyme–substrate intermediate in their respective catalytic cycles (5). The six tables are in a sense the heart of the book. They show how widespread and pervasive is the phenomenon of covalent catalysis among enzymes. And in default of a single instance to the contrary, it is argued that covalent catalysis is probably universal in enzyme action, just as it seems to be in the other forms of chemical (mass-law) catalysis.

In 1971 Bell and Koshland found "sixty cases in which there is strong evidence for covalent enzyme–substrate intermediates." Seeking to know "whether they are an occasional aberration or a significant route in enzyme mechanisms," these authors concluded from their survey that "*Thus, covalent intermediates are not essential for enzyme action*" (italics added) (6). In the present book the list of enzymes which act through covalent enzyme–substrate intermediates is greatly enlarged, and the opposite conclusion is drawn.

## Nonenzymic Chemical Catalysis: Its Resemblance to Enzymic Catalysis

Students of chemical catalysis have found it convenient to divide their large subject into three smaller and separate fields of study, known commonly as homogeneous, heterogeneous, and enzymic catalysis. Since they all have to do with speeding up the progress of chemical reactions, it is natural to seek the common thread which unites them. Enough is now known about homogeneous (7, 8) and heterogeneous (9, 10) catalysis to state confidently that *both require the transient covalent linkage of catalyst to reactant at some stage of the catalytic cycle.* Examples of such linkage are so familiar in acid–base, nucleophilic–electrophilic, and transition-metal catalysis—all in homo-

---

(5) The expression "covalent enzyme–substrate intermediate" is used indiscriminately in this book to mean the covalent combination of an enzyme with the whole of its substrate, or, more commonly, with some fragment of it. Some enzymes have an attached coenzyme. Often the holoenzyme forms a covalent bond to substrate through its coenzyme, which may or may not be covalently fixed to the apoenzyme. This last point is regarded as immaterial, since only the holoenzyme is enzymically active. The covalent bond with substrate may be to an enzymic nucleophile or to an enzymic electrophile (e.g., pyridoxal-P, oxidized flavin, a metal cation).
(6) R. M. Bell and D. E. Koshland, *Science* 172, 1253–1256 (1971).
(7) T. C. Bruice and S. J. Benkovic, *Bioorganic Mechanisms*, Benjamin, New York, 1966, Vol. 1; W. P. Jencks, *Catalysis and Enzymology*, McGraw-Hill, New York, 1969; G. W. Parshall, *Homogeneous Catalysis*, Wiley-Interscience, New York, 1980.
(8) M. L. Bender, *Mechanisms of Homogeneous Catalysis from Protons to Proteins*, Wiley-Interscience, New York, 1971.
(9) J. R. Anderson, ed., *Chemisorption and Reactions on Metallic Films*, Academic Press, New York, 1971, Vols. 1 and 2.
(10) G. C. Bond, *Heterogeneous Catalysis*, Clarendon Press, Oxford, 1974.

Physically adsorbed hydrogen molecule    Transition state    Chemisorbed hydrogen atoms

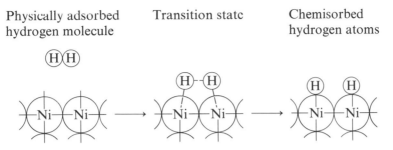

**Fig. 1.** Physisorption and chemisorption of a hydrogen molecule on a nickel surface [reproduced with permission, ref. (10), p. 21].

geneous solution—that none need be given here (7, 8). But since heterogeneous catalysis is generally less familiar to some biochemists, a few facts relating to this subject seem worth recalling.

Catalysis on a solid surface proceeds in several stages, the first two of which require the reactant to adsorb to the surface, first physically (physisorption) and then chemically (chemisorption) (Fig. 1). Chemisorption is a consequence of a mutual orbital overlap between the adsorbate and an atom (or atoms) in the surface, making use of the free valences ("dangling orbitals") of the latter. To react catalytically at a solid surface a molecule (or an atom) must first be chemisorbed. *It is an essential step in the preparation of a molecule for reaction* (10). The antecedent physical adsorption, on the other hand, is relevant to catalysis only as a prelude to chemisorption. When the reaction is consummated, desorption of the product(s) from the surface regenerates the catalyst.

It could well be asked why a pair of gas molecules—to take an extreme case—should react with each other more expeditiously while adsorbed to an appropriate solid surface than they are apt to do in the homogeneous gas phase. This question has no one single answer. Any credible answer must be an amalgam of several contributing elements, of which the following are the most prominent. While in the adsorbed condition a molecule (or atom) is in constant contact with the underlying solid, which is a source of relatively unlimited activation energy (11). Thus, curbs on the rate of transfer of activation energy, which may slow or block a reaction in the gas phase, are nonexistent in surface reactions. Moreover, two reacting molecules adsorbed on adjacent sites have more time to come into a favorable configuration than they have in the fleeting moment of a gas collision. And lastly, the surface atoms of a solid catalyst intervene actively in the chemistry by creating reaction intermediates (chemisorbed species) as components of a new reaction pathway, one with a lower activation energy than is possible in the gas phase. These considerations, as they apply to heterogeneous catalysis, have their

(11) C. N. Hinshelwood, *Kinetics of Chemical Change*, Oxford University Press, 1940, Oxford, p. 220; ref. (4), p. 405.

exact counterparts in enzymic catalysis, and highlight the natural kinship of the two kinds of catalysis.

But to be more precise, the enzymic catalysis of chemical reactions lies rather between the extremes of homogeneous and heterogeneous catalysis. It is neither one nor the other, but shares in the properties of both. As a lyophilic colloid, the protein of the enzyme is not truly in homogeneous solution, yet it is "soluble." When an enzyme is part of an insoluble membrane, its resemblance to a solid catalyst is of course greater. But even when it is "soluble," an enzyme still has a surface for the adsorption and reaction of substrate molecules, and in this sense continues to resemble the solid catalyst. The term "active center" is as familiar to heterogeneous catalysis as it is to enzymology. Specificity, too, is inseparable from the phenomenon of chemisorption. While physisorption occurs quite generally, chemisorption requires that an appropriate surface be matched by an appropriate adsorbate. Enzymes, of course, have developed the property of specificity to the ultimate degree. Inhibition of heterogeneous catalysis can be competitive or it can be irreversible (poisoning). As is well known, enzymes can also be inhibited competitively or irreversibly. Kinetic rate expressions for catalyzed chemical reactions—whether of the homogeneous, heterogeneous, or enzymic kind— exhibit the same mathematical form (12). For enzymes, this is the familiar Michaelis–Menten equation, and for solid surface catalysis it is the equally familiar Langmuir isotherm or derivative expressions. The changes in binding properties induced in enzyme molecules by substrates and other ligands are paralleled in heterogeneous catalysis. Thus, the adsorption of an atom or a molecule to a site on the surface of a solid catalyst often causes a significant change in the chemisorptive properties of a neighboring site (13, 14). It is clear from this short account that catalysis by enzymes has many ties to homogeneous and, especially, to heterogeneous catalysis. And since these latter require that a covalent bond between substrate and catalyst share in the process, it seems only natural and fitting that the same covalent principle should govern the action of enzymes. Accordingly, it is no mere happenstance that the physisorbed and chemisorbed states in heterogeneous catalysis bear a conspicuous likeness, respectively, to the Michaelis–Menten complex and the covalent enzyme–substrate intermediate of enzymic catalysis.

Yet another important, but hitherto unrecognized, point of resemblance between heterogeneous and enzymic catalysis is only now coming to light. This is the singular property of *mobility*, by which is meant the capacity of an atom or a molecular grouping to *migrate* from the site of original attachment on the catalytic surface to some other site. Such migration is a necessity in solid surface catalysis if the atoms (or groups) are to get close enough to

(12) Ref. (8), pp. 12–15.
(13) G. C. Bond, *Catalysis by Metals*, Academic Press, New York, 1962, pp. 52–53.
(14) J. T. Yates, Jr., *Chem. Eng. News* 52, 19–29 (1974).

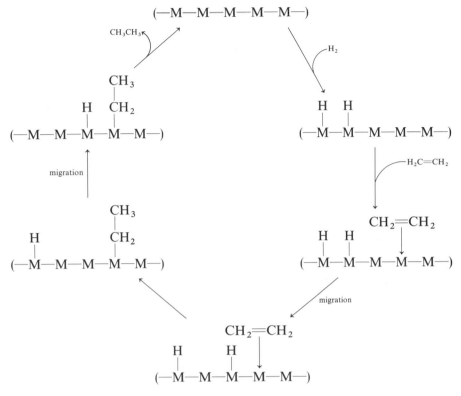

**Fig. 2.** Proposed mechanism of the hydrogenation of ethylene on a metal surface, depicting the migration of adsorbed hydrogen atoms. Adapted from (15). (—M—M—M—M—M— is a one-dimensional representation of a metal surface.)

react with each other. Consider the catalytic hydrogenation of ethylene

$$H_2 + CH_2{=}CH_2 \longrightarrow CH_3CH_3$$

on a metal surface (Fig. 2). If the hydrogen molecule alights on the surface and is dissociatively chemisorbed at a site distant from where the ethylene is adsorbed, migration must precede reaction. The activation energy for migration is much less than the activation energy for desorption, since the surface bond must break in desorption but only weakens during migration. Figure 2 portrays the migration of hydrogen (15), but ethylene is also capable of migration (16). Similar "surface walks" are possible for chemisorbed carbon monoxide, oxygen, alkyl groups, etc. (17).

(15) E. L. Mutterties, *Science* 196, 839–848 (1977).
(16) Ref. (13), p. 93.
(17) B. M. W. Trapnell, *Chemisorption*, Academic Press, New York, 1955, p. 201 ; P. M. Gundry and F. C. Tompkins, *Quart. Rev.* 14, 282–284 (1960); E. L. Mutterties, *Bull. Soc. Chim. Belg.* 84, 959–986 (1975).

Catalysis by some enzymes may also require migration of covalently fixed fragments of substrate. In the well-known case of redox reactions, electrons often react with an acceptor at a point on the enzyme surface which is distant from the site where the electron donor released them to the enzyme (Chapter 2). But even larger fragments than electrons can migrate across an enzyme surface while in the covalently bound condition. A case in point is the multi-enzyme complex, pyruvate dehydrogenase from *E. coli*. It consists of three enzymes acting in sequence to catalyze the net reaction

$$CH_3\overset{\displaystyle O}{\overset{\|}{C}}COOH + CoASH + NAD^+ \underset{FAD}{\overset{TPP}{\rightleftharpoons}}$$

$$CH_3\overset{\displaystyle O}{\overset{\|}{C}}{-}SCoA + CO_2 + NADH + H^+$$

The complex is organized around dihydrolipoyl transacetylase (EC 2.3.1.12), acting as a core, to which the other two enzymes (pyruvate dehydrogenase [EC 1.2.4.1] and dihydrolipoyl dehydrogenase [EC 1.6.4.3]) are noncovalently joined (Fig. 3). The transacetylase has 24 identical polypeptide chains, each with two molecules of lipoic acid bound by amide linkage to a lysine residue. The 48 lipoyl sulfhydryl groups of the transacetylase can interact with each other, and jointly constitute a network which can transfer an acetyl (plus a pair of electrons) among all of the lipoyl groups through thiol-disulfide reactions (18). In transferring an acetyl in this way, the lipoyl network is analogous to the "network" of metal atoms in the solid surface upon which, for instance, hydrogens are transferred (Fig. 2). In both cases, an activated fragment of substrate is enabled to migrate from its point of origin on the catalyst to some distant point where it can react conveniently with an acceptor (19). While dihydrolipoyl transacetylase is a huge multisubunit enzyme, other, smaller enzymes are showing signs that their covalently linked substrate fragments can also migrate; but in these cases the migration is from one locus to another within the same active center. Alluded to here is the possibility of a "triple-displacement" mechanism for an enzyme, wherein one and the same substrate fragment (other than an electron or a proton) joins covalently, in succession, to two different catalytic groups of the active center, in a kind of "surface walk" (see below the section on Steric Inversion and Covalent Catalysis).

---

(18) J. H. Collins and L. J. Reed, *PNAS* 74, 4223–4227 (1977).

(19) D. L. Bates, M. J. Danson, G. Hale, E. A. Hooper, and R. N. Perham, *Nature (London)* 268, 313–316 (1977); M. J. Danson, E. A. Hooper, and R. N. Perham, *Biochem. J.* 175, 193–198 (1978); G. B. Shepherd and G. G. Hammes, *B* 16, 5234–5241 (1977); R. L. Cate and T. E. Roche, *JBC* 254, 1659–1665 (1979); K. J. Angelides and G. G. Hammes, *B* 18, 1223–1229 (1979); M. C. Ambrose-Griffin, M. J. Danson, W. G. Griffin, G. Hale, and R. N. Perham, *Biochem. J.* 187, 393–401 (1980).

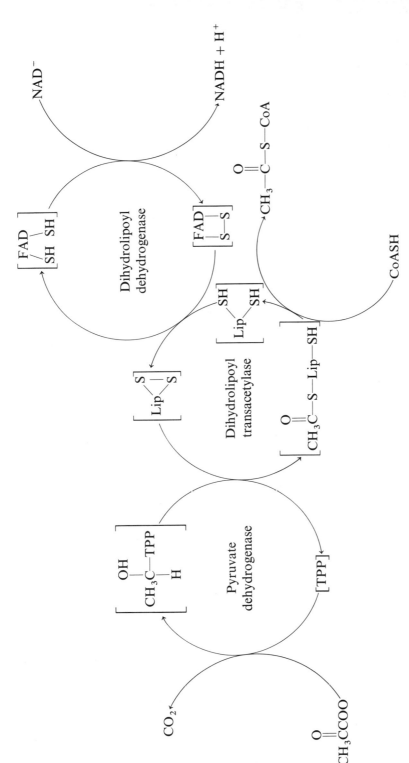

**Fig. 3.** Reaction sequence in pyruvate oxidation by the pyruvate dehydrogenase complex of enzymes (18). (TPP, thiamine pyrophosphate; Lip-$(SH)_2$ and Lip-$S_2$, reduced and oxidized lipoic acid, respectively; FAD, flavin adenine dinucleotide.)

## Enzymic Catalysis

In 1947 biochemists learned for the first time that an enzyme participates chemically in catalysis in the form of a covalent enzyme–substrate intermediate. In that year Doudoroff, Barker, and Hassid gave the first evidence that the phosphorolysis of sucrose by sucrose phosphorylase (p. 78)

$$\alpha\text{-Glucosyl-fructose} + P_i \longleftrightarrow \alpha\text{-glucosyl-1-P} + \text{fructose} \qquad (1)$$

is mediated by the glucosylated enzyme (20).

$$\alpha\text{-Glucosyl-fructose} + E \longleftrightarrow \beta\text{-glucosyl}—E + \text{fructose} \qquad (2)$$

$$\beta\text{-Glucosyl}—E + P_i \longleftrightarrow \alpha\text{-glucosyl-1-P} + E \qquad (3)$$

The reaction proceeds by a double-displacement mechanism—that is, by covalent catalysis—and not by the single displacement theretofore implied by Eq. 1 (21). Since 1947 the number of enzymes known to enter covalently into the reactions which they catalyze has grown to well over 400 (Chapter 8) (22).

While all biochemists agree that many enzymes make use of covalent catalysis, some insist that this is not true for *all* enzymes. Some reactions, it is claimed, do indeed proceed by direct displacements without covalent intervention by the enzyme. This point of view has long held sway in mechanistic enzymology, which in its early years leaned heavily on what physical organic chemistry taught about the mechanism of uncatalyzed reactions in solution. It was natural in those early times to *assume* that the enzyme fulfilled its catalytic function by providing a special surface upon which single-displacement reactions could go forward at a greatly enhanced rate. The *assumption* that the enzyme is a template upon which single displacements are accelerated became firmly fixed in biochemical thinking, and has for decades passed for truth. But the plain fact is that the foregoing assumption

---

(20) M. Doudoroff, H. A. Barker, and W. Z. Hassid, *JBC* 168, 725–732 (1947).

(21) Because of their long usage and convenience, the terms "double-displacement" and "single-displacement" are used in this book in place of more cumbersome expressions to denote, respectively, the covalent and noncovalent participation of the enzyme in a reaction. Thus, "double-displacement" in the context of this book always implies covalent catalysis by an enzyme, *irrespective of kinetics*.

(22) With just a few exceptions, these more than 400 enzymes (Chapters 2–7) do not include any that are merely protonated by substrate. When a basic group in the active center of an enzyme acquires a proton from a substrate molecule, a genuine covalent enzyme–substrate intermediate is thereby formed, since the proton is obviously a fragment of the substrate. The vast majority of enzyme reactions probably require such proton abstraction in the course of substrate transformation. On this view alone, the vast majority of enzymes should qualify as covalent enzymes. From the standpoint of covalent catalysis, however, such proton transfer to enzyme is regarded by some as "trivial." For this reason I have omitted from the tables in Chapters 2–7 any enzyme whose only known covalent bonding is to a substrate proton, *unless proton transfer is the only chemical reaction in which the enzyme participates*, as in the case of some isomerases; proline racemase, for instance (p. 188).

remains just that—an assumption, long held but seldom questioned. In all these years no unambiguous evidence has appeared in its favor. In default of such evidence, some proponents of the single-displacement hypothesis have even resorted to *negative* evidence. They assert that for any individual enzyme, the *failure* to find evidence for covalent catalysis is equivalent to positive evidence for the single-displacement mechanism for that enzyme (23). But one could with as much reason assert the converse: that negative evidence for the single displacement is equivalent to positive evidence for covalent catalysis. Both assertions are, plainly, non sequiturs. Biologists have long known that "negative facts, *when considered alone*, never teach us anything" (italics added) (24). The resort to negative evidence is rooted in the extreme difficulty of devising an experiment which can, without ambiguity, uphold the single-displacement hypothesis (25). A hypothesis so resistant to verification by experiment ought to be viewed with suspicion (26).

Covalent catalysis by the above-cited sucrose phosphorylase might have been anticipated. There is a net retention of steric configuration at C-1 of glucose which the enzyme must accommodate, and a double (Walden) displacement offers the ideal means to the accommodation (Eqs. 2 and 3) (27). But covalent catalysis is operative, too, in reactions where there is no question of chirality. To cite a few familiar instances: the enzymic transfer of such diverse chemical groups as the acyl (arylamine acetyltransferase, p. 74), the phosphoryl (acetate kinase, p. 92), the amidino (glycine amidinotransferase, p. 68), coenzyme A (coenzyme A transferase, p. 104), and the sulfur atom of thiocyanate (rhodanese, p. 103) all proceed by covalent catalysis. Thus, a chiral demand is not a precondition for covalent catalysis. When a retention of steric configuration is needed, covalent catalysis can indeed provide it. But it can also provide for *inversion* of configuration when inversion is needed.

---

(23) J. Henkin and R. H. Abeles, *B* 15, 3472–3479 (1976); P. Dimroth, R. Loyal, and H. Eggerer, *EJB* 80, 479–488 (1977).

(24) C. Bernard, *An Introduction to the Study of Experimental Medicine*, Dover Publication reprint in English, p. 174 (1865). Bernard's admonition antedates the modern catchier phrase: "the absence of evidence is not evidence of absence."

(25) I am often told: "Well, I have made every conceivable test on my enzyme in search of a covalent intermediate, and I can find none. Therefore my enzyme must be a single-displacement enzyme!" Which prompts the questions: "Have you done every conceivable test on your enzyme to prove *single-displacement* catalysis? If so, what were the tests and what were the results?" To these questions no proper reply is ever made, other than "No" to the first one.

(26) Elsewhere in this chapter and in Chapter 8, when considering the totality (that is, a statistical number) of enzymes, I contrast the negative evidence for single-displacement catalysis with the positive evidence for covalent catalysis. I regard this use of negative evidence as logically valid because positive evidence for one side of a debated issue is set against negative evidence for the other side. Such weighing of positive versus negative evidence for alternative mechanisms has been singularly lacking thus far in the discussion on some individual enzymes (23). In these cases, only negative evidence for covalent catalysis is offered because positive evidence for the single displacement is nonexistent.

(27) J. Voet and R. H. Abeles, *JBC* 245, 1020–1031 (1970).

## Steric Inversion and Covalent Catalysis

There is general agreement that retention of configuration—as in the sucrose phosphorylase reaction—is well explained by covalent (double-displacement) catalysis. Inversion of configuration, on the other hand, is said by some to be so obviously in accord with the single displacement that covalent catalysis is out of the question. The reality, however, is otherwise. In proof of this we cite the adenine phosphoribosyltransferase reaction,

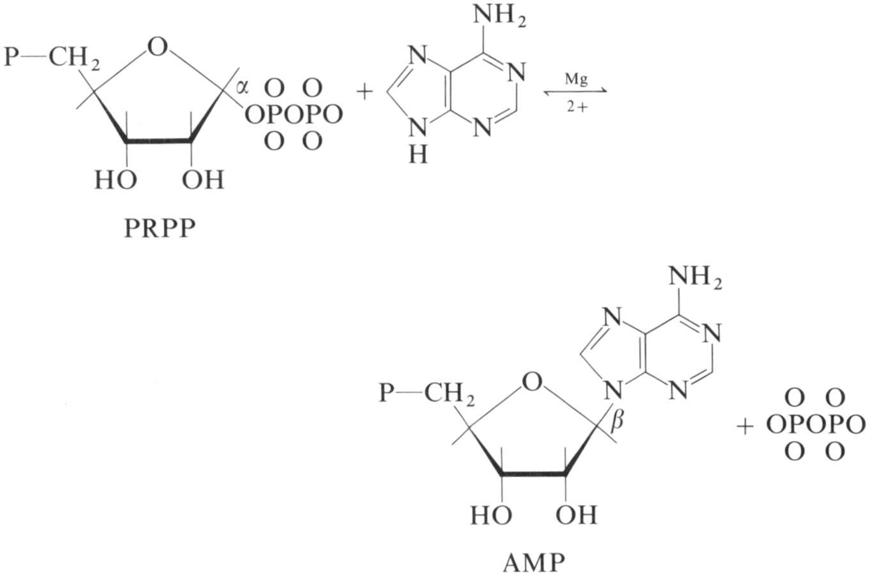

PRPP

AMP

in which an inversion occurs on C-1 of the transferring 5'-phosphoribosyl group (p. 80). The reaction is mediated by a phosphoribosyl enzyme. The enzyme is in fact isolated from tissue as the phosphoribosyl enzyme. The very existence of the latter makes the single-displacement mechanism for this enzyme impossible. Though only one phosphoribosyl-enzyme inter- mediate is so far known, the reaction is best thought of as a triple displace- ment. The mobile phosphoribosyl group must link covalently through its C-1, sequentially, to two different catalytic groups in the active center during one turn of the catalytic cycle. This "surface walk" includes three $S_N2$ re- actions (i.e., Walden inversions) on C-1, resulting in net steric inversion. From these considerations it is plain that steric inversion on substrate is no bar to covalent catalysis.

A recent surge of interest in phosphoryl-transferring enzymes has revealed that many of them catalyze their reactions with a net steric inversion on the transferred phosphorus atom (28). Acetate kinase (p. 92), hexokinase (p. 87),

(28) Configurational effects on phosphorus are observable, of course, only when the phosphorus has been made into an artificially chiral center.

and pyruvate kinase (p. 91) are among such enzymes. A net inversion on phosphorus implies, of course, an odd number of in-line (Walden-type) displacements (29). Proponents of the single-displacement mechanism hold that the odd number in question is "one," solely because it gives the "simplest interpretation" of the steric inversion. But such might have been said of the reaction catalyzed by adenine phosphoribosyltransferase before it was known to be covalently catalyzed. Acetate kinase, it is found, inverts the phosphorus atom which it transfers, but does so through at least one experimentally established phosphoenzyme intermediate (p. 92). Acetate kinase, accordingly, shows every sign of being a triple-displacement enzyme (30). Consider, too, that all of the more than 400 covalent enzymes of which we have record (Chapters 2–7) might have chosen—were the choice possible—the "simplicity" of single-displacement catalysis. Instead, all of these enzymes chose the "complexity" of covalent catalysis. There is therefore no reason whatever for believing that enzymes set a premium on simplicity. Quite the contrary. All the evidence suggests that the single-displacement pathway, *though it may look "simple" to the investigator*, is the truly toilsome pathway for the enzyme to take. No enzyme is known for certain to take it (2).

# Why Covalent Catalysis Is Favored Over Single-Displacement Catalysis

Why then are so many enzymic reactions fragmented into two (or more) partial chemical reactions, with a covalent enzyme–substrate complex as a linking intermediate (31)? And, conversely, why is there no unambiguous evidence for even a single instance of a single-displacement reaction? The simplest answer is, as was said above, that it must be easier for an enzyme (sucrose phosphorylase, for instance) to catalyze its reaction by dividing it into a suite of partial reactions (Eqs. 2 and 3). For all of its seemingly greater

---

(29) A role for pseudorotation in enzymic phosphoryltransfer is generally discounted [W. A. Blättler and J. R. Knowles, *B* 18, 3927–3933 (1979); B. M. Dunn, C. DiBello, and C. B. Anfinsen, *JBC* 248, 4769–4774 (1973); K.-F. R. Sheu, J. P. Richard, and P. A. Frey, *B* 18, 5548–5556 (1979); F. H. Westheimer, in *Molecular Rearrangements*, P. de Mayo, ed., Interscience, New York and London, in press].

(30) L. B. Spector, *PNAS* 77, 2626–2630 (1980).

(31) In the context of this book the expression "partial reaction" refers to a reaction in which the enzyme, covalently linked to substrate (or a fragment thereof), is a participant. The fragment of substrate in question is *never* a proton, except in the case of a few isomerases in which proton transfer is the sole chemical reaction taking place (Chapter 6). When the net reaction requires abstraction of a substrate proton by a basic group of the enzyme, such proton abstraction is, to be sure, a partial chemical reaction [cf. ref. (22)]. It is however separate from—but in reality additional to—the partial reactions under consideration in this book.

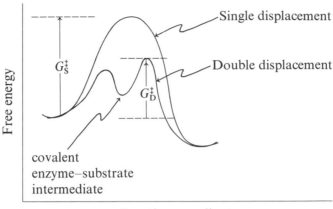

Reaction coordinate

**Fig. 4.** Hypothetical free energy relationships for enzymic catalysis by the single- and double-displacement pathways. The reaction coordinate will have a different physical significance for the two pathways.

complexity, the double-displacement pathway opposes a lower net energy barrier ($G_D^{\ddagger}$, Fig. 4) to the reaction than does the single-displacement pathway $G_S^{\ddagger}$, Fig. 4). But why should this be so? In this section I offer some reasons why the covalent pathway is the easier one for the enzyme to take.

One advantage the double-displacement has over the single-displacement pathway for sucrose phosphorylase—to take a concrete example—is evident in the different ways the enzyme would handle its substrates on the two pathways (32). In the single-displacement pathway the enzyme is obliged *simultaneously* to align *both* substrates—sucrose and phosphate— for reaction with each other (Eq. 1). A correct alignment of *both* substrates must be arranged by the enzyme in order to maximize orbital overlap of a phosphate oxygen with C-1 of the glucosyl portion of sucrose prior to the displacement of fructose from that carbon. By contrast, the double-displacement pathway is much less constraining. It allows the enzyme to manipulate its substrates *one at a time*. First, the enzyme aligns sucrose with a carboxyl oxygen in the active center in order to form the glucosyl bond to this enzyme carboxyl (Eq. 2) (27). Then the second substrate, phosphate, is correctly aligned by the enzyme so that a phosphate oxygen atom can bond easily to C-1 of the glucosyl portion of the glucosyl-enzyme intermediate, displacing the enzyme carboxyl with concurrent formation of α-glucose-1-P (Eq. 3). Having to adjust but one substrate at a time into productive alignment gives

---

(32) L. B. Spector, *Bioorg. Chem.* 2, 311–321 (1973).

the double-displacement pathway an entropic advantage over the single-displacement pathway (33).

Such considerations ought to hold generally for enzymes, whether they follow "ping-pong" or sequential kinetics. In ping-pong kinetics the second substrate is absent altogether from the active center when the first substrate reacts with the catalytic group of the enzyme. In sequential kinetics the second substrate is also present in the active center, but primarily for conformational reasons. It stands by while the first substrate reacts with the catalytic group of the enzyme. When the covalent enzyme–substrate intermediate is formed, the second substrate is brought into productive alignment with it for completion of the reaction cycle.

It might be thought a disadvantage that the enzyme, in double displacement, must arrange two successive orbital alignments in place of only the one required by single displacement. But we note in the latter case that the two molecules to be aligned are both external to the enzyme. On the other hand, each of the two alignments necessary to the double displacement involves the enzyme itself as one of the partners in the alignment. Alignments in which the enzyme itself is a partner must be easier to arrange for the same (anthropomorphic) reason that "aligning one's own left hand for insertion into its glove is easier than aligning someone else's left hand for insertion into the same glove" (32).

The foregoing considerations have to do with what we may term the time-related, or temporal, aspect of enzymic events. In a somewhat different vein, we may speak of the spatial aspect of those same events. The single-displacement mode of enzyme action demands, as we know, that the two substrates be jammed together in correct orbital alignment for reaction with each other, at one unique position on the enzyme's surface. There is in this something of the character of a three-body interaction, an event of some improbability. It follows that single-displacement catalysis must of necessity impose a high degree of order upon its substrates, which is of course expensive in activation energy (30). Compared with this are the two-body interactions of covalent catalysis, with their intrinsically greater probability. Implicit in these notions is the possibility—offered only by covalent catalysis—that substrates can prepare for reaction by binding in the enzyme's active center at a comfortable, *nontouching* distance from each other. They need not be jammed together for maximal orbital overlap of their reacting parts. The catalytic group of the enzyme will negotiate the distance between them, and see to it that a transferring fragment of substrate will get easily from donor

---

(33) Professor W. P. Jencks has voiced a similar point of view on the entropic virtue of acting upon substrates one at a time [W. P. Jencks and M. I. Page, *FEBS Symp.* 29, 45–58 (1972)]. But despite this, he holds firmly to the conviction that covalent catalysis is not universal, and that many enzymes make use of single-displacement catalysis (34).
(34) W. P. Jencks, *Adv. Enzymol.* 43, 219–410 (1975).

to acceptor. In conformity with this, we note that all of the common catalytic groups of enzymes—cysteinyl sulfhydryl, seryl hydroxyl, the carboxyl of aspartate or glutamate, the ε-amino of lysine, the imidazole ring of histidine, and the coenzyme prosthetic groups—all possess in greater or lesser degree the properties of a "swinging arm." Within the limits of their molecular dimensions some can swing further than others, but all are capable of some degree of swing. In any case, a spatial requirement less exacting than that of a three-body interaction ought surely to benefit the entropy of activation (35).

While donor and acceptor substrates are bound in their noncontiguous loci, their reacting parts need not necessarily line up opposite to each other for "direct" transfer by simple double displacement. The reacting parts may be askew to each other, further reducing the degree of order imposed in the binding process. Yet, thanks to covalent catalysis (and the "swinging arm"), group transfer is easily managed through triple displacements, or perhaps even longer "surface walks" (30). Such roundabout transfer of migrating groups recalls the surface walks of heterogeneous catalysis and of the acetyl group in the pyruvate dehydrogenase complex. Some enzymes, in their evolutionary development, may have found it easier to arrange for a surface walk than to fix the relative orientation of substrates for "direct" transfer by simple double displacement. Such may indeed have happened in the case of acetate kinase (36) (p. 92).

Covalent catalysis can also benefit the enthalpy of activation. After a pair of substrate atoms has been correctly aligned for reaction, the enzyme must work to press them into close enough contact to form a covalent bond between them. Such work ought to be easier when one of the two atoms in question is a part of the enzyme itself than when both atoms are external to the enzyme. Pursuing our "hand-in-glove" analogy; after the fingers of the hand are aligned with the fingers of the glove, the two sets of fingers must be pressed together to complete the gloving process. This, clearly, is more

---

(35) The phenomenon of the "swinging arm" is one of several important distinctions between enzymic and solid surface catalysis. Relative to the enzyme, the solid catalyst operates under a handicap, since reactants on the metal surface must get into direct orbital contact with each other for reaction to occur (Fig. 2). This may explain in part why enzymes can catalyze a chemistry which, in complexity and sophistication, is unmatched in heterogeneous catalysis.

Other conspicuous distinctions between enzymic and solid surface catalysts are: (a) the conformational flexibility of enzymes compared with the rigidity of solid catalysts; (b) the greater chemical diversity of catalytic groups found in the active center of an enzyme than is found at the active sites of a given solid catalyst; (c) the three-dimensional character of the enzymic active center compared with the generally two-dimensional character of the solid surface.

(36) No guess is ventured as to an optimal distance between substrates for covalent catalysis. It probably varies from enzyme to enzyme. But even if no distance at all separates the substrates (that is, if they are in van der Waal's contact with each other), the enzyme, through covalent catalysis, is still spared the entropy-expensive alignment of their reacting parts for direct group transfer by single-displacement catalysis. For, however far apart the reacting parts of substrates in contact may be, group transfer between them remains possible through covalent catalysis via the roundabout "surface walk."

easily done with one's own hand (a part of one's total structure) than with someone else's hand. Though this is intuitively obvious, it is not easily explained in physical terms.

In short, covalent catalysis endows the enzyme with entropic and enthalpic advantages which are unavailable from single-displacement catalysis. Entropic benefits accrue from the possibility, through covalent catalysis, of binding substrates noncontiguously in the active center and of acting upon them one at a time. And, for "anthropomorphic" reasons, covalent catalysis eases the task of orbital alignment and the work of forcing atoms together to make them react.

# Is Covalent Catalysis a Means of Stabilizing Very Reactive (Hypothetical) Intermediates?

The foregoing considerations on the temporal and spatial features of catalysis have, obviously, a special relevance to catalysis by enzymes. They are, however, less relevant, if at all, in many instances of nonenzymic mass-law catalysis. Yet these latter, as is recounted earlier in this chapter, make universal use of the covalent principle. There must, therefore, be something about covalent catalysis which makes it indispensable in all kinds of mass-law catalysis, enzymic and nonenzymic alike. This something, it has been suggested, may be the stabilization of very reactive (hypothetical) intermediates (37).

For any established pathway (enzymic or nonenzymic) it is possible to picture a corresponding "uncatalyzed" pathway, which often involves a very reactive intermediate. Comparing the two pathways reveals that the catalyst stabilizes the reactive intermediate of the uncatalyzed pathway by combining with it. In illustration, consider the nonenzymic oxidation of molecular hydrogen by thallium(III) in homogeneous aqueous solution

$$H_2 + Tl(III) \xrightarrow{Cu(II)} 2 H^+ + Tl(I)$$

The reaction is catalyzed by Cu(II) and its complexes (37). Catalysis takes the form

$$H_2 + Cu(II) \longrightarrow CuH^+ + H^+ \qquad (\Delta H = 26 \text{ kcal/mole}) \qquad (4)$$

$$CuH^+ + Tl(III) \longrightarrow Cu(II) + Tl(I) + H^+$$

wherein molecular hydrogen reacts with catalytic Cu(II) in the rate-determining step to make copper hydride (Eq. 4), which then transfers two electrons to

(37) J. Halpern, *Discuss. Faraday Soc.* 46, 7–19 (1968); J. Chatt and J. Halpern, in *Catalysis*, F. Basolo and R. L. Burwell, Jr., eds., Plenum Press, New York, 1973, pp. 107–129.

Tl(III) to complete the oxidation. The corresponding uncatalyzed reaction

$$H_2 \longrightarrow II^+ + H^- \qquad \text{(estimated } \Delta H = 35 \text{ kcal/mole)} \qquad (5)$$

$$H^- + Tl(III) \longrightarrow H^+ + Tl(I)$$

begins with the heterolytic dissociation of molecular hydrogen into a proton and a free hydride ion (Eq. 5). The reaction of Eq. 5 is clearly disfavored thermodynamically relative to that of Eq. 4, and it results moreover in the extremely unstable free hydride ion. Both strictures are removed by catalytic Cu(II) when it stabilizes a latent hydride ion in the form of copper hydride.

This way of thinking about mass-law catalysis finds support in numberless instances drawn from all fields of enzymic and nonenzymic chemistry. From organic chemistry there is the long-known benzoin condensation catalyzed by cyanide ion.

The established catalytic pathway (Eq. 6)

$$(6)$$

proceeds over an intermediate (shown in brackets) in which the catalytic cyanide ion is linked to a substrate molecule (38). This catalyst–substrate complex may be regarded as a stabilized form of the hypothetical aldehyde-anion of the uncatalyzed benzoin condensation

Aldehyde-anion

(38) A. Lapworth, *J. Chem. Soc.* 83, 995–1005 (1903); E. Stern, *Z. Phys. Chem.* 50, 512–559 (1905).

The aldehyde-anion is surely an extremely reactive entity, difficult to prepare and preserve for the intended reaction other than in the cryptic form of the cyanide complex of Eq. 6.

Heterogeneous catalysis, too, affords countless instances of the stabilization of hypothetical "energy-rich" intermediates through covalent linkage to the catalytic surface. A case in point is the familiar hydrogenation of ethylene depicted in Fig. 2 of the present chapter. If this reaction were possible in the absence of catalyst it would require the dissociation of molecular hydrogen into free hydrogen atoms at a great cost of activation energy. A nickel surface, however, can dissociate a hydrogen molecule into a pair of hydrogen–nickel complexes at a vastly diminished energy cost, while conserving most of the chemical reactivity of the hypothetical free hydrogen atoms.

In much the same way, enzymes may be said to stabilize reactive intermediates which would otherwise be virtually inaccessible. We instance here just a few well-known reactions. The uncatalyzed decarboxylation of acetoacetate at neutral pH

$$CH_3\overset{O}{\overset{\|}{C}}CH_2COO^- \longrightarrow CO_2 + \left[CH_3\overset{O}{\overset{|}{C}}-\overset{-}{C}H_2\right]$$

demands the separation of carbon dioxide from the highly basic enolate ion. The enzyme, acetoacetate decarboxylase, circumvents this difficulty by forming a Schiff base between acetoacetate and a lysine residue in the active center. Loss of carbon dioxide leaves the stabilized eneamine form of enolic acetone (p. 157). Similarly, the uncatalyzed decarboxylation of amino acids

$$R \quad CH_2\underset{NH_2}{\overset{}{C}H}COO^- \longrightarrow CO_2 + \left[R-CH_2-\underset{NH_2}{\overset{H}{\underset{|}{\overset{|}{C}}}}:^-\right]$$

would lead to a very reactive carbanion. Enzymatically, however, this is stabilized by Schiff base formation with pyridoxal-P or a pyruvyl residue in the active center of amino acid decarboxylases. The uncatalyzed decarboxylation of pyruvate would yield the unstable acetaldehyde-anion

$$CH_3\overset{O}{\overset{\|}{C}}COO^- \longrightarrow CO_2 + \left[CH_3\overset{O}{\overset{\|}{C}}:^-\right]$$

which is easily stabilized by pyruvate decarboxylase as the α-hydroxyethyl-thiamine-pyrophosphate–enzyme intermediate (p. 155).

In the same vein, one can image the uncatalyzed acetyl transfer

$$CH_3\overset{O}{\overset{\|}{C}}-CoA + arylamine \longrightarrow N\text{-acetylarylamine} + CoA$$

as taking place in stages each of which involves the free acetylium ion

$$CH_3\overset{\overset{O}{\|}}{C}-CoA \longrightarrow CoA + \left[CH_3\overset{\overset{O}{\|}}{C}^+\right]$$

$$\left[CH_3\overset{\overset{O}{\|}}{C}^+\right] + \text{arylamine} \longrightarrow \textit{N}\text{-acetylarylamine}$$

Enzymatically, arylamine acetyltransferase catalyzes this acetyl transfer via an acetyl-enzyme intermediate in which the acetyl group is joined to a sulfhydryl in the enzyme's active center (p. 74). The acetyl enzyme may thus be regarded as a stabilized form of the acetylium ion.

The uncatalyzed reaction of two molecules of hydrogen peroxide to yield dioxygen and water

$$H_2O_2 + H_2O_2 \longrightarrow O_2 + 2\,H_2O$$

can be imagined to take place in two stages

$$HO-OH \qquad HO^- + [HO^+]$$

$$[HO^+] + HO-OH \longrightarrow O_2 + H_2O + H^+$$

First a molecule of hydrogen peroxide dissociates into a hydroxide ion and a free hydroxonium ion. The latter then oxidizes a second molecule of hydrogen peroxide to dioxygen and water. In the presence of catalase, however, the hypothetical free hydroxonium ion is stabilized by the enzyme in the form of Compound I (p. 43).

In the reaction catalyzed by sucrose phosphorylase, the glucosyl-enzyme intermediate of Eqs. 2 and 3 is the stabilized form of the glucosyl-cation of Eq. 7.

$$(7)$$

β-Glucosylenzyme                Glucosyl-cation

In some glycosyl-transferring reactions (e.g., lysozyme, β-galactosidase) such glycosyl-cations are thought by some investigators to be sufficiently stable (as part of an ion-pair) to be real intermediates. But even if this were so, it seems hardly possible that such carbonium ions can escape altogether from a substantial degree of collapse into covalent linkage with the enzyme, as in the reversal of Eq. 7.

Corresponding to an enzymic phosphoryl transfer

$$\text{Donor}\text{—}PO_3^- + \text{acceptor} \longleftrightarrow \text{donor} + \text{acceptor}\text{—}PO_3^-$$

one can also imagine an uncatalyzed phosphoryl transfer as proceeding via monomeric metaphosphate ion as an intermediate.

$$\text{Donor}\text{—}PO_3^- \longleftrightarrow \text{donor} + [PO_3^-]$$

$$[PO_3^-] + \text{acceptor} \longleftrightarrow \text{acceptor}\text{—}PO_3^-$$

Monomeric metaphosphate, though never isolated, is believed to be an exceedingly active and unselective phosphorylating agent. Yet its participation in enzymic reactions by single-displacement catalysis has been speculated upon (39). Metaphosphate has been invoked particularly in the creatine kinase reaction (40).

$$\text{Creatine} + \text{ATP} \longleftrightarrow \text{creatine}\text{—}PO_3 + \text{ADP}$$

Here it is thought that metaphosphate is locked firmly between the donor and acceptor substrates aligned optimally in a presumed transition-state complex for *direct* phosphoryl transfer. This trapping of metaphosphate [designated in (40) as the "trigonal planar form of the transferable phosphoryl group"] in the presumed transition state has, of course, the intention of stabilizing it. But to impose such orbital constraints simultaneously on creatine, ATP, and enzyme in order to accommodate metaphosphate in a transition state is to forfeit the entropic gains we spoke of in the previous section of this chapter. Through covalent catalysis, however, metaphosphate is stabilized as phosphoenzyme, and all entropic gains can be realized (Eqs. 8 and 9) (41)

$$\text{ATP} + \text{E} \longleftrightarrow \text{E}\text{—}PO_3 + \text{ADP} \qquad (8)$$

$$\text{E}\text{—}PO_3 + \text{creatine} \longleftrightarrow \text{creatine}\text{—}PO_3 + \text{E} \qquad (9)$$

---

(39) S. J. Benkovic and K. J. Schray, in *The Enzymes*, 3rd ed., P. D. Boyer, ed., Academic Press, New York, 1973, Vol. 8, pp. 201–238; G. Lowe and B. S. Sproat, *Chem. Commun.* 783–785 (1978); *idem. J. Chem. Soc. Perkin Trans. 1*, 1622–1630 (1978); A. S. Mildvan, *Adv. Enzymol.* 49, 103–126 (1979); J. R. Knowles, *Ann. Rev. Biochem.* 49, 877–919 (1980); A. Satterthwait and F. H. Westheimer, *JACS* 103, 1177–1180 (1981).

(40) E. J. Milner-White and D. C. Watts, *Biochem. J.* 122, 727–740 (1971); A. C. McLaughlin, J. S. Leigh, Jr., and M. Cohn, *JBC* 251, 2777–2787 (1976).

(41) At the time of writing creatine kinase has not been established as a covalent enzyme. Equations 8 and 9 are for this reason hypothetical, but no more so than the metaphosphate-containing transition-state remarked upon in the text. Equations 8 and 9 are written as they are not to imply that a phosphorylated creatine kinase will necessarily have a free existence, but merely to keep the argument simple.

From the few examples given here one can sense how appealing it is to think of enzymes (and other mass-law catalysts) as stabilizers of reactive (hypothetical) intermediates. But, all things considered, it is doubtful whether such stabilization is the "purpose" of covalent catalysis, or is accountable for its universality. The argument for stabilization is at best a subtle one, and somewhat artificial. For one thing, it is doubtful whether many of the con-jectured reactive intermediates have any real existence in uncatalyzed reac-tions in aqueous solution. It is hard to believe, for instance, that free acetylium ion could be an intermediate in an uncatalyzed acetyl transfer in aqueous medium. It is far more probable that catalyzed and uncatalyzed reactions proceed by altogether different chemical pathways. Thus, to say that enzymes stabilize reactive intermediates is to say, in different words, that covalent catalysis provides an energetically easier pathway. It is nonetheless significant that two unstable entities—the glycosyl-cation and the metaphosphate ion— have been seriously proposed as participants in single-displacement catalysis by enzymes. In such cases, covalent catalysis does indeed claim the advantage.

# Enzymes as Phase Transfer Catalysts and Energy Transducers

One of the ways in which an enzyme promotes a chemical reaction is to remove its substrate from the aqueous phase and transfer it into one which is essentially solvent-free. Hexokinase, for instance, binds a glucose molecule in its active site cleft and then closes its two great lobes around the glucose, squeezing out the water and enveloping the glucose molecule in a largely solvent-free space (42). Dehydrating the substrate and the catalytic groups renders them "naked," and therefore more reactive. Nevertheless, an activa-tion energy must be supplied to the reaction.

Just as in heterogeneous catalysis the adsorbed molecules (or atoms) can exchange energy with the underlying solid, so can the substrates liganded to an enzyme exchange energy with the underlying enzyme. Thus, the energy of binding given up to the enzyme by reactants can be conserved for such later uses as desolvation, desorption of products, or activation of reactants. Since the reaction within the active center occurs in a mostly water-free environment, activation energy for the reaction cannot come directly from collisions of the reactants with water molecules. The activation energy comes instead from the enzyme as the immediate source. Energy conserved in the enzyme from the binding of reactants (34), or derived from collisions of the enzyme with water molecules of the medium, is transmitted into the active center for the activation of substrate. In all this the enzyme acts the part of a

---

(42) C. M. Anderson, F. H. Zucker, and T. A. Steitz, *Science* 204, 375–380 (1979).

transducer (or reservoir) of energy. The catalytic group in the active center is of course an integral part of the covalent structure of the enzyme. As such, it ought to have easier access to the enzyme's energy store than has a reactant molecule which is external to the structure. When the catalytic group makes a covalent bond to a fragment of substrate, that fragment becomes, literally, an extension of the enzyme's covalent structure. And, like the catalytic group to which it is bound, it ought to be more easily infused with activation energy from the enzyme's store than a molecule adsorbed externally to the enzyme. Such considerations suggest yet another advantage which covalent catalysis has over single-displacement catalysis.

# A Definition of Catalysis

It is often said that a catalyst acts to lower the energy of activation of a reaction. It is becoming more accurate, however, to say that a catalyst alters the (uncatalyzed) chemical pathway of a reaction to one with a lower activation energy (Fig. 4). This accords better with Hinshelwood's encompassing and widely accepted definition of catalysis as a process providing "an alternative and more speedy reaction route" (4). That so many enzymes—through covalent catalysis—do indeed provide an alternative chemical route is in harmony with this definition. Notably in discord with it, however, is the single-displacement mechanism. Hard to imagine is how the single-displacement reaction on the surface of an enzyme and the same single displacement proceeding (uncatalyzed) free in solution can follow chemical pathways that are truly different (43).

While covalent linkage to substrate is considered here to be a constant in all mass-law catalysis, it is recognized that still other factors contribute to the special success of enzymes. Such success is possible only in close conjunction with the effects of binding, strain, general acid–base (covalent) catalysis, conformational change, etc., each with a part of its own to play in the orchestration of enzymic catalysis. Apart from their mere mention here, nothing further will be said of these effects, since they fall outside the restricted theme of this book.

---

(43) In harmonizing covalent catalysis with Hinshelwood's definition we inevitably lapse into circular reasoning, since the definition is itself built upon the existence of covalent catalysis. Oddly enough, circular reasoning which proves in the end to accord with the facts is not new to the annals of science. In formulating his laws of motion, Isaac Newton felt obliged to define mass as the product of density and volume—an outright circularity since the concept of mass is needed to define density. And, on mass, thus vaguely defined, he proceeded to erect the concept of force. Yet, despite the circularity and vagueness in their development, Newton's laws of motion soon came to underlie much of physical science [I. Newton, *The Mathematical Principles of Natural Philosophy and His System of the World*, translated by Andrew Motte, 1729; revised by F. Cajori, University of California Press, Berkeley, 1934, p. 1].

## Enzymes as Transferases

Of the six major classes into which enzymes have been sorted by the Enzyme Commission of the International Union of Biochemistry, only one is officially designated as "transferases." Yet a little reflection compels the view that *all* enzymes are in fact transferases (32). The division into six major classes is merely a matter of classificatory convenience. Enzymes generally transfer something from a donor to an acceptor. In the case of isomerases the donor and acceptor can be different parts of the same molecule, or even the very same atom. The something being transferred may be an electron, a proton, or a polyatomic entity such as a methyl, a phosphoryl, an acetyl, a glycosyl group, etc. At some stage in its transit from donor to acceptor the transferring fragment of substrate links covalently to the enzyme. The next six chapters tabulate, and illustrate with examples, over 400 enzymes for which the foregoing statement is believed to be true.

# Chapter 2
# Oxidoreductases

Oxidoreductases are transferases whose special object of transfer is the electron or "hydride ion." They generally possess one or more prosthetic groups—NAD, flavin, heme, pterin, a metal ion—or a built-in, redox-active disulfide function. These groups engage in the detailed process of electron transport between the donor and acceptor substrates. Mounting evidence upholds the view that the transfer of electrons between the holoenzyme, on the one hand, and the donor and acceptor, on the other, is effected by covalent union of the holoenzyme with these substrate molecules, the point of union being a prosthetic group. Examples of such covalent unions form the substance of the present chapter. For ease of presentation, the redox enzymes chosen for illustration are grouped together according to the prosthetic group(s) which they bear. The discussions are brief, and highlight the covalent aspects of redox activity.

## Flavoenzymes

A covalent bond between substrate and prosthetic group figures prominently in the electron transfer catalyzed by the flavoenzyme, D-amino-acid oxidase [EC 1.4.3.3]. This enzyme catalyzes the oxidative deamination of D-amino acids (e.g., Eq. 1).

$$\text{D-Alanine} + H_2O + O_2 \longrightarrow \text{pyruvate} + NH_3 + H_2O_2 \qquad (1)$$

In addition to D-amino acids, the enzyme is active on a number of pseudosubstrates. This fortunate fact permits the fruitful study of reaction mechanism. One such pseudosubstrate is nitroethane, which, in the presence of enzyme, is oxidized by oxygen to acetaldehyde and nitrite ion (1).

$$CH_3CH_2NO_2 + O_2 + H_2O \longrightarrow CH_3CHO + NO_2^- + H_2O_2 + H^+$$

The special utility of nitroethane lies in its being completely ionized at pH 8.3 to the carbanion

$$CH_3CH_2NO_2 \longrightarrow CH_3-\overset{..}{\underset{H}{C}}{}^- -NO_2 + H^+$$

and it is the carbanion which is the immediate substrate of the enzyme, being oxidized at a rate 100-fold faster than neutral nitroethane.

Oxidation begins with the covalent fixation of the nitroethane carbanion on N-5 of the isoalloxazine ring of the coenzyme (Fig. 1). This has the effect of reducing FAD to the oxidation level of $FADH_2$, as revealed by anaerobic spectrophotometry. There follows the rapid elimination of nitrite ion resulting in the pivotal cationic imine 1. The latter is a powerful electrophile and reacts rapidly with water or hydroxide ion to yield the hydroxyethyl derivative 2, which promptly eliminates the holoenzyme to give acetaldehyde 3. The cationic imine 1 can also react with cyanide ion, which competes effectively with water and inhibits the enzyme by forming the stable structure 4. Enzyme, so inhibited, can be resolved into apoenzyme and the flavin-substrate-cyanide adduct. Comparison of the physical and chemical properties of the isolated adduct with those of flavin derivatives of established structure makes it a virtual certainty that the adduct has the structure of 5-cyanoethyl-1,5-dihydro FAD (5) (1).

It is clear from the foregoing that the transfer of electrons from pseudosubstrate to flavoenzyme entails covalent bond formation between the anion of the pseudosubstrate and the prosthetic group of the enzyme. That such would prove to be generally true of enzymic redox reactions was foretold in earlier speculations on the subject (2). It accords, moreover, with general chemical experience in which sufficient orbital overlap (that is, a substantial degree of covalent bond formation) is a precondition of electron transfer between donor and acceptor molecules. In order for an amino acid to react covalently with a flavoenzyme a pair of electrons on the $\alpha$-carbon must first be laid bare by removal of the $\alpha$-proton. From nitroethane the $\alpha$-proton ionizes away nonenzymatically. But from amino acids the proton is removed by the enzyme. Consonant with such removal is the action of D-amino-acid oxidase on D-$\beta$-chloroalanine—in the absence of oxygen—to effect the $\alpha,\beta$-elimination of hydrogen chloride (3–5).

(1) D. J. T. Porter, J. G. Voet, and H. J. Bright, *JBC* 248, 4400–4416 (1973).

(2) P. Hemmerich, G. Nagelschneider, and C. Veeger, *FEBS lett.* 8, 69–83 (1970); P. Hemmerich, *Chimia* 26, 149–150 (1972).

(3) C. T. Walsh, A. Schonbrunn, and R. H. Abeles, *JBC* 246, 6855–6866 (1971).

(4) D. J. T. Porter, J. G. Voet, and H. J. Bright, *Biochem. Biophys. Res. Commun.* 49, 257–263 (1972).

(5) C. T. Walsh, E. Krodel, V. Massey, and R. H. Abeles, *JBC* 248, 1946–1955 (1973).

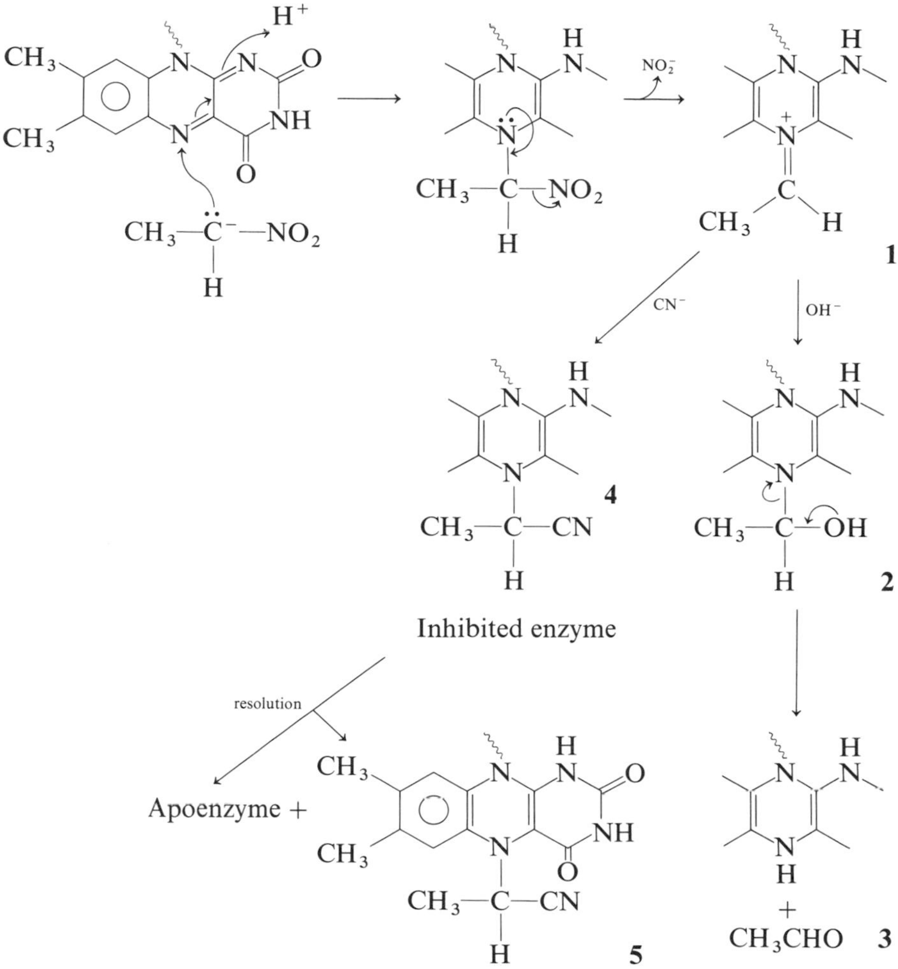

**Fig. 1.** Mechanism of oxidation of nitroethane anion by D-amino-acid oxidase and inhibition of the enzyme by cyanide ion (1).

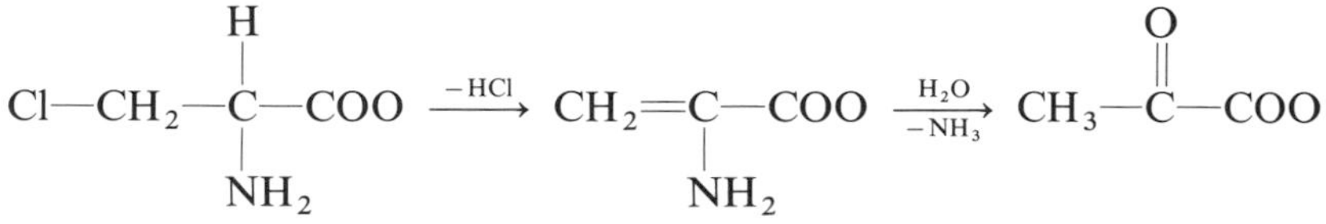

If the α-proton is replaced by a deuteron a large isotope effect is observed (3). And with tritium in the α-position some 20–40% of it is recovered in the methyl group of pyruvate (5). These observations speak for the removal of the α-proton of substrate by a base in the active center of the enzyme and its transient preservation there in substantial isolation from the medium (6).

(6) D. J. T. Porter, J. G. Voet, and H. J. Bright, *JBC* **252**, 4464–4473 (1977).

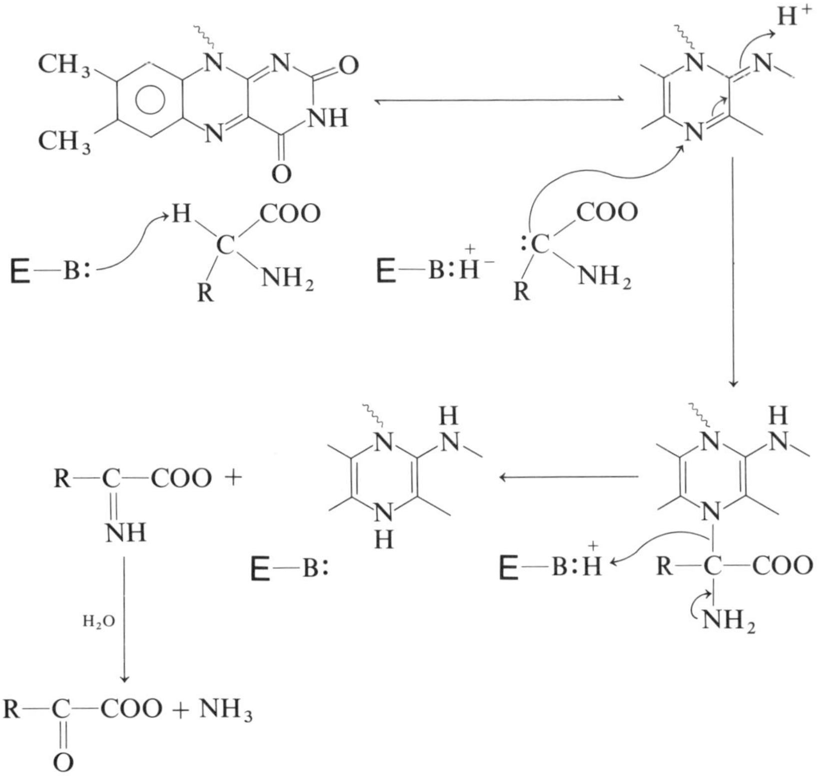

**Fig. 2.** Proposed mechanism of the oxidation of an amino acid by D-amino-acid oxidase (1).

α-Deprotonation of an amino acid generates the α-carbanion which, like the carbanion of nitroethane, doubtless transfers a pair of electrons to flavoenzyme after first binding covalently to it (Fig. 2).

Also in accord with these conceptions are the observations made on reconstituted D-amino-acid oxidase in which FAD is replaced by 5-deaza FAD. The reconstituted enzyme cannot catalyze the oxidation of D-alanine

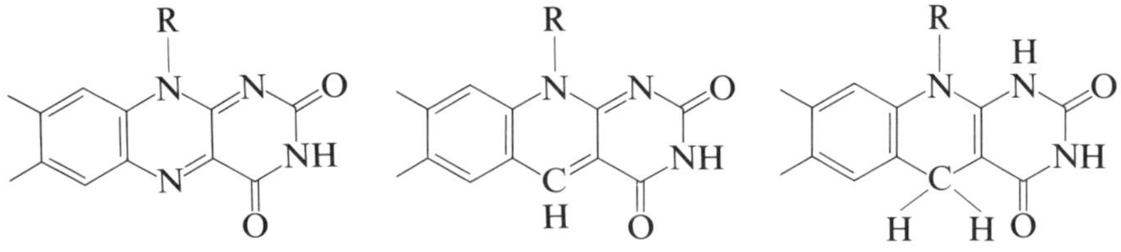

FAD                    5-Deaza FAD                 5-Deaza FADH$_2$
                       (R = ribityl-ADP)

according to Eq. 1 because reduced enzyme (that is, E—deaza $FADH_2$) is unreactive with oxygen and cannot complete the catalytic cycle. Despite this, the reconstituted enzyme acts on D-alanine to effect its oxidative de-amination to pyruvate plus ammonia with stoichiometric reduction of the enzyme to E—deaza $FADH_2$ (7, 8). With DL-[$\alpha$-$^3$H]alanine as substrate, a tritium ion and a pair of electrons are transferred to the C-5 of deaza FAD during reduction of the enzyme (7). The reduced holoenzyme, after isolation, can act upon pyruvate plus ammonia to give back [$\alpha$-$^3$H]alanine and completely detritiated E—deaza FAD. Reduced deaza FAD has two hy-drogens on C-5 which are potentially removable during reoxidation. The complete transfer of tritium to alanine points to a stereospecific transfer of the $\alpha$-proton between substrate and deaza FAD. The same hydrogen transfer to normal E—FAD seems a certainty, but it cannot be visualized because the N-5 proton of reduced FAD exchanges rapidly with the solvent. Cited above was evidence that one of the early steps in D-amino acid oxidase action is the abstraction from the substrate of its $\alpha$-hydrogen as a proton, mediated by a basic group in the active center. This proton must be the one which comes ultimately to rest on C-5 of deaza FAD, and presumably on N-5 of FAD (of the normal holoenzyme) before exchanging with solvent.

It is a matter of some interest that while deaza FAD is a structural analogue of FAD, the central ring of deaza FAD and deaza $FADH_2$ is at the same time an analogue of the pyridine and reduced pyridine rings of NAD and NADH, respectively (9). Any reduction of C-5 of deaza FAD is, therefore, at the same time a reduction at C-4 of the molecule in its character of NAD analo-gue. It follows that if the reduction at C-5 (of the FAD analogue) takes the form of two electrons and a separated proton, the same may be true for reduction at C-4 (of the NAD analogue). Such separation of two electrons and their attendant proton in the reduction of NAD by alcohol dehydro-genase is inquired into later in this chapter.

The mechanistic findings made on D-amino-acid oxidase are echoed in the similar findings made on other flavoenzymes. Both L-amino-acid oxidase [EC 1.4.3.2] and glucose oxidase [EC 1.1.3.4] (which catalyzes the oxidation of $\beta$-D-glucose to D-glucono-$\delta$-lactone)

(7) L. B. Hersh and M. S. Jorns, *JBC* 250, 8728–8734 (1975).
(8) J. Fisher, R. Spencer, and C. Walsh, *B* 15, 1054–1064 (1976).
(9) P. Hemmerich and M. S. Jorns, *FEBS Symp.* 29, 95–188 (1973); G. Blankenhorn, *B* 14, 3172–3176 (1975); P. Hemmerich, V. Massey, and H. Fenner, *FEBS lett.* 84, 5–21 (1977).

use the nitroethane anion as pseudosubstrate (10). Direct tritium transfer from [1-³H] glucose to glucose oxidase reconstituted with deaza FAD is also an established fact (8).

Lactate monooxygenase [EC 1.13.12.4], which catalyzes the oxidative decarboxylation of L-lactate

$$\underset{\underset{OH}{|}}{\overset{\overset{H}{|}}{CH_3-C-COO}} + O_2 \longrightarrow CH_3-COO + CO_2 + H_2O$$

can, under anaerobiasis, catalyze the elimination of hydrogen chloride from β-chloro-L-lactate.

$$\underset{\underset{OH}{|}}{\overset{\overset{H}{|}}{Cl-CH_2-C-COO}} \longrightarrow \overset{\overset{O}{\parallel}}{CH_3-C-COO} + Cl^- + H^+$$

With β-chloro-α-[³H]lactate as substrate, 30% of the tritium is recovered in the methyl group of pyruvate, testifying to a considerable shielding of the α-proton from the solvent after removal from the substrate (11). Besides, lactate monooxygenase reconstituted with deaza FMN is reduced by L-[α-³H]lactate with tritium transfer to enzyme-bound deaza FMN (12). In its action on glycolate

$$HOCH_2COO + O_2 \longrightarrow \overset{\overset{O}{\parallel}}{HC-COO} + H_2O_2$$

lactate monooxygenase acts through a covalent intermediate—an N(5)-glycolyl adduct—formed when the substrate (as carbanion) links to the N-5 position of the flavin (Fig. 3) (13).

The N-5 position of oxidized flavin is clearly an electrophile of considerable potency (14). Besides binding to the nitroalkane and glycolate anions, it is also the site of covalent attachment of sulfite ion during the oxidation of sulfite to adenylylsulfate by adenylylsulfate reductase [EC 1.8.99.2] (15).

(10) D. J. T. Porter and H. J. Bright, *JBC* 252, 4361–4370 (1977).
(11) C. Walsh, O. Lockridge, V. Massey, and R. H. Abeles, *JBC* 248, 7049–7054 (1973).
(12) B. A. Averill, A. Schonbrunn, R. H. Abeles, L. T. Weinstock, C. C. Cheng, J. Fisher, R. Spencer, and C. Walsh, *JBC* 250, 1603–1605 (1975).
(13) V. Massey and S. Ghisla, *Proc. 10th FEBS Meet.* (1975) pp. 145–158, North-Holland, Amsterdam; S. Ghisla, V. Massey, and Y. S. Choong, *JBC* 254, 10662–10669 (1979); V. Massey, S. Ghisla, and K. Kieschke, *JBC* 255, 2796–2806 (1980); S. Ghisla and V. Massey, *JBC* 255, 5688–5696 (1980).
(14) B. E. P. Swoboda and V. Massey, *JBC* 241, 3409–3416 (1966); F. Müller and V. Massey, *JBC* 244, 4007–4016 (1969); L. Hevesi and T. C. Bruice, *B* 12, 290–297 (1973).
(15) G. B. Michaels, J. T. Davidson, and H. D. Peck, Jr. *Biochem. Biophys. Res. Commun.* 39, 321–328 (1970); G. B. Michaels, J. T. Davidson, and H. D. Peck, Jr. in *Flavins and Flavoproteins*, H. Kamin, ed., University Park Press, Baltimore, 1971, pp. 555–580.

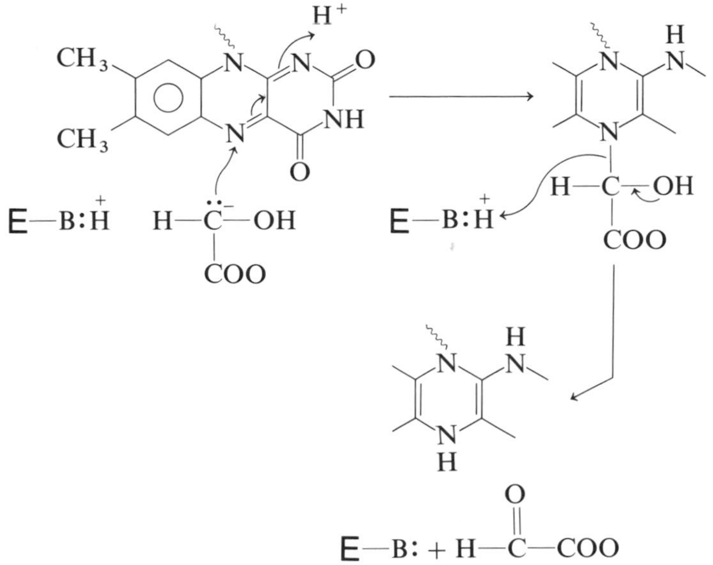

**Fig. 3.** The oxidation of glycolate by lactate monooxygenase via carbanion and covalent intermediates (13).

$$AMP^{2-} + SO_3^{2-} + 2\ Fe(CN)_6^{3-} \xrightarrow[Fe/S]{FAD} AMP-SO_3^{2-} + 2\ Fe(CN)_6^{4-}$$

While binding covalently at N-5 of flavin, sulfite reduces the flavin and raises itself to the sulfonyl level of oxidation. Eventually it transfers as a sulfonyl group to a phosphoryl oxygen of AMP (Fig. 4).

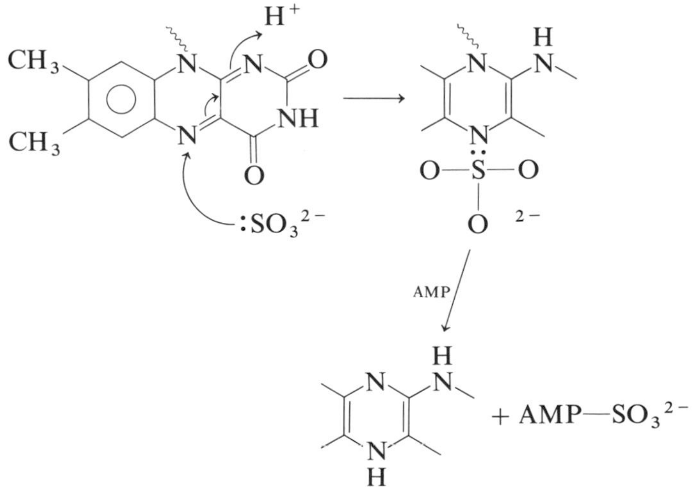

**Fig. 4.** Mechanism of the oxidation of sulfite to the oxidation level of sulfate by adenylyl-sulfate reductase.

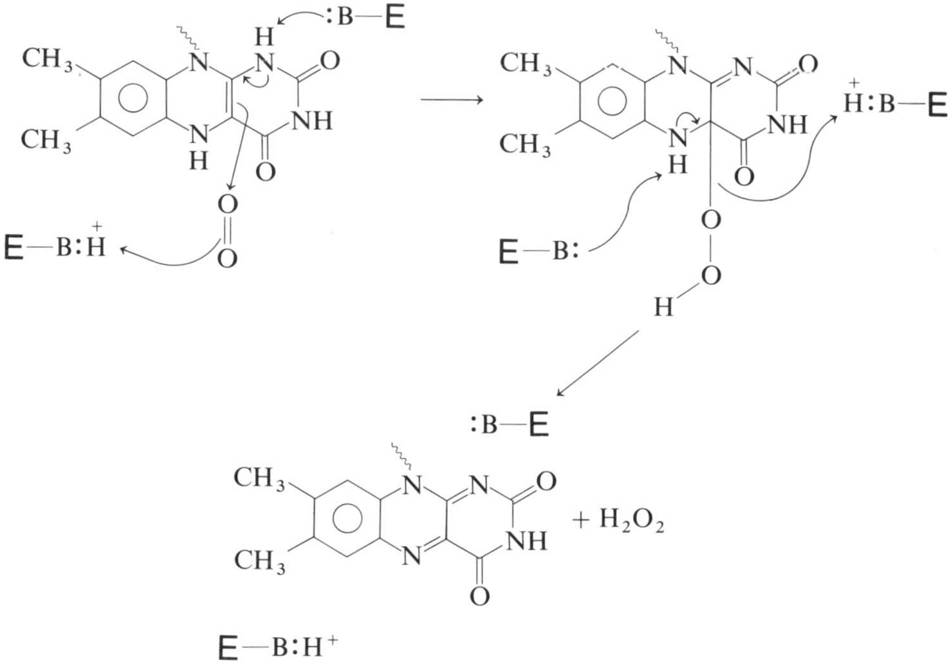

**Fig. 5.** Oxidation of reduced flavoenzyme by molecular oxygen through a C(4a)-peroxyflavin intermediate (17).

The reduced flavoenzymes formed in oxidase reactions complete their catalytic cycle by surrendering two electrons to molecular oxygen. Observations on several flavoenzymes (16–18) and on model flavins (19) have bred the conviction that reduced flavoenzyme forms a covalent adduct with molecular oxygen as a preliminary to electron release (Fig. 5). The flavin-oxygen adduct has the structure of a $C$(4a)-peroxyflavin (17). Upon elimination of hydrogen peroxide from the adduct the oxidized flavin is restored.

(16) V. Massey, F. Müller, R. Feldberg, M. Schumann, P. A. Sullivan, L. G. Howell, S. G. Mayhew, R. G. Matthews, and G. P. Foust, *JBC* 244, 3999–4006 (1969); T. Spector and V. Massey, *JBC* 247, 5632–5636 (1972); S. Strickland and V. Massey, *JBC* 248, 2953–2962 (1973); V. Massey, G. Palmer, and D. Ballou, in *Oxidases and Related Redox Systems*, T. E. King, H. S. Mason, and M. Morrison, eds., University Park Press, Baltimore, 1973, Vol. 1, pp. 25–43.
(17) B. Entsch, D. P. Ballou, and V. Massey *JBC* 251, 2550–2563 (1976); L. M. Schopfer and V. Massey, *JBC* 255, 5355–5363 (1980).
(18) B. Entsch, M. Husain, D. P. Ballou, V. Massey, and C. Walsh, *JBC* 255, 1420–1429 (1980); N. B. Beaty and D. P. Ballou, *JBC* 255, 3817–3819 (1980).
(19) P. Hemmerich, A. P. Bhaduri, G. Blankenhorn, M. Brüstlein, W. Haas, and W. R. Knappe, in *Oxidases and Related Redox Systems*, T. E. King, H. S. Mason, and M. Morrison, eds., University Park Press, Baltimore, 1973, Vol. 1, pp. 3–24; S. Ghisla, B. Entsch, V. Massey, and M. Husain, *EJB* 76, 139–148 (1977).

Such flavin-oxygen adducts mediate the action of monooxygenases (e.g., p-hydroxybenzoate hydroxylase [EC 1.14.13.2]) (17, 18) and of bacterial luciferase [EC 1.2.3.–] (20). In the reaction catalyzed by the latter enzyme, the reduced flavin (FMNH$_2$) reacts with oxygen to form an adduct which is actually isolable at low temperature. Spectroscopically, the luciferase-oxygen adduct is very similar to the peroxyflavin intermediate of the monooxygenases.

In xanthine oxidase [EC 1.2.3.2] FAD shares its prosthetic duties with a molybdenum atom, a reduced pterin (21), and a pair of iron–sulfur centers. It is virtually certain that during the catalytic reaction

$$\text{Xanthine} + \text{H} + O_2 + H_2O \longrightarrow \text{Uric acid} \cdot =O + H_2O_2$$

xanthine and molecular oxygen make covalent contact with the holoenzyme at separated locales within the active center (22, 23). The two locales are connected by an electron transport chain along which electrons are channeled from xanthine, coordinated to Mo at the input end, to molecular oxygen at the exit. The chain is composed of five redox elements ranged in order of their respective redox potentials (24).

$$e^- \longrightarrow \text{Mo(VI) and reduced pterin} \longrightarrow \text{Fe/S (1)}$$
$$\longrightarrow \text{Fe/S (2)} \longrightarrow \text{FAD} \longrightarrow O_2$$

Molecular oxygen reacts at the exit terminal with reduced FAD (25). Before it can insert electrons into molybdenum xanthine must, according to the current conception, combine covalently with a nucleophile of the enzyme (X in Fig. 6) (22, 26). Linkage of the enzymic nucleophile to C-8 of xanthine induces the departure of the proton from this position with release of two electrons to Mo(VI). Electron transfer is assumed to require the tight coordination of xanthine to molybdenum. The hydrogen on C-8 is transferred

(20) J. W. Hastings, C. Balny, C. Le Peuch, and P. Douzou, *PNAS* 70, 3468–3472 (1973); J. W. Hastings and C. Balny, *JBC* 250, 7288–7293 (1975); S. C. Tu, *B* 18, 5940–5945 (1979).
(21) J. L. Johnson, B. E. Hainline, and K. V. Rajagopalan, *JBC* 255, 1783–1786 (1980).
(22) J. S. Olson, D. P. Ballou, G. Palmer, and V. Massey, *JBC* 249, 4363–4382 (1974).
(23) R. C. Bray, *The Enzymes*, 3rd ed., P. D. Boyer, ed., Academic Press, New York, 1975, Vol. 12, pp. 299–419.
(24) R. C. Bray and P. F. Knowles, *Proc. R. Soc. London, Ser. A* 302, 351 (1968); R. C. Bray and J. C. Swann, *Struct. Bonding* 11, 107–144 (1972); D. Edmondson, D. Ballou, D. van Heuvelen, G. Palmer, and V. Massey *JBC* 248, 6135–6144 (1973).
(25) J. S. Olson, D. P. Ballou, G. Palmer, and V. Massey, *JBC* 249, 4350–4362 (1974).
(26) R. C. Bray, S. Gutteridge, D. A. Stotter, and S. J. Tanner, *Biochem. J.* 177, 357–360 (1979).

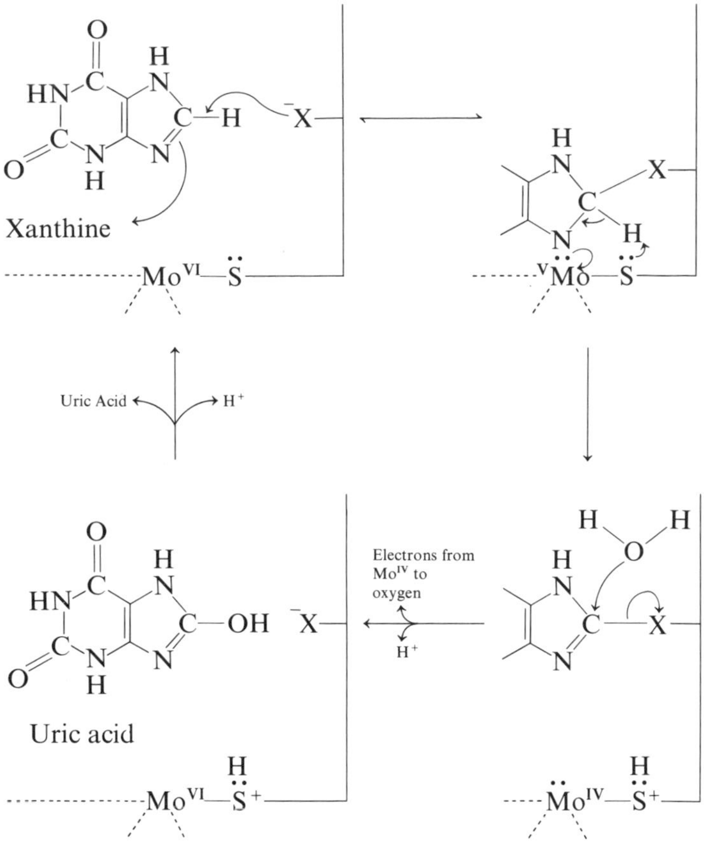

**Fig. 6.** A chemical representation of the oxidation of xanthine by xanthine oxidase [adapted from (22) and (26)].

as a proton to a basic group of the enzyme, perhaps to a sulfur liganded to molybdenum (27). Oxidation of substrate is completed when a water molecule displaces the enzymic nucleophile at C-8 to yield the enolic form of uric acid. The oxidation of xanthine to uric acid is seen thus to be a hydroxylative process, with water supplying the oxygen. To complete the catalytic cycle the electrons given up by xanthine are channeled from molybdenum to oxygen as indicated above.

Very like xanthine oxidase are the enzymes aldehyde oxidase [EC 1.2.3.1] (28) and xanthine dehydrogenase [EC 1.2.1.37] (29). Each possesses the

(27) E. I. Stiefel, *PNAS* 70, 988–992 (1973).

(28) K. V. Rajagopalan and P. Handler, in *Biological Oxidations*, T. P. Singer, ed., Interscience, New York, 1968, pp. 301–337; U. Branzoli and V. Massey, *JBC* 249, 4346–4349 (1974).

(29) H. Dalton, D. J. Lowe, R. T. Pawlik, and R. C. Bray, *Biochem. J.* 153, 287–295 (1976).

same set of prosthetic groups as xanthine oxidase and is thought to act by an analogous chemical mechanism.

## NAD Enzymes

These enzymes confront us with a curious problem in definition. Is NAD a prosthetic group or a substrate? Alcohol dehydrogenase [EC 1.1.1.1], for instance, catalyzes the reaction

$$\text{Ethanol} + \text{NAD}^+ \longleftrightarrow \text{acetaldehyde} + \text{NADH} + \text{H}^+ \qquad (2)$$

from which it appears that NAD is not, strictly speaking, a prosthetic group of the enzyme, but rather a substrate, like ethanol. If this is so, then alcohol dehydrogenase catalyzes the *direct* transfer of electrons between donor and acceptor. Yet it may after all be more accurate to think of NAD as a prosthetic group in such reactions, a conclusion which emerges from the considerations that follow.

Lactate-malate transhydrogenase [EC 1.1.99.7] catalyzes the following redox reaction:

$$\text{L-Lactate} + \text{oxaloacetate} \xrightarrow{\text{NAD}} \text{pyruvate} + \text{L-malate} \qquad (3)$$

NAD participates in the reaction as a tightly bound prosthetic group which undergoes alternate oxidation and reduction (30). The NAD cannot be separated from the enzyme without denaturing the protein. When the enzyme is treated with $[2\text{-}^3\text{H}]$lactate, in the absence of oxaloacetate, tritium is transferred to the holoenzyme with concurrent increase in absorbance at 345 nm and of fluorescence at 440 nm, both characteristic of reduced NAD. The radioactive enzyme can be separated from substrates by gel filtration. Upon treatment of the tritiated holoenzyme with oxaloacetate the optical manifestations are reversed. Ninety-eight percent of the radioactivity is lost from the holoenzyme and is recovered in the malate. It is clear that in the action of lactate-malate transhydrogenase a pair of electrons and a proton are passed from the donor to the holoenzyme, and then on to the acceptor.

Reverting to alcohol dehydrogenase, we recall that this enzyme has a quite broad specificity, being active on a diversity of substrates. It is found that the rate of ethanol oxidation by the horse liver enzyme is much enhanced in the presence of lactaldehyde.

$$\text{CH}_3\text{CH}_2\text{OH} + \text{CH}_3 \overset{\overset{\text{H}}{|}}{\underset{\underset{\text{OH}}{|}}{\text{C}}} \overset{\overset{\text{O}}{\|}}{\text{CH}} \xrightarrow{\text{NAD}} \text{CH}_3 \overset{\overset{\text{O}}{\|}}{\text{CH}} + \text{CH}_3 \overset{\overset{\text{H}}{|}}{\underset{\underset{\text{OH}}{|}}{\text{C}}} \text{CH}_2\text{OH} \qquad (4)$$

---

(30) S. H. G. Allen, *JBC* 241, 5266–5275 (1966); S. H. G. Allen and J. R. Patil, *JBC* 247, 909–916 (1972).

Relative to the rate of reaction 2, a 7- to 22-fold enhancement, depending on the aldehyde, can be achieved (31, 32). And only a *catalytic* amount of NAD is needed. In the coupled reaction, the NADH formed in the oxidation of ethanol seems not to dissociate from the enzyme, but is directly reoxidized by the aldehyde. Operating in this fashion, alcohol dehydrogenase mimics the action of lactate-malate transhydrogenase (Eq. 3). In the presence of lactaldehyde the oxidation of ethanol even follows ping-pong kinetics under some conditions (32). Besides, a strong deuterium isotope effect on the velocity is observed, there being no such effect in the absence of acceptor aldehyde. Evidently, when the rate-limiting dissociation of NADH is no longer a factor (reaction 2), the H-transfer step becomes rate limiting (reaction 4). It seems, then, that alcohol dehydrogenase—and perhaps other dehydrogenases too—can operate more efficiently when harnessed into a coupled reaction, as exemplified by reaction 4. In this context, reaction 2 is but a partial reaction, which depends on the dissociability of the prosthetic group in the absence of an alternative electron acceptor.

Further in keeping with this view is the aldehyde dismutase activity of horse liver alcohol dehydrogenase. Formaldehyde (33), acetaldehyde, and butyraldehyde (34) undergo dismutation to the corresponding alcohols and acids in a reaction catalyzed by the enzyme and a *catalytic* amount of NAD. In these reactions one molecule of aldehyde is the electron donor and a second molecule is the acceptor. It is generally agreed that the dismutase activity of alcohol dehydrogenase accords best with the view that NAD shares in the reaction as a prosthetic group (33, 34). An analogous dismutation of glyoxylate to oxalate and glycolate is catalyzed by lactate dehydrogenase [EC 1.1.1.27] and a *catalytic* amount of NAD or NADH (35).

Apart from their specificities, alcohol dehydrogenase, lactate dehydrogenase, and lactate-malate transhydrogenase are clearly alike in the chemistry of their action on the hydroxyl and carbonyl functions of substrates, and hardly different in this respect from the other NAD enzymes listed by the Enzyme Commission. And since the three illustrated dehydrogenases can use NAD catalytically—as a prosthetic group—in redox reactions, I hazard the guess that many other NAD enzymes can do the same.

Horse liver alcohol dehydrogenase is a metalloenzyme bearing two zinc ions in each of its two identical subunits (36). One of the zinc ions is so bound by protein ligands as to be inaccessible to chelating agents, and

(31) N. K. Gupta and W. G. Robinson, *BBA* 118, 431–434 (1966).

(32) H. Gershman and R. H. Abeles, *Arch. Biochem. Biophys.* 154, 659–674 (1973).

(33) R. H. Abeles and H. A. Lee, *JBC* 235, 1499–1503 (1960); N. K. Gupta, *Arch. Biochem. Biophys.* 141, 632–640 (1970).

(34) K. Dalziel and F. M. Dickenson, *Nature (London)* 206, 255–257 (1965).

(35) M. Romano and M. Cerra, *BBA* 177, 421–426 (1969); R. J. S. Duncan and K. F. Tipton, *EJB* 11, 58–61 (1969); W. A. Warren, *JBC* 245, 1675–1681 (1970); R. J. S. Duncan, *Arch. Biochem. Biophys.* 201, 128–136 (1980).

(36) C. I. Brändén, H. Eklund, B. Nordström, T. Boiwe, G. Söderlund, E. Zeppezauer, I. Ohlsson, and Å. Åkeson *PNAS* 70, 2439–2442 (1973).

its function is regarded as structural. But the second zinc ion of the subunit is fixed in the bottom of the active site cleft where it is reactive with the chelating agent, 1,10-phenanthroline, as well as with substrate. One such substrate is the chromophoric *trans*-4-*N*,*N*-dimethylaminocinnamaldehyde

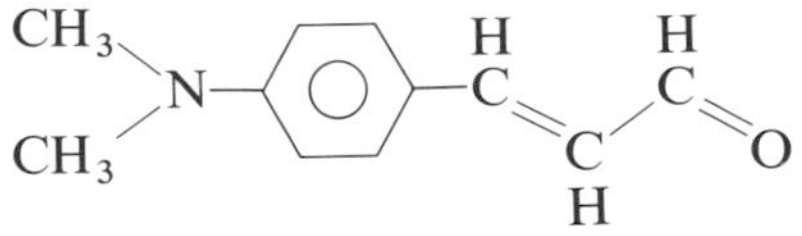

which is reducible by alcohol dehydrogenase to the corresponding primary alcohol. But prior to the redox stage of the reaction, the chromophore links with the enzyme–NADH-Zn complex to form a chemical intermediate which, from its chemical, spectroscopic, and kinetic properties, is identified as a *coordination complex formed between the zinc ion and the carbonyl oxygen of substrate* (37). As a prosthetic group of the enzyme, zinc ion is thought to act the part of a Lewis acid, polarizing the carbonyl function of the substrate to enhance its electrophilicity. The substrate molecule is gripped by the zinc ion in such wise as to bring the carbonyl carbon within van der Waals distance of C-4 of NADH, preparatory to the passage of electrons between them (38).

Some tentative new light was shone on the obscure process of electron transfer by alcohol dehydrogenase through studies made on yet another substrate, the intense chromophore, *p*-nitroso-*N*,*N*-dimethylaniline, **6**. With the horse liver enzyme this compound has a turnover number 2.4-fold greater than that of acetaldehyde (39). The nitroso compound is irreversibly reduced to the corresponding hydroxylamine, **7**, in presumed accordance with the stoichiometry of Eq. 5 (40).

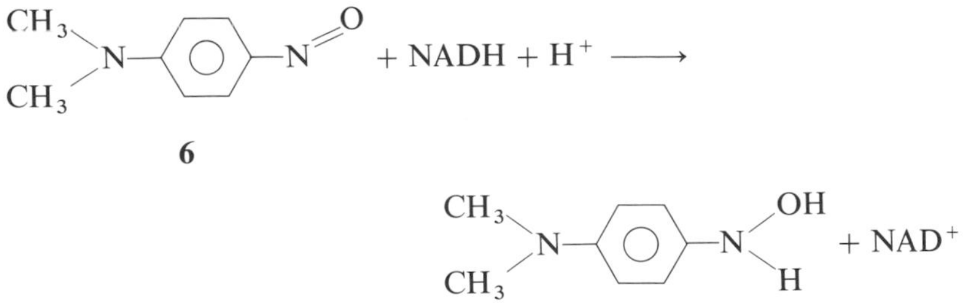

**6**

$$(5)$$

**7**

(37) M. F. Dunn and J. S. Hutchison, *B* 12, 4882–4892 (1973); C. T. Angelis, M. F. Dunn, D. C. Muchmore, and R. M. Wing, *B* 16, 2992–2931 (1977); H. Dietrich, W. Maret, L. Wallén, and M. Zeppezauer, *EJB* 100, 267–270 (1979); S. A. Evans and J. D. Shore, *JBC* 255, 1509–1514 (1980).

(38) M. F. Dunn, J.-F. Biellmann, and G. Branlant, *B* 14, 3176–3182 (1975).

(39) M. F. Dunn and S. A. Bernhard, *B* 10, 4569  4575 (1971).

(40) The hydroxylamine (structure **7**) undergoes further nonenzymic transformations unconnected with the enzymic redox process [S. C. Koerber, P. Schack, A. M. J. Au, and M. F. Dunn, *B* 19, 731–738 (1980)].

Rapid-kinetic measurements of the reaction, made under a variety of conditions, reveals that the decreases of optical density at 440 nm (due to nitroso reduction) and at 330 nm (due to NADH oxidation) are *not* concurrent processes. The disappearance of the 440 nm absorbance is essentially complete before noticeable change occurs at 330 nm; that is to say, reduction of the nitroso group is apparently complete before the oxidation of NADH has begun. From these observations is inferred the transient existence of an obligatory *covalent adduct* between the nitroso substrate and NADH, having an extinction coefficient at 330 nm similar to that of NADH itself (41). A reasonable structure for the adduct is shown as **8** (42), which collapses to NAD and the hydroxylamine.

8                                                                                              (6)

7

The hydroxylamine suffers further chemical changes not indicated here (40). It follows from these considerations that alcohol dehydrogenase may catalyze electron transfer not as a hydride ion but as a pair of electrons and a separated proton.

   If covalent linkage of substrate with the coenzyme of alcohol dehydrogenase is a factor in electron transfer, then the reversible oxidation of a primary alcohol to aldehyde can be represented by the sequence of stages in Fig. 7, which attempts to portray only the bare chemistry of the reaction. After coordination of the alcohol oxygen to the zinc ion in the active center (stages $2 \rightarrow 3$), removal of a proton from the $\alpha$-carbon of the alcohol (stages $3 \rightarrow 4$) generates a carbanion (stage 4) which joins covalently at C-4

---

(41) These optical observations were later confirmed in another laboratory by rapid-scanning spectrophotometry, but without interpretative comment [C. H. Suelter, R. B. Coolen, N. Papadakis, and J. L. Dye, *Anal. Biochem.* 69, 155–163 (1975)].

(42) Structure **8** will be recognized as the product formed by the (hypothetical) addition of the substituted hydroxylamine (structure **7**) to NAD in (hypothetical) reversal of reaction 6. Such addition of nitrogen nucleophiles to NAD has long been known [J. van Eys, *JBC* 233, 1203–1210 (1958)]. Here I suggest its use by the enzyme as a means of electron transfer.

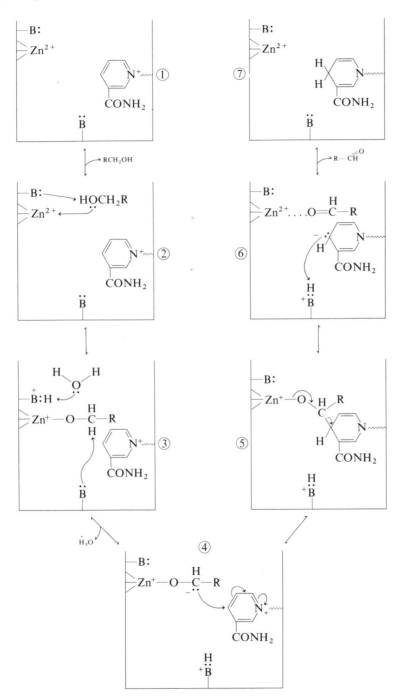

**Fig. 7.** A proposed mechanism of the alcohol dehydrogenase reaction showing the mode of electron transfer through covalent union of substrate with holoenzyme.

of the positively charged pyridine ring of NAD (stage 5) (43). The inter-mediate so formed fragments into aldehyde and the carbanionic form of NADH (stage 6). The latter is deemed to possess some inherent stability because the C-4 protons have some intrinsic acidity (44). To complete the reaction, carbanionic NADH retrieves the proton which was earlier removed from the substrate (stage 6 → 7). The enzyme base shown near the opening of the active site cleft can exchange protons easily with the medium. But the base located in the innermost region of the active site is totally seques-tered from the medium, and is the point of transfer for the unexchangeable proton which moves reversibly between substrate and NAD (45).

The elements of a hydride ion are seen on this view to be transferred as a pair of electrons and a separated proton. The formation of a covalent intermediate from the alcoholic substrate and the holoenzyme requires prior deprotonation of the substrate with ultimate retrieval of the removed proton by reduced NAD. The detailed mode of electron transfer between substrate and pyridine nucleotide proposed in Fig. 7 rests admittedly on slim experimental grounds; namely, the rapid-kinetic measurements reported in (39) and (41) and described above. Yet this way of transferring electrons conforms closely with the analogous mechanism for the flavoenzymes, D-amino-acid oxidase and lactate monooxygenase, wherein electron transfer is conducted through covalent union of substrate with prosthetic flavin. It conforms, too, with other "hydride" transfers well known to enzymology. In the intramolecular redox reaction catalyzed by glucose-6-P isomerase [EC 5.3.1.9]

Glucose-6-P                          Fructose-6-P

---

(43) The addition of carbanions to the pyridine ring of NAD was observed years ago [R. M. Burton and N. O. Kaplan, *JBC* 206, 283–297 (1954); R. M. Burton, A. San Pietro, and N. O. Kaplan, *Arch. Biochem. Biophys.* 70, 87–106 (1957)]. The only new element here is its use as a device for electron transfer.

(44) F. Fowler, *JACS* 94, 5926–5927 (1972).

(45) F. H. Westheimer, H. F. Fisher, E. E. Conn, and B. Vennesland, *JACS* 73, 2403 (1951).

a proton and a pair of electrons move reversibly between C-1 and C-2 of the sugar via an ene-diol mechanism, wherein the proton traverses a pathway which includes a basic group in the active center of the enzyme (46). The same may be said of triosephosphate isomerase [EC 5.3.1.1] (47). Glyoxalase I [EC 4.4.1.5] also catalyzes an intramolecular redox reaction.

$$CH_3-\overset{\overset{O}{\|}}{C}-\overset{\overset{O}{\|}}{C}-H \ + \ HS-G \ \longrightarrow \ CH_3-\overset{\overset{OH}{|}}{\underset{H}{C}}-\overset{\overset{O}{\|}}{C}-S-G$$

   Methylglyoxal    Glutathione       S-Lactylglutathione

Here, too, a "hydride ion" transfer is the essence of the reaction. But the proton moves by way of a basic group in a protected (from solvent) region of the active center (48). In these intramolecular redox reactions it is clear that the proton, which has only to shift to the adjacent carbon atom, is forced by the enzyme to take the long way round, to disconnect itself from its electron pair, to go first to a basic group of the enzyme, and only then to rejoin an electron pair on the atom adjacent to the atom of origin. Whether enzymic redox reactions be *intra*molecular (sugar isomerases) or *inter*molecular (flavoenzymes), the transfer of "hydride ion" seems always to demand a separation of the migrating proton from its electrons. On grounds of chemical analogy, the same principle ought to govern the action of NAD enzymes.

## Copper Enzymes

Among the most extensively studied metalloenzymes is the copper- and zinc-containing superoxide dismutase [EC 1.15.1.1] of eukaryotic cytosols, which catalyzes the disproportionation of two superoxide radical anions into a molecule each of oxygen and hydrogen peroxide (49).

$$O_2^- + O_2^- + 2\,H^+ \longrightarrow O_2 + H_2O_2$$

Each subunit of the enzyme contains one copper and one zinc ion, but only the copper participates in the chemistry of the reaction, oscillating between

(46) I. A. Rose and E. L. O'Connell, *JBC* 236, 3086–3091 (1961).
(47) J. M. Herlihy, S. G. Maister, W. J. Albery, and J. R. Knowles, *B* 15, 5601–5607 (1976).
(48) S. S. Hall, A. M. Doweyko, and F. Jordan, *JACS* 98, 7460–7461 (1976) *idem*., *JACS* 100, 5934–5939 (1978).
(49) I. Fridovich, *Adv. Enzymol.* 41, 35–97 (1974); I. Fridovich, *Ann. Rev. Biochem.* 44, 147–159 (1975).

the $Cu^{2+}$ and $Cu^{+}$ valence states (50). Cyanide ion is a competitive inhibitor of superoxide dismutase (51), forming a coordinate bond to copper with its carbon atom (52, 53). Superoxide ion binds to copper at its water co-ordination site, which is also where the inhibition by cyanide takes place (51). Hydrogen peroxide, a product of the dismutation, also coordinates directly to the copper of the enzyme (52, 54). It seems therefore safe to say that all electron transfers between copper and the oxygen atoms of substrates occur when the latter are directly coordinated to the metal, and the mechanism of superoxide dismutase is as follows (55):

$$ECu^{2+} + O_2^{-} \longrightarrow ECu^{2+}{-}O_2^{-}$$

$$ECu^{2+}{-}O_2^{-} \longrightarrow ECu^{+} + O_2$$

$$ECu^{+} + O_2^{-} \longrightarrow ECu^{+}{-}O_2^{-}$$

$$ECu^{+}{-}O_2^{-} + 2\,H^{+} \longrightarrow ECu^{2+} + H_2O_2$$

Laccase [EC 1.10.3.2], one of the "blue oxidases," is a copper-containing enzyme catalyzing the oxidation of aryl diamines and diphenols by a process in which both atoms of molecular oxygen are reduced to water.

Like xanthine oxidase, laccase is an asymmetric enzyme in the sense that the reducing substrate and the oxygen react at different regions within the active center (56). For prosthetic groups, laccase contains only copper atoms, four per molecule, distributed in three distinct sites with distinctive spectral and chemical properties. The copper atoms of type 1 (absorbing at 614 nm) and type 2 (no absorption) are paramagnetic; and laccase possesses one atom in each of these categories. Type 3 copper is diamagnetic and absorbs at 330 nm. A molecule of enzyme has two type 3 copper atoms, which together function as a two-electron acceptor. The two type 3 atoms are

(50) G. Rotilio, L. Morpurgo, L. Calabrese, and B. Mondovi, *BBA* 302, 229–235 (1973); D. Klug-Roth, I. Fridovich, and J. Rabani, *JACS* 95, 2786–2790 (1973).
(51) A. Rigo, P. Viglino, and G. Rotilio, *Biochem. Biophys. Res. Commun.* 63, 1013–1018 (1975).
(52) G. Rotilio, L. Morpurgo, C. Giovagnoli, L. Calabrese, and B. Mondovi, *B* 11, 2187–2192 (1972).
(53) P. H. Haffner and J. E. Coleman, *JBC* 248, 6626–6629 (1973).
(54) G. Rotilio, L. Calabrese, F. Bossa, D. Barra, A. Finazzi Agro, and B. Mondovi, *B* 11, 2182–2187 (1972).
(55) D. Klug, J. Rabani, and I Fridovich, *JBC* 247, 4839–4842 (1972).
(56) B. G. Malmström, in *Symmetry and Function of Biological Systems at the Macromolecular Level*, A. Engström and B. Strandberg, eds., Almquist and Wiksell, Uppsala, 1969, pp. 153–163.

spin-coupled; and, as a binuclear pair, are considered to be the site of oxygen ligation (57).

The reduction of oxygen to water by laccase is a four-electron transaction which occurs maximally after all the $Cu^{2+}$ ions in the enzyme are reduced (58). Reduction of laccase begins with the formation of an inner-sphere complex between phenolate ion of the reducing substrate and type 1 $Cu^{2+}$, which serves as the entry port for electrons into the enzyme (57, 59). A succession of four one-electron transfers from two molecules of substrate follows, with the intermediary appearance of free radical forms of the substrate (60). From the site of entry the electrons spread over the various copper centers, whereupon the reduction of oxygen can proceed. Passage of electrons to dioxygen is thought to begin with a two-electron transfer, the dioxygen being reduced thereby to the oxidation level of peroxide as an intermediary stage (61, 62). Such an intermediate has been identified optically and magnetically (63). It has properties spectrally similar to those of a hydrogen peroxide complex with native laccase (64). It parallels, moreover, the findings made on the copper-containing protein hemocyanin, the oxygenation of which entails the reduction of dioxygen to peroxide coordinated to the diamagnetic, binuclear copper pair in this protein (65).

$$Cu^+ \cdots\cdots\cdots Cu^+ + O_2 \longrightarrow Cu^{2+}-O^--O^--Cu^{2+}$$

A reasonable conjecture, therefore, is that oxygen reacts with reduced laccase at its diamagnetic pair of type 3 copper ions (57, 62). A subsequent transfer of two more electrons completes the reduction of the peroxide intermediate to water (66).

Ceruloplasmin [EC 1.16.3.1], the plasma enzyme which catalyzes the oxidation of ferrous ion to ferric ion,

$$4\,Fe^{2+} + 4\,H^+ + O_2 \longrightarrow 4\,Fe^{3+} + 2\,H_2O$$

(57) R. Malkin and B. G. Malmström, *Adv. Enzymol.* 33, 177–244 (1970).

(58) B. G. Malmström, A. Finazzi Agro, and E. Antonini, *EJB* 9, 383–391 (1969).

(59) B. Reinhammar and Y. Oda, *J. Inorg. Biochem.* 11, 115–127 (1979).

(60) L. Broman, B. G. Malmström, R. Aasa, and T. Vänngård, *BBA* 75, 365–376 (1963).

(61) R. Malkin, B. G. Malmström, and T. Vänngård, *EJB* 10, 324–329 (1969).

(62) B. Reinhammar, *BBA* 275, 245–259 (1972); O. Farver, M. Goldberg, D. Lancet and I. Pecht, *Biochem. Biophys. Res. Commun.* 73, 494–500 (1976); M. Goldberg, O. Farver, and I. Pecht, *JBC* 255, 7353–7361 (1980).

(63) L. E. Andréasson, R. Brändén, B. G. Malmström, and T. Vänngård, *FEBS lett.* 32, 187–189 (1973).

(64) R. Brändén, B. G. Malmström, and T. Vänngård, *EJB* 18, 238–241 (1971); O. Farver and I. Pecht, *FEBS Lett.* 108, 436–438 (1979).

(65) J. S. Loehr, T. B. Freedman, and T. M. Loehr, *Biochem. Biophys. Res. Commun.* 56, 510–515 (1974).

(66) R. Aasa, R. Brändén, J. Deinum, B. G. Malmström, B. Reinhammar, and T. Vänngård, *FEBS lett.* 61, 115–119 (1976); *idem.*, *Biochem. Biophys. Res. Commun.* 70, 1204–1209 (1976).

is also a "blue oxidase" with the same three types of copper as are found in laccase (67). Upon oxidation of reduced ceruloplasmin with oxygen, an intermediate is formed with optical properties very similar to those of the corresponding peroxide intermediate formed in the oxygenation of reduced laccase (68). Parallel observations are reported for ascorbate oxidase [EC 1.10.3.3], which is also a "blue oxidase" (69).

## Copper-Heme Enzymes

Cytochrome $c$ oxidase [EC 1.9.3.1], whose activity is so vital to aerobic life, conveys electrons from reduced cytochrome $c$ to molecular oxygen.

$$4 \text{ Ferrocytochrome } c + O_2 + 4 H^+ \longrightarrow 4 \text{ ferricytochrome } c + 2 H_2O$$

As with laccase and ceruloplasmin, both atoms of the oxygen molecule are reduced to water in a four-electron transaction. Electrons enter the enzyme one at a time and are delivered up to oxygen in one two-electron transfer followed by two one-electron transfers (70). Cytochrome $c$ oxidase has four redox centers: cytochromes $a$ and $a_3$ and a pair of copper atoms ($Cu_A$ and $Cu_B$). Cytochrome $a_3$ and $Cu_B$ are thought to form a coupled binuclear center, like the two type 3 copper ions of the "blue oxidases," which constitutes the dioxygen-reducing unit. Electrons probably enter the enzyme at cytochrome $a$. They make their way via $Cu_A$ to the cytochrome $a_3$–$Cu_B$ unit, which after a two-electron reduction, can react with dioxygen, reducing the latter to the level of peroxide. Compounds of dioxygen with cytochrome $a_3$ have been detected (71, 72, 73). Dioxygen coordinates to the iron of cytochrome $a_3$ much as it does to hemoglobin (74), or as carbon monoxide does to both of these hemes (71). Indications are that a peroxide bridge may then form between cytochrome $a_3$ and $Cu_B$ (72). Two further one-electron transfers reduces the peroxo intermediate to water. Though many mysteries still surround the action of cytochrome $c$ oxidase it is nonetheless clear that at

(67) R. J. Carrico, B. G. Malmström, and T. Vänngård, *EJB* 22, 127–133 (1971).
(68) T. Manabe, N. Manabe, K. Hiromi, and H. Hatano, *FEBS lett.* 23, 268–270 (1972); T. Manabe, H. Hatano, and K. Hiromi, *J. Biochem. (Tokyo)* 73, 1169–1174 (1973).
(69) R. E. Strothkamp and C. R. Dawson, *Biochem. Biophys. Res. Commun.* 85, 655–661 (1978).
(70) B. G. Malmström, *BBA* 549, 281–303 (1979).
(71) B. Chance, C. Saronio, and J. S. Leigh, Jr., *JBC* 250, 9226–9237 (1975).
(72) G. M. Clore and E. M. Chance, *Biochem. J.* 177, 613–621 (1979).
(73) G. M. Clore, L. E. Andréasson, B. Karlsson, R. Aasa, and B. G. Malmström, *Biochem. J.* 185, 139–154, 155–167 (1980).
(74) W. S. Caughey, C. H. Barlow, J. C. Maxwell, J. A. Volpe, and W. J. Wallace, *Ann. N.Y. Acad. Sci.* 244, 1–8 (1975).

least one substrate (oxygen) forms a coordinate (covalent) bond to the enzyme during catalysis.

## Heme Enzymes

In reactions catalyzed by the heme enzymes catalase [EC 1.11.1.6], peroxidase [EC 1.11.1.7], and chloroperoxidase [EC 1.11.1.10], the oxidizing equivalents furnished by the oxidant (hydrogen peroxide) are conveyed to the reductant by an enzyme–substrate intermediate known for many years as Compound I (75). In the special case of catalase the reductant is a second molecule of hydrogen peroxide, and molecular oxygen and water are the products.

$$H_2O_2 + enzyme \longrightarrow Compound\ I + H_2O$$

$$Compound\ I + H_2O_2 \longrightarrow enzyme + O_2 + H_2O$$

$$Sum:\quad 2\ H_2O_2 \longrightarrow O_2 + 2\ H_2O$$

A combination of incisive isotope experiments on chloroperoxidase (76, 77) and of titration studies on horse radish peroxidase (78) point to a structure for Compound I in which one oxygen atom from the oxidant (hydrogen peroxide or a monosubstituted hydrogen peroxide) coordinates axially to the ferric atom of the heme (77). How a molecule of hydrogen peroxide might react with the heme iron of these enzymes to transform the two oxidizing equivalents of peroxide into those of Compound I is shown in Fig. 8. For

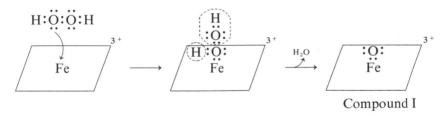

Compound I

**Fig. 8.** How Compound I and water might be formed from the reaction of enzyme with hydrogen peroxide.

(75) H. Theorell, *Enzymologia* 10, 250–252 (1941); B. Chance, *Arch. Biochem. Biophys.* 22, 224–252 (1949).
(76) L. P. Hager, D. L. Doubek, R. M. Silverstein, J. H. Hargis, and J. C. Martin, *JACS* 94, 4364–4366 (1972).
(77) L. P. Hager, D. Doubek, and P. Hollenberg, in *Molecular Basis of Electron Transport*, J. Schultz and B. F. Cameron, eds., Academic Press, New York, 1972, pp. 347–364.
(78) G. R. Schonbrunn and S. Lo, *JBC* 247, 3353–3360 (1972).

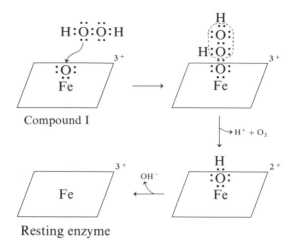

**Fig. 9.** How Compound I of catalase might react with a (second) molecule of hydrogen peroxide to form oxygen and water.

simplicity Compound I is represented here as an oxene, since it is formally the product of the addition of a hydroxonium ion ($H:\overset{..}{\underset{..}{O}}^+$ with an open sextet of electrons) to the ferric iron of heme, with ultimate loss of the proton. The electronic structure of Compound I probably differs from the one shown here (79). As an oxene, Compound I is a potent oxidant. Accordingly, Compound I (of catalase) converts a second molecule of hydrogen peroxide into molecular oxygen as depicted in Fig. 9. The Compound I of peroxidase, however, is slow to react with a second molecule of hydrogen peroxide, preferring instead to act upon one of a number of oxidizable substrates, including phenols, aminophenols, amines, leucodyes, ascorbic acid, and amino acids.

Chloroperoxidase has as its prime activity the chlorination of organic compounds using chloride ion as the source of chlorine and hydrogen peroxide as oxidant (80).

$$RH + Cl^- + H_2O_2 \longrightarrow RCl + H_2O + OH^-$$

(79) The oxidizing equivalents (i.e., the absence of electrons) shown residing on the oxygen atom of Compound I (Fig. 8) are probably delocalized over the iron atom and the porphyrin ring of the heme, resulting in contributions from tetravalent iron and a $\pi$-cation radical to the electronic structure of Compound I [D. Dolphin and R. H. Felton, *Acc. Chem. Res.* 7, 26–32 (1974)].

(80) L. P. Hager, D. R. Morris, F. S. Brown, and H. Eberwein, *JBC* 241, 1769–1777 (1966).

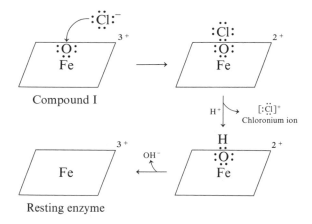

Fig. 10. How Compound I of chloroperoxidase might oxidize chloride ion.

Here again, Compound I is the immediate source of oxidizing equivalents to raise the oxidation level of chloride to that of chloronium ion which, in a formal sense, may be said to supply the requisite chlorinating power (Fig. 10) (81).

Camphor 5-monooxygenase [EC 1.14.15.1] is a cytochrome $P$-450 enzyme which uses dioxygen (and a pair of electrons) in the hydroxylation of a methylene carbon of camphor to give 5-hydroxycamphor.

$+ O_2 + 2$ reduced putidaredoxin $+ 2 H^+ \longrightarrow$

Camphor

HO $+ H_2O + 2$ oxidized putidaredoxin
H

5-Hydroxycamphor

The chemical mechanism of this and the other cytochrome $P$-450 enzymes differs materially from that of catalase and the peroxidases. But there is a resemblance to these latter enzymes in the important sense that anaerobically

(81) R. Chiang, T. Rand-Meir, R. Makino, and L. P. Hager, *JBC* 251, 6340–6346 (1976).

reduced cytochrome *P*-450 combines with dioxygen to a reactive oxygenated intermediate (82). Mössbauer spectra of this intermediate resemble closely that of oxyhemoglobin (83), in which dioxygen is an axial ligand of the heme iron (84). Further reduction is thought to induce the heterolytic rupture of the oxygen–oxygen bond in the dioxygen-heme–enzyme intermediate, generating thus a monooxygen species equivalent to Compound I of the peroxidases (85). As an oxene, the form of cytochrome *P*-450 equivalent to Compound I inserts an oxygen atom into the C—H bond of camphor. Analogous oxygen insertions into their respective substrates are presumably catalyzed by the many other cytochrome *P*-450 enzymes which have been discovered.


# Disulfide Enzymes

In place of, or in addition to, the more familiar prosthetic groups, some oxidoreductases make use of a built-in, redox-active disulfide function as a focal point of catalysis. One such enzyme is glutathione-cystine transhydrogenase [EC 1.8.4.4], which catalyzes the oxidation of glutathione by cystine.

$$2 \text{ Glutathione} + \text{cystine} \longleftrightarrow \text{oxidized glutathione} + 2 \text{ cysteine}$$

In its mode of action the enzyme has more the character of a thioltransferase than of a transhydrogenase (86). Thus, enzyme isolated from rat liver has glutathione covalently fixed to it (87). In its oxidized (disulfide) form, the

(82) Y. Oshimura, V. Ullrich, and J. A. Peterson, *Biochem. Biophys. Res. Commun.* 42, 140–146 (1971); J. A. Peterson, Y. Oshimura, J. Baron, and R. W. Estabrook, in *Oxidases and Related Redox Systems*, T. E. King, H. S. Mason, and M. Morrison, eds., University Park Press, Baltimore, 1973, Vol. 2, pp. 565–577; L. Eisenstein, P. Debey, and P. Douzou, *Biochem. Biophys. Res. Commun.* 77, 1377–1383 (1977); C. Bonfils, P. Debey, and P. Maurel, *Biochem. Biophys. Res. Commun.* 88, 1301–1307 (1979); C. Larroque and J. E. van Lier, *FEBS lett.* 115, 175–177 (1980).
(83) M. Sharrock, E. Münck, P. G. Debrunner, V. Marshall, J. D. Lipscomb, and I. C. Gunsalus, *B* 12, 258–265 (1973).
(84) C. H. Barlow, J. C. Maxwell, W. J. Wallace, and W. S. Caughey, *Biochem. Biophys. Res. Commun.* 55, 91–95 (1973).
(85) E. G. Hrycay, J.-Å. Gustafsson, M. Ingelman-Sundberg, and L. Ernster, *EJB* 61, 43–52 (1976); F. Lichtenberger, W. Nastainczyk, and V. Ullrich, *Biochem. Biophys. Res. Commun.* 70, 939–946 (1976). But the notion that cytochrome *P*-450 activity is mediated by a Compound I-like intermediate has come under some criticism [R. C. Blake and M. J. Coon, *JBC* 255, 4100–4111 (1980)].
(86) P. Askelöf, K. Axelsson, S. Eriksson, and B. Mannervik, *FEBS lett.* 38, 263–267 (1974).
(87) S. Eriksson, P. Askelöf, K. Axelsson, and B. Mannervik, *Acta Chem. Scand.* B28, 931–936 (1974).

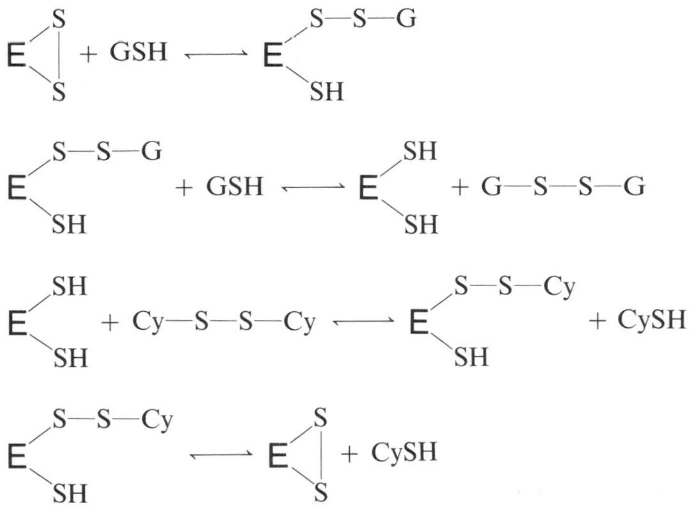

$$E\begin{smallmatrix}S\\|\\S\end{smallmatrix} + GSH \longrightarrow E\begin{smallmatrix}S-S-G\\ \\SH\end{smallmatrix}$$

$$E\begin{smallmatrix}S-S-G\\ \\SH\end{smallmatrix} + GSH \longrightarrow E\begin{smallmatrix}SH\\ \\SH\end{smallmatrix} + G-S-S-G$$

$$E\begin{smallmatrix}SH\\ \\SH\end{smallmatrix} + Cy-S-S-Cy \longrightarrow E\begin{smallmatrix}S-S-Cy\\ \\SH\end{smallmatrix} + CySH$$

$$E\begin{smallmatrix}S-S-Cy\\ \\SH\end{smallmatrix} \longrightarrow E\begin{smallmatrix}S\\|\\S\end{smallmatrix} + CySH$$

**Fig. 11.** Mechanism of the thioltransferase action of glutathione-cystine transhydrogenase.

enzyme reacts with [$^{35}$S]glutathione to give the glutathione-containing enzyme, which can be isolated by gel and ion exchange chromatography. The enzyme probably operates by the chemical mechanism shown in Fig. 11.

A disulfide "prosthetic" group, acting in conjunction with FAD, also figures in the reduction of oxidized glutathione, as catalyzed by glutathione reductase [EC 1.6.4.2].

Oxidized glutathione + NADPH + H$^+$ $\longrightarrow$ 2 glutathione + NADP$^+$

Electrons supplied to the enzyme by NADPH are first shared between FAD and the adjacent disulfide function in the active center (Fig. 12) (88). Ultimately these electrons are delivered to oxidized glutathione through thiol–disulfide interchange between substrate and the reduced disulfide of the enzyme. Though the intermediary glutathione-enzyme has not so far been isolated, it is a virtual certainty that it lies on the reaction pathway; this because of all that is known about enzymic (see above) and nonenzymic (89) thiol–disulfide interchanges. Mechanisms analogous to that of Fig. 12 for glutathione reductase are thought to hold good for the reactions catalyzed by

(88) V. Massey and C. H. Williams, Jr., *JBC* 240, 4470–4480 (1965); G. E. Schultz, R. H. Schirmer, W. Sachsenheimer, and E. F. Pai, *Nature (London)* 273, 120–124 (1978).
(89) A. J. Parker and N. Kharasch, *Chem. Rev.* 59, 583–628 (1959); R. Cecil, in *The Proteins*, H. Neurath, ed., Academic Press, New York, 1963, Vol. 1, p. 379.

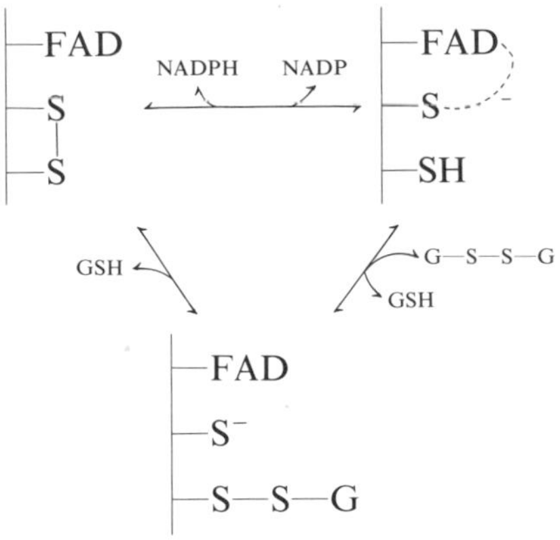

**Fig. 12.** Mechanism proposed for glutathione reductase (88).

dihydrolipoamide reductase [EC 1.6.4.3] (90) and thioredoxin reductase [EC 1.6.4.5] (91).

## Summary

This chapter illustrates, with enzymes from the major classes of oxidoreductases, some of the evidence for covalent union of substrate and holoenzyme during oxidoreduction. Few will deny that a redox enzyme, as an agent of electron transfer, receives electrons from donor molecules and holds them transiently before releasing them to acceptors. This fact alone establishes the redox enzyme as a covalent catalyst, since the transferring electron must share an orbital with the holoenzyme during catalysis. Beyond this are the augmenting signs that the process of electron transfer engenders covalent linkages between the enzyme, on the one hand, and the electron donor and acceptor, on the other. On grounds of chemical analogy we are right to believe that what holds good for one flavoenzyme ought to hold good for the next; and what is true for one heme enzyme should be true for the next; and so on. Table 2.1 lists 139 oxidoreductases for which evidence exists that, during oxidoreduction, electrons pass from the donor to the holoenzyme and thence to the acceptor. This number of enzymes amounts to 24% of all the redox enzymes recognized by the Enzyme Commission. Chapter 8 explains how a fair extrapolation of the 24% figure to 93% may reasonably be claimed on grounds of chemical analogy, leaving only 7% of the recognized redox enzymes as unrepresented in Table 2.1.

(90) R. G. Matthews, D. P. Ballou, C. Thorpe, and C. H. Williams, Jr., *JBC* 252, 3199–3207 (1977); R. G. Matthews, D. P. Ballou, and C. H. Williams, Jr., *JBC* 254, 4974–4981 (1979).
(91) C. H. Williams, Jr., in *The Enzymes*, 3rd ed., P. D. Boyer, ed., Academic Press, New York, 1976, Vol. 13, pp. 90–171.

**Table 2.1.** Oxidoreductases Known to Act by Covalent Catalysis

| EC no. | Familiar name of enzyme[a] | Criteria[b] | References |
|--------|---------------------------|-------------|------------|
| 1.1.1.1 | Alcohol dehydrogenase [NAD, Zn] | K, M | (92) |
| 1.1.1.22 | UDPglucose dehydrogenase [NAD] | M, S | (93) |
| 1.1.1.42 | Isocitrate dehydrogenase (NADP) [Mn, NADP] | Ma | (94) |
| 1.1.2.3 | Lactate dehydrogenase (cytochrome) [FMN, heme] | R, M | (95) |
| 1.1.2.4 | D-Lactate dehydrogenase (cytochrome) [FAD] | R, M | (96) |
| 1.1.3.1 | Glycolate oxidase [FMN] | R, M | (97) |
| 1.1.3.4 | Glucose oxidase [FAD] | R, K, M | (98) |
| 1.1.3.6 | Cholesterol oxidase [FAD] | R | (99) |
| 1.1.3.8 | L-Gulonolactone oxidase [flavin] | R | (100) |
| 1.1.3.15 | L-2-Hydroxyacid oxidase [FMN] | R, M | (101) |
| 1.1.3.17 | Choline oxidase [FAD] | R | (102) |
| 1.1.99.5 | Glycerol-3-P dehydrogenase [FMN, Fe] | R | (103) |
| 1.1.99.6 | D-2-Hydroxyacid dehydrogenase [FMN] | R, K, M | (104) |
| 1.1.99.7 | Lactate–malate transhydrogenase [NAD] | I, K | (105) |
| 1.1.99.10 | Glucose dehydrogenase [FAD] | R, K | (106) |
| 1.1.99.13 | D-Glucoside 3-dehydrogenase [FAD] | R | (107) |
| 1.1.99.15 | 5,10-Methylenetetrahydrofolate reductase (FADH$_2$) | R | (108) |
| 1.1.99 | Cellobiose: guinone oxidoreductase [FAD] | R | (109) |
| 1.2.1.2 | Formate dehydrogenase [FMN, Fe—S] | R | (110) |
| 1.2.1.2 | Formate dehydrogenase [heme, Se, Fe—S, Mo] | R | (111) |
| 1.2.1.3 | Aldehyde dehydrogenase [NAD] | B, M | (112) |
| 1.2.1.11 | Aspartate-semialdehyde dehydrogenase [NAD] | I, M | (113) |
| 1.2.1.12 | Glyceraldehyde-3-P dehydrogenase [NAD] | I[c], E, K, B | (114) |
| 1.2.1.37 | Xanthine dehydrogenase [FAD, Mo, Fe—S, —S—S—H] | R, K | (115) |
| 1.2.2.2 | Pyruvate dehydrogenase (cytochrome) [FAD, TPP] | I, R | (116) |
| 1.2.3.1 | Aldehyde oxidase [FAD, Mo, Fe—S, —S—S—H] | R | (117) |
| 1.2.3.2 | Xanthine oxidase [FAD, Mo, Fe—S, —S—S—H] | R, K | (118) |
| 1.2.3.3 | Pyruvate oxidase [FAD, TPP] | R | (119) |
| 1.2.3. | Bacterial luciferase [FMN] | I | (120) |
| 1.2.4.1 | Pyruvate dehydrogenase (lipoate) [TPP] | I, K | (121) |
| 1.2.7.1 | Pyruvate synthase [TPP, Fe—S, chromophore] | R | (122) |
| 1.3.1.14 | Orotate reductase (NAD) [FAD, FMN] | R, M | (123) |
| 1.3.1.15 | Orotate reductase (NADPH) [FAD, FMN] | R, M | (124) |
| 1.3.1.26 | Dihydropicolinate reductase [FMN] | R, K | (125) |
| 1.3.3.1 | Dihydroorotate oxidase [FAD, FMN, Fe—S] | R, Ma | (126) |
| 1.3.99.1 | Succinate dehydrogenase [FAD, Fe—S] | R, Ma, M | (127) |
| 1.3.99.2 | Butyryl-CoA dehydrogenase [FAD, Cu] | R | (128) |
| 1.3.99.3 | Acyl-CoA dehydrogenase [FAD] | R, Ma | (129) |
| 1.3.99.6 | 3-Oxo-5$\beta$-steroid $\Delta^4$-dehydrogenase [FMN] | M | (130) |
| 1.3.99.9 | $\beta$-Cyclopiazonate oxidocyclase [FAD] | R, K | (131) |
| 1.4.1.13 | Glutamate synthase [FAD, FMN, Fe—S] | G, R | (132) |
| 1.4.3.1 | D-Aspartate oxidase [FAD] | R | (133) |
| 1.4.3.2 | L-Amino-acid oxidase [FAD] | R, M | (134) |

**Table 2.1.** Oxidoreductases Known to Act by Covalent Catalysis  (*Continued*)

| EC no. | Familiar name of enzyme[a] | Criteria[b] | References |
|---|---|---|---|
| 1.4.3.3 | D-Amino-acid oxidase [FAD] | R, M | (135) |
| 1.4.3.4 | Amine oxidase [FAD] | R, K | (136) |
| 1.4.3.6 | Amine oxidase [PLP?, Cu] | R, K, M | (137) |
| 1.4.3.9 | Tyramine oxidase [FAD] | R, K | (138) |
| 1.4.3.10 | Putrescine oxidase [FAD] | R, K | (139) |
| 1.4.3.12 | Cyclohexylamine oxidase [FAD] | R | (140) |
| 1.4.99.1 | D-Amino-acid dehydrogenase [FAD] | R | (141) |
| 1.5.3.1 | Sarcosine oxidase [flavin] | R | (142) |
| 1.5.3.5 | 6-Hydroxy-L-nicotine oxidase [FAD] | R | (143) |
| 1.5.3.6 | 6-Hydroxy-D-nicotine oxidase [FAD] | R | (144) |
| 1.5.99.1 | Sarcosine dehydrogenase [flavin] | R | (145) |
| 1.5.99.4 | Nicotine dehydrogenase [FMN] | R | (146) |
| 1.5.99.5 | N-Methylglutamate dehydrogenase [flavin] | R, K | (147) |
| 1.5.99.6 | Spermidine dehydrogenase [FAD, heme] | R | (148) |
| 1.5.99.7 | Trimethylamine dehydrogenase [FMN, Fe—S] | R, Ma | (149) |
| 1.6.1.1 | NAD (P) transhydrogenase [FAD] | R, K, M | (150) |
| 1.6.2.2 | Cytochrome $b_5$ reductase [FAD] | R | (151) |
| 1.6.2.4 | NADPH-cytochrome reductase [FAD, FMN] | R, K | (152) |
| 1.6.4.2 | Glutathione reductase (NAD(P)H) [FAD, —S—S—] | R, K, E | (153) |
| 1.6.4.3 | Lipoamide reductase (NADH) [FAD, —S—S—] | R, K, M | (154) |
| 1.6.4.5 | Thioredoxin reductase (NADPH) [FAD, —S—S—] | R | (155) |
| 1.6.6.1 | Nitrate reductase (NADH) [FAD, heme, Mo] | R, Ma | (156) |
| 1.6.6.2 | Nitrate reductase (NAD(P)H) [FAD, Mo] | R | (157) |
| 1.6.6.3 | Nitrate reductase (NADPH) [FAD, cyt, Mo] | R | (158) |
| 1.6.6.4 | Nitrite reductase (NAD(P)H) [FAD, siroheme] | I[d], R | (159) |
| 1.6.99.1 | NADPH dehydrogenase [FMN] | R, K, M | (160) |
| 1.6.99.2 | Quinone reductase [FAD] | R, K | (161) |
| 1.6.99.3 | NADH dehydrogenase [FMN, Fe—S, Mo] | R, Ma, K | (162) |
| 1.6.99.5 | NADH dehydrogenase (quinone) [FAD] | R | (163) |
| 1.7.2.1 | Nitrite reductase (cytochrome) [Cu] | R, I[d] | (164) |
| 1.7.3.1 | Nitroethane oxidase [FAD, Fe] | R | (165) |
| 1.7.3.4 | Hydroxylamine oxidase [heme] | R | (166) |
| 1.7.7.1 | Ferredoxin–nitrite reductase [Fe—S, siroheme] | I[d], Ma | (167) |
| 1.7.99.1 | Hydroxylamine reductase [FAD] | R | (168) |
| 1.7.99.3 | Nitrite reductase [hemes c and d] | I[d], R | (169) |
| 1.7.99.3 | Nitrite reductase [FAD, Cu] | R | (170) |
| 1.7.99.4 | Nitrate reductase [Mo, Fe—S] | R, Ma | (171) |
| 1.8.1.2 | Sulfite reductase (NADPH) [FAD, FMN, Fe—S, siroheme] | I[d], R, K | (172) |
| 1.8.3.1 | Sulfite oxidase [Mo, heme] | R, K | (173) |
| 1.8.4.2 | Protein–disulfide reductase [—S—S—] | R, M | (174) |
| 1.8.4.4 | Glutathione–cystine transhydrogenase [—S—S—] | I[e] | (175) |
| 1.8.99.1 | Sulfite reductase [Fe] | R | (176) |
| 1.8.99.2 | Adenylylsulfate reductase [FAD, Fe—S] | I[d], K, Ma | (177) |

**Table 2.1.** Oxidoreductases Known to Act by Covalent Catalysis   (*Continued*)

| EC no. | Familiar name of enzyme[a] | Criteria[b] | References |
|---|---|---|---|
| 1.9.3.1 | Cytochrome oxidase [Cu, heme] | $I^d$, R | (178) |
| 1.9.6.1 | Nitrate reductase (cytochrome) [heme] | R | (179) |
| 1.10.3.1 | Catechol oxidase [Cu] | R | (180) |
| 1.10.3.2 | Laccase [Cu] | $I^d$, R | (181) |
| 1.10.3.3 | Ascorbate oxidase [Cu] | $I^d$, R, K | (182) |
| 1.10.3.4 | Isophenoxazine synthase [FMN, Mn, —S—S—] | R | (183) |
| 1.11.1.1 | NAD peroxidase [FAD, —S—S—] | R | (184) |
| 1.11.1.5 | Cytochrome peroxidase [heme] | $I^d$, Ma, M | (185) |
| 1.11.1.6 | Catalase [heme] | $I^d$, M | (186) |
| 1.11.1.7 | Peroxidase [heme] | $I^d$, M | (187) |
| 1.11.1.9 | Glutathione peroxidase [Se] | R, K, M | (188) |
| 1.11.1.10 | Chloroperoxidase [heme] | $I^d$, M | (189) |
| 1.12.1.2 | Hydrogen dehydrogenase [Fe, FMN] | E, K | (190) |
| 1.12.2.1 | Cytochrome $c_3$ hydrogenase [Fe] | E, R | (191) |
| 1.13.11.1 | Pyrocatechase [Fe] | Ma | (192) |
| 1.13.11.3 | Protocatechuate 3,4-dioxygenase [Fe] | Ma | (193) |
| 1.13.11.11 | Tryptophan 2,3-dioxygenase [heme] | $I^d$, R | (194) |
| 1.13.11.12 | Lipoxygenase [Fe] | R, Ma | (195) |
| 1.13.11. | Indoleamine 2,3-dioxygenase [heme] | $I^d$ | (196) |
| 1.13.12.1 | Arginine 2-monooxygenase [FAD] | R | (197) |
| 1.13.12.2 | Lysine 2-monooxygenase [FAD] | R | (198) |
| 1.13.12.4 | Lactate 2-monooxygenase [FMN] | $I^d$, R, K, M | (199) |
| 1.14.12.3 | Benzene 1,2-dioxygenase [FAD, Fe] | R | (200) |
| 1.14.12.4 | Methylhydroxypyridine-carboxylate dioxygenase [FAD] | R | (201) |
| 1.14.13.1 | Salicylate hydroxylase [FAD] | R | (202) |
| 1.14.13.2 | p-Hydroxybenzoate hydroxylase [FAD] | $I^d$, R | (203) |
| 1.14.13.4 | Melilotate hydroxylase [FAD] | $I^d$, R | (204) |
| 1.14.13.5 | Imidazoleacetate oxygenase [FAD] | R | (205) |
| 1.14.13.6 | Orcinol hydroxylase [FAD] | R | (206) |
| 1.14.13.7 | Phenol 2-monooxygenase [FAD] | $I^d$, R | (207) |
| 1.14.13.8 | Dimethylaniline oxidase [FAD] | $I^d$, R | (208) |
| 1.14.13.10 | 2,6-Dihydroxypyridine oxidase [flavin] | R | (209) |
| 1.14.13.16 | Cyclopentanone oxygenase [FAD] | R | (210) |
| 1.14.13 | Cyclohexanone oxygenase [FAD] | R | (211) |
| 1.14.13 | Ketone monooxygenase [FAD] | R | (212) |
| 1.14.13 | Resorcinol hydroxylase [FAD] | R | (213) |
| 1.14.13 | 4-Hydroxyisophthalate hydroxylase [FAD] | R | (214) |
| 1.14.14.1 | Mixed-function oxidase [cyt *P*-450] | $I^d$, R | (215) |
| 1.14.15.1 | Camphor 5-monooxygenase [cyt *P*-450] | I, $I^d$, R, M | (216) |
| 1.14.15.3 | Alkane 1-monooxygenase [cyt *P*-450] | R | (217) |
| 1.14.15.4 | Steroid 11β-monooxygenase [cyt *P*-450] | $I^d$, R, M | (218) |
| 1.14.17.1 | Dopamine β-monooxygenase [Cu] | I, R, Ma | (219) |
| 1.14.18.1 | Tyrosinase [Cu] | $I^d$, R, K | (220) |

**Table 2.1.** Oxidoreductases Known to Act by Covalent Catalysis   (*Continued*)

| EC no. | Familiar name of enzyme[a] | Criteria[b] | References |
|---|---|---|---|
| 1.14.99.15 | 4-Methoxybenzoate O-demethylase [FMN, Fe—S] | R | (221) |
| 1.15.1.1 | Superoxide dismutase [Cu] | R, M | (222) |
| 1.16.3.1 | Ceruloplasmin [Cu] | I[d], R | (223) |
| 1.17.1.1 | CDP-4-keto-6-deoxy-D-glucose reductase [PMP] | S | (224) |
| 1.17.4.1 | Ribonucleoside-diphosphate reductase [—S—S—, Fe] | R, K, M | (225) |
| 1.17.4.2 | Ribonucleoside-triphosphate reductase [—S—S—, cobalamin] | R | (226) |
| 1.18.1.1 | Rubredoxin-NAD reductase [FAD] | R | (227) |
| 1.18.1.2 | Ferredoxin-NADP reductase [FAD] | R | (228) |
| 1.18.2.1 | Nitrogenase [Mo, Fe—S] | R, Ma, M | (229) |
| 1.18.3.1 | Hydrogenase [Fe—S] | E, R, Ma, M | (230) |

[a] Any parenthetical expression is part of the official name of the enzyme. Prosthetic groups are indicated in brackets. Abbreviations: TPP, thiamine pyrophosphate; PLP, pyridoxal-5′-P; cyt, cytochrome; —S—S—H, persulfide; —S—S—, disulfide; Fe—S, a nonheme iron labile–sulfur center.

[b] The symbols mean the following:

I, the holoenzyme links covalently with a fragment of its substrate (other than electrons) to form a chemically competent intermediate;

R, the oxidized holoenzyme accepts electrons directly from reduced substrate, or the reduced holoenzyme gives up electrons to oxidized substrate, or both;

K, the enzyme exhibits kinetic properties consistent with the participation of a reduced holoenzyme in the reaction;

B, the enzyme displays "burst" kinetics;

E, the enzyme catalyzes one or more exchange reactions consistent with the participation of a covalent enzyme–substrate intermediate;

M, miscellaneous data and derivative arguments which are peculiar to the enzyme in question;

Ma, magnetic resonance or other physical studies show that this enzyme binds a substrate molecule by coordinate linkage to its prosthetic group (a metal ion) or that the metal undergoes a valence change;

S, the holoenzyme forms a Schiff base with substrate;

G, the enzyme is irreversibly inactivated for glutamine utilization by stoichiometric alkylation with a glutamine analogue; inferred from this is the participation of a glutamyl-enzyme intermediate.

[c] This enzyme has been isolated from tissue as the 3-P-glyceryl-enzyme.

[d] Identified spectroscopically.

[e] This enzyme is isolated from tissue with a glutathione residue bound to it by a disulfide bond.

(92) N. K. Gupta and W. G. Robinson, *BBA* 118, 431–434 (1966); H. Gershman and R. H. Abeles, *Arch. Biochem. Biophys.* 154, 659–674 (1973); Refs. (37, 39); J. D. Shore and D. Santiago, *JBC* 250, 2008–2012 (1975).

(93) W. P. Ridley, J. P. Houchins, and S. Kirkwood, *JBC* 250, 8761–8767 (1975); A. O. Ordman and S. Kirkwood, *JBC* 252, 1320–1326 (1977); H. S. Prihar and D. S. Feingold, *FEBS lett.* 99, 106–108 (1979).

(94) J. J. Villafranca and R. F. Colman, *B* 13, 1152–1160 (1974); M. H. O'Leary and J. A. Limburg, *B* 16, 1129–1135 (1977); R. S. Levy and J. J. Villafranca, *B* 16, 3293–3309 (1977).

(95) C. A. Appleby and R. K. Morton, *Biochem. J.* 73, 539–550 (1959); R. K. Morton and J. M. Sturtevant, *JBC* 239, 1614–1624 (1964); F. Lederer, *EJB* 46, 393–399 (1974); C. Capeillère-

Blandin, R. C. Bray, M. Iwatsubo, and F. Labeyrie, *EJB* 54, 549–566 (1975); D. Pompon, M. Iwatsubo, and F. Lederer, *EJB* 104, 479–488 (1980).

(96) C. Gregolin and T. P. Singer, *BBA* 67, 201–218 (1963); M. Futai, *B* 12, 2468–2474 (1973); L. D. Kohn and H. R. Kaback, *JBC* 248, 7012–7017 (1973); S. Ghisla, J.-M. Lhoste, S. Olson, C. J. Whitfield, and V. Massey, in *Flavins and Flavoproteins*, K. Yagi and T. Yamano, eds., University Park Press, Baltimore, 1980, pp. 55–66.

(97) N. A. Frigerio and H. A. Harbury *JBC* 231, 135–157 (1958); M. S. Jorns and L. B. Hersh, *JBC* 251, 4872–4881 (1976).

(98) D. Keilin and E. F. Hartree, *Biochem. J.* 42, 221–229 (1948); Q. H. Gibson, B. E. P. Swoboda, and V. Massey, *JBC* 239, 3927–3934 (1964); D. J. T. Porter, J. G. Voet, and H. J. Bright, *Z. Naturforsch.* 27B, 1052–1053 (1972); Refs. (8, 10).

(99) M. Fukuyama and Y. Miyake, *J. Biochem.* (*Tokyo*) 85, 1183–1193 (1979).

(100) H. Nakagawa and A. Asano, *J. Biochem.* (*Tokyo*) 68, 737–746 (1970); H. Nakagawa, A. Asano, and R. Sato, *J. Biochem.* (*Tokyo*) 77, 221–232 (1975).

(101) J. C. Robinson, L. Keay, R. Molinari, and I. W. Sizer, *JBC* 237, 2001–2010 (1962); T. H. Cromartie and C. T. Walsh, *B* 14, 3482–3490 (1975).

(102) S. Ikuta, S. Imamura, H. Misaki, and Y. Horiuti, *J. Biochem.* (*Tokyo*) 82, 1741–1749 (1977); M. Ohta-Fukuyama, Y. Miyake, S. Emi, and T. Yamano, *J. Biochem.* (*Tokyo*) 88, 197–203 (1980).

(103) R. L. Ringler, *JBC* 236, 1192–1198 (1961); N. Sone, *J. Biochem.* (*Tokyo*) 74, 297–305 (1973).

(104) E. Boeri, T. Cremona, and T. P. Singer, *Biochem. Biophys. Res. Commun.* 2, 298–302 (1960); A. P. Nygaard, *JBC* 236, 920–925 (1961); P. K. Tubbs, *Biochem. J.* 82, 36–42 (1962); S. Ghisla, S. T. Olson, V. Massey, and J.-M. Lhoste, *B* 18, 4733–4742 (1979).

(105) Ref. (30); M. I. Dolin, *JBC* 244, 5273–5285 (1969).

(106) T. G. Bak, *BBA* 139, 277–293 (1967); *Idem. BBA* 146, 317–327 (1967).

(107) K. Hayano and S. Fukui, *JBC* 242, 3665–3672 (1967).

(108) R. E. Cathou and J. M. Buchanan, *JBC* 238, 1746–1751 (1963); R. G. Matthews and S. Kaufman, *JBC* 255, 6014–6017 (1980).

(109) U. Westermark and K. E. Eriksson, *Acta Chem. Scand.* B29, 419–424 (1975).

(110) T. Höpner and A. Trautwein, *Z. Naturforsch.* 27B, 1075–1076 (1972).

(111) H. G. Enoch and R. L. Lester, *JBC* 250, 6693–6705 (1975).

(112) H. Weiner, J. H. J. Hu, and C. G. Sanny, *JBC* 251, 3853–3855 (1976); A. K. H. MacGibbon, S. J. Haylock, P. D. Buckley, and L. F. Blackwell, *Biochem. J.* 171, 533–538 (1978); T. M. Kitson, *Biochem J.* 175, 83–90 (1978); R. J. S. Duncan, *Biochem. J.* 183, 459–462 (1979).

(113) M. J. Holland and E. W. Westhead, *B* 12, 2276–2281 (1973); J. F. Biellmann, P. Eid, C. Hirth, and H. Jornvall, *EJB* 104, 53–58 (1980).

(114) J. Harting and S. Velick, *Fed. Proc.* 11, 226 (1952); S. Velick and J. E. Hayes, Jr., *JBC* 203, 545–562 (1953); H. L. Segal and P. D. Boyer, *JBC* 204, 265–281 (1953); P. Oesper, *JBC* 207, 421–429 (1954); I. Krimsky and E. Racker, *Science* 122, 319–321 (1955); J. H. Park, B. P. Meriwether, P. Clodfelder, and L. W. Cunningham, *JBC* 236, 136–141 (1961); I. Harris, B. P. Meriwether, and J. H. Park, *Nature* (*London*) 198, 154–157 (1963); W. Bloch, R. A. MacQuarrie, and S. A. Bernhard, *JBC* 246, 780–790 (1971); F. Seydoux, S. Bernhard, O. Pfenninger, M. Payne, and O. P. Malhotra, *B* 12, 4290–4300 (1973).

(115) K. V. Rajagopalan and P. Handler *JBC* 242, 4097–4107 (1967); K. V. Rajagopalan, in *Flavins and Flavoproteins*, H. Kamin, ed., University Park Press, Baltimore, 1971, p. 291; Ref. (29).

(116) L. P. Hager and L. O. Krampitz, *Fed. Proc.* 22, 536 (1963); F. R. Williams and L. P. Hager, *Arch. Biochem. Biophys.* 116, 168–176 (1966).

(117) H. R. Mahler, B. Mackler, and D. E. Green, *JBC* 210, 465–480 (1954); K. V. Rajagopalan and P. Handler *JBC* 239, 2027–2035 (1964); Ref. (28).

(118) Refs. (22, 24, 26); D. Edmondson, V. Massey, G. Palmer, L. M. Beachum, III, and G. B. Elion, *JBC* 247, 1597–1604 (1972).

(119) L. P. Hager, D. M. Geller, and F. Lipmann, *Fed. Proc.* 13, 734–738 (1954).

(120) Ref. (20).

(121) H. Holzer, H. W. Goedde, K.-H. Göggel, and B. Ulrich, *Biochem. Biophys. Res. Commun.* 3, 599–602 (1960); H. W. Goedde, H. Inouye, and H. Holzer, *BBA* 50, 41–44 (1961); C. S. Tsai, M. W. Burgett, and L. J. Reed, *JBC* 248, 8348–8352 (1973); L. S. Khailova, R. Bernhardt, and G. Huebner, *Biokhimiya* 42, 93–96 (1977) (Engl. transl.); R. L. Cate and T. E. Roche, *JBC* 254, 1659–1665 (1979).

(122) K. Uyeda and J. C. Rabinowitz, *JBC* 246, 3120–3125 (1971).

(123) J. L. Graves and B. Vennesland, *JBC* 226, 307–316 (1957); H. C. Friedmann and B. Vennesland, *JBC* 233, 1398–1406 (1958).

(124) S. Udaka and B. Vennesland, *JBC* 237, 2018–2024 (1962); G. Krakow, J. Ludowieg, J. H. Mather, W. M. Normore, L. Tosi, S. Udaka, and B. Vennesland, *B* 2, 1009–1014 (1963).

(125) K. Kimura and T. Goto, *J. Biochem. (Tokyo)* 77, 415–420 (1975).

(126) K. V. Rajagopalan, V. Aleman, P. Handler, W. Heinen, G. Palmer, and H. Beinert, *Biochem. Biophys. Res. Commun.* 8, 220–226 (1962); R. W. Miller and V. Massey, *JBC* 240, 1453–1465 (1965); P. Blattmann and J. Rétey, *EJB* 30, 130–137 (1972).

(127) T. P. Singer, E. B. Kearney, and P. Bernath, *JBC* 223, 599–613 (1956); D. V. Dervartanian, W. P. Zeylemaker, and C. Veeger, in *Flavins and Flavoproteins*, E. P. Slater, ed., Elsevier, Amsterdam, 1966, pp. 183–199; D. V. Dervartanian, C. Veeger, W. H. Orme-Johnson, and H. Beinert, *BBA* 191, 22–37 (1969); M. Conjalka and T. C. Hollocher, in *Structure and Function of Oxidation-Reduction Enzymes*, Å. Åkeson and A. Ehrenberg, eds., Pergamon Press, New York, 1972, pp. 509–518; T. A. Alston, L. Mela, and H. J. Bright, *PNAS* 74, 3767–3771 (1977); C. J. Coles, D. E. Edmondson, and T. P. Singer, *JBC* 254, 5161–5167 (1979).

(128) H. R. Mahler, *JBC* 206, 13–26 (1954); H. Beinert and R. H. Sands, in *Free Radicals in Biological Systems*, M. S. Blois, H. M. Brown, R. M. Lemmon, R. O. Lindblom, and M. Weissbluth, eds., Academic Press, New York, 1961, p. 26.

(129) H. Beinert, *JBC* 225, 465–478 (1957); C. Thorpe, R. G. Matthews, and C. H. Williams, Jr., *B* 18, 331–337 (1979); F. E. Freman, J.-J. Park Kim, K. Hubta, and M. C. McKean, *JBC* 255, 2195–2198 (1980); C. L. Hall and J. D. Lambeth, *JBC* 255, 3591–3595 (1980).

(130) S. J. Davidson and P. Talalay, *JBC* 241, 906–915 (1966).

(131) D. J. Steenkamp, J. C. Schabort, and N. P. Ferreira *BBA* 309, 440–456 (1973); D. J. Steenkamp and J. C. Schabort, *EJB* 40, 163–170 (1973); C. W. Holzapfel and N. P. Ferreira, *BBA* 358, 126–143 (1974); J. C. Schabort and M. Marx, *Int. J. Biochem.* 10, 61–65 (1979).

(132) R. E. Miller and E. R. Stadtman, *JBC* 247, 7407–7410 (1972); D. W. Tempest, J. L. Meers, and C. M. Brown, in *The Enzymes of Glutamine Metabolism*, S. Prusiner and E. R. Stadtman, eds., Academic Press, New York, 1973, p. 167; P. P. Trotta, K. E. B. Platzer, R. H. Haschemeyer, and A. Meister *PNAS* 71, 4607–4611 (1974); A. R. Rendina and W. H. Orme-Johnson, *B* 17, 5388–5393 (1978).

(133) E. Rocca, *Boll. Soc. Ital. Biol. Sper.* 40, 390–391 (1964).

(134) D. Wellner and A. Meister, *JBC* 235, PC12 (1960); M. Nakano and T. S. Danowski, *JBC* 241, 2075–2083 (1963); D. J. T. Porter and H. J. Bright, *Biochem. Biophys. Res. Commun.* 36, 209–214 (1969); C. T. Walsh, A. Schonbrunn, and R. H. Abeles, *JBC* 246, 6855–6866 (1971).

(135) V. Massey, G. Palmer, and R. Bennett, *BBA* 48, 1–9 (1961); Refs. (1, 3, 5, 7 and 8).

(136) H. Yamada and K. T. Yasunobu, *JBC* 237, 1511–1516 (1962); V. G. Erwin and L. Hellerman, *JBC* 242, 4230–4238 (1967); S. Oi, K. Shimada, M. Inamasu, and K. T. Yasunobu, *Arch. Biochem. Biophys.* 139, 28–37 (1970); M. D. Houslay and K. F. Tipton, *Biochem. J.* 135, 735–750 (1973).

(137) F. Buffoni, in *Pyridoxal Catalysis: Enzymes and Model Systems*, E. E. Snell, A. E. Braunstein, E. S. Severin, and Y. M. Torchinsky, eds., Interscience, New York, 1968, p. 363; H. Kumagai, T. Nagate, H. Yamada, and H. Fukami, *BBA* 185, 242–244 (1969); B. Mondovi, G. Rotilio, A. Finazzi Agro, M. P. Vallogini, B. G. Malmström, and E. Antonini, *FEBS lett.* 2, 182–184 (1969); A. Finazzi Agro, G. Rotilio, M. T. Costa, and B. Mondovi, *FEBS lett.* 4, 31–32 (1969); S. Oi, M. Inamasu, and K. T. Yasunobu, *B* 9, 3378–3383 (1970); W. G. Bardsley,

M. J. C. Crabbe, and J. S. Shindler, *Biochem. J.* 131, 459–469 (1973); R. Neumann, R. Hevey, and R. H. Abeles, *JBC* 250, 6362–6367 (1975); A. Lindström, B. Olsson, and G. Petterson, *EJB* 42, 377–381 (1974); R. H. Suva and R. H. Abeles, *B* 17, 3538–3545 (1978).

(138) H. Kumagai, H. Matsui, K. Ogata, and H. Yamada, *BBA* 171, 1–8 (1969); H. Kumagai, H. Yamada, H. Suzuki, and Y. Ogura, *J. Biochem. (Tokyo)* 69, 137–144 (1971).

(139) R. J. DeSa, *JBC* 247, 5527–5534 (1972).

(140) T. Tokieda, T. Niimura, F. Takamura, and T. Yamaha, *J. Biochem. (Tokyo)* 81, 851–858 (1977).

(141) K. Tsukada, *JBC* 241, 4522–4528 (1966).

(142) W. R. Frisell, *Arch. Biochem. Biophys.* 142, 213–222 (1971).

(143) K. Decker and V. D. Dai, *EJB* 3, 132–138 (1967).

(144) M. Brühmüller, H. Möhler, and K. Decker, *EJB* 29, 143–151 (1972).

(145) W. R. Frisell and C. G. Mackenzie, *JBC* 237, 94–98 (1962); Ref. (142).

(146) L. I. Hochstein and B. P. Dalton, *BBA* 139, 56–68 (1967).

(147) L. B. Hersh, M. J. Stark, S. Worthen, and M. K. Fiero, *Arch. Biochem. Biophys.* 150, 219–226 (1972); C. W. Bamforth and P. J. Large, *Biochem. J.* 161, 357–370 (1977).

(148) C. W. Tabor and P. D. Kellogg, *JBC* 245, 5424–5433 (1970).

(149) D. J. Steenkamp, T. P. Singer, and H. Beinert, *Biochem. J.* 169, 361–369 (1978); T. P. Singer, D. J. Steenkamp, W. C. Kenney, and H. Beinert, in *Flavins and Flavoproteins*, K. Yagi and T. Yamano, eds., University Park Press, Baltimore, 1980, pp. 277–287.

(150) P. T. Cohen and N. O. Kaplan *JBC* 245, 2825–2836 (1970); *idem. JBC* 245, 4666–4672 (1970); D. D. Louie and N. O. Kaplan, *JBC* 245, 5691–5698 (1970); E. Jacobs and R. R. Fisher, *B* 18, 4315–4322 (1979).

(151) P. Strittmatter and S. F. Velick *JBC* 228, 785–799 (1957); H. R. Mahler, I. Raw, R. Molinari, and D. F. de Amaral, *JBC* 233, 230–239 (1958); P. Strittmatter, *JBC* 233, 748–753 (1958); G. R. Drysdale, M. J. Spiegel, and P. Strittmatter, *JBC* 236, 2323–2328 (1961); P. Strittmatter, *JBC* 237; 3250–3254 (1962).

(152) E. Haas, B. L. Horecker, and T. R. Hogness, *JBC* 136, 747–774 (1940); B. L. Horecker, *JBC* 183, 593–605 (1950); B. S. S. Masters, H. Kamin, Q. H. Gibson, and C. H. Williams, Jr., *JBC* 240, 921–931 (1965); T. Iyanagi and H. S. Mason, *B* 12, 2297–2308 (1973); Y. Yasukochi and B. S. S. Masters, *JBC* 251, 5337–5344 (1976); J. D. Dignam and H. W. Strobel, *B* 16, 1116–1123 (1977); T. Iyanagi, F. K. Anan, Y. Imai, and H. S. Mason, *B* 17, 2224–2230 (1978); Y. Yasukochi, J. A. Peterson, and B. S. S. Masters, *JBC* 254, 7097–7104 (1979).

(153) S. Black and B. Hudson, *Biochem. Biophys. Res. Commun.* 5, 135–138 (1961); L. W. Mapson and F. A. Isherwood, *Biochem. J.* 86, 173–191 (1963); Ref. (88); I. Carlberg and B. Mannervik, *JBC* 250, 5475–5480 (1975); G. Moroff, R. S. Ochs, and K. G. Brandt, *Arch. Biochem. Biophys.* 173, 42–49 (1976); B. Mannervik, V. Boggaram, I. Carlberg, and K. Larson, in *Flavins and Flavoproteins*, K. Yagi and T. Yamano, eds., University Park Press, Baltimore, 1980, pp. 173–187.

(154) V. Massey, Q. H. Gibson, and C. Veeger, *Biochem. J.* 77, 341–351 (1960); V. Massey and C. Veeger, *BBA* 67, 679–681 (1963); J. K. Reed, *JBC* 248, 4834–4839 (1973); C. Thorpe and C. H. Williams, Jr., *JBC* 251, 7726–7728 (1976); Ref. (90); D. M. Templeton, B. R. Hollebone, and C. S. Tsai, *B* 19, 3868–3873 (1980).

(155) E. C. Moore, P. Reichard, and L. Thelander, *JBC* 239, 3445–3452 (1964); V. Massey, R. G. Matthews, G. P. Foust, L. G. Howell, C. H. Williams, Jr., G. Zanetti, and S. Ronchi, in *Pyridine Nucleotide-dependent Dehydrogenases*, H. Sund, ed., Springer, Berlin, 1970, pp. 393–409; Ref. (91).

(156) C. A. Fewson and D. J. D. Nicholas, *BBA* 49, 335–349 (1961); L. P. Solomonson, G. H. Lorimer, R. L. Hall, R. Borchers, and J. L. Bailey, *JBC* 250, 4120–4127 (1975).

(157) D. J. D. Nicholas and A. Nason, *Plant Physiol.* 30, 135–138 (1955); M. A. De La Rosa, J. Diez, J. M. Vega, and M. Losada, *EJB* 106, 249–256 (1980).

(158) D. J. D. Nicholas and A. Nason, *JBC* 211, 183–197 (1954); R. H. Garrett and A. Nason, *JBC* 244, 2870–2882 (1969).

(159) O. Prakash and J. C. Sadana, *Arch. Biochem. Biophys.* 148, 614–632 (1972); J. M. Vega, R. H. Garrett, and L. M. Siegel, *JBC* 250, 7980–7989 (1975).

(160) T. Nakamura, J. Yoshimura, and Y. Ogura, *J. Biochem. (Tokyo)* 57, 554–564 (1965); Ref. (8); D. J. T. Porter and H. J. Bright, *JBC* 255, 7362–7370 (1980).

(161) F. Märki and C. Martius, *Biochem. Z.* 333, 111–135 (1960); S. Hosoda, W. Nakamura, and K. Hayashi, *JBC* 249, 6416–6423 (1974); E. P. Titovets and G. G. Petrovskii, *Biokhimiya* 41, 1241–1248 (1976) (Engl. transl.).

(162) D. V. DerVartanian, *Z. Naturforsch.* 27B, 1082–1084 (1972); T. Imagawa and T. Nakamura, in *Flavins and Flavoproteins*, K. Yagi and T. Yamano, eds., University Park Press, Baltimore, 1980, pp. 199–207.

(163) E. Misaka and K. Nakanishi, *J. Biochem. (Tokyo)* 53, 465–471 (1963).

(164) H. Iwasaki and T. Matsubara, *J. Biochem. (Tokyo)* 71, 645–652 (1972); J. LeGall, W. J. Payne, T. V. Morgan, and D. DerVartanian, *Biochem. Biophys. Res. Commun.* 87, 355–362 (1979).

(165) T. Kido, K. Soda, T. Suzuki, and K. Asada, *JBC* 251, 6994–7000 (1976).

(166) M. K. Rees, *B* 7, 366–372 (1968); G. A. F. Ritchie and D. J. D. Nicholas, *Biochem. J.* 138, 471–480 (1974); T. Yamanaka, M. Shinra, K. Takahashi, and M. Shibasaka, *J. Biochem. (Tokyo)* 86, 1101–1108 (1979).

(167) P. J. Aparicio, D. B. Knaff, and R. Malkin, *Arch. Biochem. Biophys.* 169, 102–107 (1975); J. M. Vega and H. Kamin, *JBC* 252, 896–909 (1977); M. C. Liu, D. V. DerVartanian, and H. D. Peck, Jr., *Biochem. Biophys. Res. Commun.* 96, 278–285 (1980).

(168) G. C. Walker and D. J. D. Nicholas *BBA* 49, 361–368 (1961).

(169) H. Shimada and Y. Orii, *FEBS lett.* 54, 237–240 (1975); H. Shimada and Y. Orii, *J. Biochem. (Tokyo)* 80, 135–140 (1976).

(170) G. C. Walker and D. J. D. Nicholas *BBA* 49, 350–360 (1961).

(171) P. Forget, *EJB* 42, 325–332 (1974); R. C. Bray, S. P. Vincent, D. J. Lowe, R. A. Clegg, and P. B. Garland, *Biochem. J.* 155, 201–203 (1976); S. P. Vincent and R. C. Bray, *Biochem. J.* 171, 639–647 (1978).

(172) A. Yoshimoto and R. Sato, *BBA* 153, 555–575 (1968); L. M. Siegel, H. Kamin, D. C. Rueger, R. P. Presswood, and Q. H. Gibson, in *Flavins and Flavoproteins*, H. Kamin, ed., University Park Press, Baltimore, 1971, pp. 523–554; L. M. Siegel, P. S. Davis, and H. Kamin. *JBC* 249, 1572–1580 (1974).

(173) H. J. Cohen, I. Fridovich, and K. V. Rajagopalan, *JBC* 246, 374–382 (1971); J. L. Johnson and K. V. Rajagopalan, *JBC* 251, 5505–5511 (1976).

(174) H. M. Katzen and F. Tietze, *JBC* 241, 3561–3570 (1966); D. F. Carmichael, M. Keefe, M. Pace, and J. E. Dixon, *JBC* 254, 8386–8390 (1979).

(175) Ref. (87).

(176) K. Asada, G. Tamura, and R. S. Bandurski, *JBC* 244, 4904–4915 (1969).

(177) Ref. (15); H. D. Peck, Jr., R. N. Bramlett, and D. V. DerVartanian, *Z. Naturforsch.* 27B, 1084–1085 (1972); K. Adachi and I. Suzuki, *Can. J. Biochem.* 55, 91–98 (1977).

(178) H. Beinert and G. Palmer, *JBC* 239, 1221–1227 (1964); Q. H. Gibson, C. Greenwood, D. C. Wharton, and G. Palmer, *JBC* 240, 888–894 (1965); C. Greenwood, M. T. Wilson, and M. Brunori, *Biochem. J.* 137, 205–215 (1974); Refs. (70–73).

(179) P. A. Straat and A. Nason *JBC* 240, 1412–1416 (1965).

(180) S. E. Bresler, E. N. Kazbekov, A. T. Sukhodolova, and V. N. Shadrin, *Biokhimiya* 44, 583–588 (1979) (Engl. transl.).

(181) B. Reinhammar, *Adv. Inorg. Biochem.* 1, 91–118 (1979); Refs. (61, 62).

(182) F. J. Dunn and C. R. Dawson, *JBC* 189, 485–497 (1951); I. Yamazaki and L. H. Piette, *BBA* 50, 62–69 (1961); Ref. (69); A. Marchesini and P. M. H. Kroneck, *EJB* 101, 65–76 (1979).

(183) P. M. Nair and L. C. Vining, *BBA* 96, 318–327 (1965).

(184) M. I. Dolin, *JBC* 225, 557–573 (1957); *idem.*, *JBC* 250, 310–317 (1975).

(185) T. Yonetani, *Adv. Enzymol.* 33, 309–335 (1970); A. F. W. Coulson, J. E. Erman, and

T. Yonetani, *JBC* 246, 917–924 (1971); G. Lang, K. Spartalian, and T. Yonetani, *BBA* 451, 250–258 (1976); T. L. Poulos and J. Kraut, *JBC* 255, 8199–8206 (1980).

(186) B. Chance, *Nature (London)* 161, 914–917 (1948); B. Chance, *Arch. Biochem. Biophys.* 41, 404–415 (1952); Ref. (77).

(187) H. Theorell, *Enzymologia* 10, 250–252 (1941); second citation of ref. (186); Ref. (77); P. Douzou, R. Sireix, and F. Travers, *PNAS* 66, 787–792 (1970); T. Araiso, K. Miyoshi, and I. Yamazaki, *B* 15, 3059–3063 (1976).

(188) L. Flohe, G. Loschen, W. A. Günzler, and E. Eichele, *Z. Physiol. Chem.* 353, 987–999 (1972); C. Little, R. Olinescu, K. G. Reid, and P. J. O'Brien, *JBC* 245, 3632–3636 (1970); A. Wendel, W. Pilz, R. Ladenstein, G. Sawatzki, and U. Weser, *BBA* 377, 211–215 (1975); J. W. Forstrom and A. L. Tappel, *JBC* 254, 2888–2891 (1979). R. J. Kraus and H. E. Ganther, *Biochem. Biophys. Res. Commun.* 96, 1116–1122 (1980).

(189) J. A. Thomas, D. R. Morris, and L. P. Hager, *JBC* 245, 3135–3142 (1970); L. P. Hager, D. L. Doubek, R. M. Silverstein, T. T. Lee, J. A. Thomas, J. H. Hargis, and J. C. Martin, in *Oxidases and Related Redox Systems*, T. E. King, H. S. Mason, and M. Morrison, eds., University Park Press, Baltimore, 1973, Vol. 1, pp. 310–327; Ref. (81); M. M. Palcic, R. Rutter, T. Araiso, L. P. Hager, and H. B. Dunford, *Biochem. Biophys. Res. Commun.* 94, 1123–1127 (1980).

(190) D. H. Bone, *BBA* 67, 589–598 (1963).

(191) T. Yagi, M. Tsuda, Y. Mori, and H. Inokuchi, *JACS* 91, 2801 (1969); T. Yagi, *J. Biochem. (Tokyo)* 68, 649–657 (1970); T. Yagi, K. Kimura, H. Daidoji, F. Sakai, S. Tamura, and H. Inokuchi, *J. Biochem. (Tokyo)* 79, 661–672 (1976); K. Kumura, A. Suzuki, H. Inokuchi, and T. Yagi, *BBA* 567, 96–105 (1979).

(192) L. Que, Jr., R. H. Heistand, II, R. Mayer, and A. L. Roe, *B* 19, 2588–2593 (1980).

(193) L. Que, Jr., J. D. Lipscomb, E. Munck, and J. M. Wood, *BBA* 485, 60–74 (1977); R. H. Felton, L. D. Cheung, R. S. Phillips, and S. W. May, *Biochem. Biophys. Res. Commun.* 85, 844–850 (1978); W. E. Keyes, T. M. Loehr, M. L. Taylor, and J. S. Loehr, *Biochem. Biophys. Res. Commun.* 89, 420–427 (1979).

(194) Y. Ishimura, M. Nozaki, O. Hayaishi, T. Nakamura, M. Tamura, and I. Yamazaki, *JBC* 245, 3593–3602 (1970); F. O. Brady, P. Feigelson, and K. V. Rajagopalan, *Arch. Biochem. Biophys.* 157, 36–72 (1973); F. O. Brady and P. Feigelson, *JBC* 250, 5041–5048 (1975).

(195) J. J. M. C. de Groot, G. A. Veldink, J. F. G. Vliegenhart, J. Boldingh, R. Wener, and B. F. van Gelder, *BBA* 377, 71–79 (1975); M. R. Egmond, A. Finazzi Agro, P. M. Fasella, G. A. Veldink, and J. F. G. Vliegenhart, *BBA* 397, 43–49 (1975); M. J. Gibian and R. A. Galaway, in *Bioorganic Chemistry*, E. E. van Tamelen, ed., Academic Press, New York, 1977, Vol. 1, p. 132.

(196) F. Hirata, T. Ohnishi, and O. Hayaishi, *JBC* 252, 4637–4642 (1977).

(197) A. Olomucki, D. B. Pho, R. Lebar, L. Delcambe, and N. V. Thoai, *BBA* 151, 353–366 (1968).

(198) H. Takeda and O. Hayaishi, *JBC* 241, 2733–2736 (1966).

(199) W. B. Sutton, *JBC* 226, 395–405 (1957); O. Lockridge, V. Massey, and P. A. Sullivan, *JBC* 247, 8097–8106 (1972); S. Takemori, Y. Nakai, K. Nakazawa, M. Katagiri, and T. Nakamura, *Arch. Biochem. Biophys.* 154, 137–146 (1973); Refs. (11–13).

(200) B. C. Axcell and P. J. Geary, *Biochem. J.* 146, 173–183 (1975).

(201) L. G. Sparrow, P. P. K. Ho, T. K. Sundaram, D. Zach, E. J. Nyns, and E. E. Snell, *JBC* 244, 2590–2600 (1969).

(202) S. Takemori, H. Yasuda, K. Mihara, K. Suzuki, and M. Katagiri, *BBA* 191, 58–76 (1969); R. H. White-Stevens, H. Kamin, and Q. H. Gibson, *JBC* 247, 2371–2381 (1972).

(203) T. Spector and V. Massey, *JBC* 247, 5632–5636 (1972); T. Spector and V. Massey, *JBC* 247, 7123–7127 (1972); Refs. (17, 18).

(204) S. Strickland and V. Massey, *JBC* 248, 2953–2962 (1973); S. Strickland, L. M. Schropfer, and V. Massey, *B* 14, 2230–2235 (1975); L. M. Schopfer and V. Massey, *JBC* 255, 5355–5363 (1980).

(205) Y. Maki, S. Yamamoto, M. Nozaki, and O. Hayaishi, *JBC* 244, 2942–2950 (1969).
(206) Y. Ohta and D. W. Ribbons, *FEBS lett.* 11, 189–192 (1970); Y. Ohta, I. J. Higgins, and
D. W. Ribbons, *JBC* 250, 3814–3825 (1975); Y. Ohta and D. W. Ribbons, *EJB* 61, 259–269
(1976).
(207) V. Massey and P. Hemmerich, in *The Enzymes*, 3rd ed., P. D. Boyer, ed., Academic Press,
New York, 1975, Vol. 12, pp. 221–223; H. Y. Neujahr and K. G. Kjellen, *JBC* 253, 8835–8841
(1978).
(208) D. M. Ziegler and C. H. Mitchell, *Arch. Biochem. Biophys.* 150, 116–125 (1972); L. L.
Poulson and D. M. Ziegler, *JBC* 254, 6449–6455 (1979).
(209) P. E. Holmes and S. C. Rittenberg, *JBC* 247, 7622–7627 (1972).
(210) M. Griffin and P. W. Trudgill, *EJB* 63, 199–209 (1976).
(211) D. B. Morris and P. W. Trudgill, *Biochem. J.* 130, 30P (1972); N. A. Donoghue, D. B.
Morris, and P. W. Trudgill, *EJB* 63, 175–192 (1976).
(212) L. N. Britton and A. J. Markovetz, *JBC* 252, 8561–8566 (1977).
(213) Y. Ohta and D. W. Ribbons, *EJB* 61, 259–269 (1976).
(214) E. A. Elmorsi and D. J. Hopper, *EJB* 76, 197–208 (1977).
(215) E. G. Hrycay, J. A. Gustafsson, M. Ingelmann-Sundberg, and L. Ernster, *EJB* 61, 43–52
(1976); R. W. Estabrook, A. G. Hildebrandt, J. Baron, K. J. Netter, and K. Liebman, *Biochem.
Biophys. Res. Commun.* 42, 132–139 (1971); V. Ullrich and K. H. Schnabel, *Arch. Biochem.
Biophys.* 159, 240–248 (1973); F. P. Guengerich, D. P. Ballou, and M. J. Coon, *Biochem.
Biophys. Res. Commun.* 70, 951–956 (1976); E. Begard, P. Debey, and P. Douzou, *FEBS lett.* 75,
52–54 (1977); C. Larroque and J. E. van Lier, *FEBS lett.* 115, 175–177 (1980).
(216) Refs. (82, 83).
(217) G. Cardini and P. Jurtshuk, *JBC* 245, 2789–2796 (1970).
(218) D. Y. Cooper, H. Schleyer, and O. Rosenthal, *Ann. N.Y. Acad. Sci.* 174 (art. 1), 205–217
(1970); H. Schleyer, D. Y. Cooper, and O. Rosenthal, in *Oxidases and Related Redox Systems*,
T. E. King, H. S. Mason, and M. Morrison, eds., University Park Press, Baltimore, 1973, Vol. 2,
pp. 469–484.
(219) S. Friedman and S. Kaufman, *JBC* 240, 4763–4773 (1965); W. Blumberg, M. Goldstein,
E. Lauber, and J. Peisach, *BBA* 99, 187–188 (1965); T. Ljones, T. Flatmark, T. Skotland, L.
Pctersson, D. Bäckström, and A. Ehrenberg, *FEBS lett.* 92, 81–84 (1978); T. Skotland, L.
Petersson, B. Bäckström, T. Ljones, T. Flatmark, and A. Ehrenberg, *EJB* 103, 5–11 (1980).
(220) R. P. F. Gregory and D. S. Bendall, *Biochem. J.* 101, 569–581 (1966); N. Makino and
H. S. Mason, *JBC* 248, 5731–5735 (1973).
(221) F.-H. Bernhardt, H. Pachowsky, and H. Staudinger, *EJB* 57, 241–256 (1975).
(222) D. Klug, J. Rabani, and I. Fridovich, *JBC* 247, 4839–4844 (1972); G. Rotilio, L. Mopurgo,
C. Giovagnoli, L. Calabrese, and B. Mondovi, *B* 11, 2187–2192 (1972); D. Klug-Roth, I.
Fridovich, and J. Rabani, *JACS* 95, 2786–2790 (1973); J. A. Fee and R. L. Ward, *Biochem.
Biophys. Res. Commun.* 71, 427–437 (1976); A. Rigo, R. Stevanato, P. Viglino, and G. Rotilio,
*Biochem. Biophys. Res. Commun.* 79, 776–783 (1977).
(223) L. Broman, B. G. Malmström, R. Aasa, and T. Vänngård, *BBA* 75, 365–376 (1963);
S. Osaki and O. Walaas, *JBC* 242, 2653–2657 (1967); Ref. (68).
(224) P. A. Rubenstein and J. L. Strominger, *JBC* 249, 3776–3781 (1974).
(225) L. Thelander, *JBC* 249, 4858–4862 (1974); L. Petersson, A. Gräslund, A. Ehrenberg,
B.-M. Sjöberg, and P. Reichard, *JBC* 255, 6706–6712 (1980); B.-M. Sjöberg, A. Gräslund,
J. S. Loehr, and T. M. Loehr, *Biochem. Biophys. Res. Commun.* 94, 793–799 (1980).
(226) E. Vitols, H. P. C. Hogenkamp, C. Brownson, R. L. Blakley, and J. Connellan, *Biochem. J.*
104, 58C–60C (1967); L. Thelander and P. Reichard, *Ann. Rev. Biochem.* 48, 133–158 (1979).
(227) T. Ueda and M. J. Coon, *JBC* 247, 5010–5016 (1972).
(228) M. Shin and D. I. Arnon, *JBC* 240, 1405–1411 (1965); M. Shin, in *Flavins and
Flavoproteins*, K. Yagi, ed., University Park Press, Baltimore, 1968, pp. 1–14; J. W. Chu and
T. Kimura, *JBC* 248, 2089–2094 (1973); J. D. Lambeth and H. Kamin, *JBC* 254, 2766–2774
(1979).
(229) B. E. Smith, D. J. Lowe, and R. C. Bray, *Biochem. J.* 130, 641–643 (1972); W. G. Zumft,

W. C. Cretney, T. C. Huang, L. E. Mortenson, and G. Palmer, *Biochem. Biophys. Res. Commun.* 48, 1525–1532 (1972); W. H. Orme-Johnson, M. D. Hamilton, T. L. Jones, M. Y. Tso, R. H. Burris, V. K. Shah, and W. J. Brill, *PNAS* 69, 3142–3145 (1972); R. N. F. Thorneley, R. R. Eady, and D. J. Lowe, *Nature (London)* 272, 557–558 (1978).

(230) A. I. Krasna and D. Rittenberg, *JACS* 76, 3015–3020 (1954); N. Tamiya and S. L. Miller, *JBC* 238, 2194–2198 (1963); A. I. Krasna, *B* 14, 2561–2568 (1975); D. L. Erbes, R. H. Burris, and W. H. Orme-Johnson, *PNAS* 72, 4795–4799 (1975); B. A. Averill and W. H. Orme-Johnson, *JACS* 100, 5234–5236 (1978).

# Chapter 3

# Transferases

It was earlier asserted that all enzymes can rightly be regarded as transferases, in the sense that they all catalyze the transfer of some fragment—an electron, a proton, a group, or even the whole—of the donor substrate to an acceptor (1). For convenience of classification, however, the Enzyme Commission of the International Union of Biochemistry has sorted the enzymes into six major classes, only one of which is officially termed "transferases" (2). These (official) transferases are group transferases. The present chapter presents a sampling of such transferases, with examples drawn from nearly all of the subsubclasses (categories) of these enzymes. The presentations are concise, and highlight mainly those aspects of each enzyme which have to do with covalent catalysis. It will be seen that group transfer is never a direct one between donor and acceptor. Always the enzyme acts to receive a group from the donor and pass it on to the acceptor. Some of the enzymes chosen for exposition have a long-familiar and accepted mode of action, and are included for the sake of completeness. But others are, at the time of writing, frankly controversial, there being no consensus as to their chemical mechanism. They are designedly included here in order to stress the issues which are in dispute.

Table 3.1 at the end of the present chapter lists all of the transferases for which some positive evidence affirms that they engage in covalent catalysis. The table includes 105 enzymes. This number comes to 18% of all the enzymes

---

(1) L. B. Spector, *Bioorg. Chem.* 2, 311–321 (1973).

(2) Enzyme Nomenclature. Recommendations (1978) of the nomenclature committee of The International Union of Biochemistry, Academic Press, New York, 1979.

listed officially by the Enzyme Commission as transferases. On grounds of chemical analogy, a reasonable extrapolation of the 18% figure to 94% can be made, leaving only 6% of the official transferases as unrepresented in Table 3.1 (see also Chapter 8).

# Methionine Synthase [EC 2.1.1.13]

In a methyl group transfer of some complexity, 5-methyl tetrahydrofolate yields up its methyl to homocysteine to form methionine, in a reaction catalyzed by methionine synthase.

$$5\text{-Methyl THF} + \text{homocysteine} \longrightarrow \text{methionine} + \text{THF}$$

The enzyme has cobalamin as prosthetic group. The cobalt atom of the latter is the locus of methyl attachment in its migration from donor to acceptor. Native methionine synthase is actually isolated from tissue in the methylated state (3). Homocysteine easily demethylates the methylated enzyme, whose chemical competence as a mediator of methyl transfer is thus established. Leading to the same conclusion are the results of kinetic investigations on methionine synthase from different sources, these showing that the enzyme follows ping-pong kinetics (4, 5). The methylated enzyme is also isolable from incubation mixtures in which the enzyme is treated with 5-[$^{14}$C]methyl THF in the absence of homocysteine (6). The [$^{14}$C]methyl-enzyme reacts quantitatively with homocysteine to give labeled methionine. Methylation of the enzyme with 5-methyl THF requires that the cobalamin be in the lowest state of oxidation; that is, cob(I)-alamin (7).

---

(3) H. Rüdiger and L. Jaenicke, *EJB* 10, 557–560 (1969).

(4) H. Rüdiger and L. Jaenicke, *FEBS lett.* 1, 293–294 (1968); G. T. Burke, J. H. Mangum, and J. D. Brodie, *B* 10, 3079–3085 (1971).

(5) The term "ping-pong" refers rightly only to the *kinetic* mechanism of an enzyme reaction, and not to its chemical mechanism. A transferase exhibiting authentic ping-pong kinetics has, to be sure, a covalent enzyme–substrate intermediate as a component of its catalytic cycle. But enzymes exhibiting non-ping-pong kinetics can also act through a covalent enzyme–substrate intermediate. Some writers have not always been clear on this point. They often call ping-pong an enzyme that has been shown to link covalently to substrate, irrespective of the kinetics actually followed. For clarity, the term ping-pong ought to be used only in a kinetic sense.

(6) J. Stavrianopoulos and L. Jaenicke, *EJB* 3, 95–106 (1967); R. T. Taylor and H. Weissbach, *Arch. Biochem. Biophys.* 119, 572–579 (1967); G. T. Burke, J. H. Mangum, and J. D. Brodie, *B* 9, 4297–4302 (1970).

(7) H. Rüdiger, *EJB* 21, 264–268 (1971).

# *N*-Methylglutamate Synthase [EC 2.1.1.21]

An unusual but profoundly interesting methyltransferase is *N*-methylgluta-
mate transferase. It has a flavin for prosthetic group, and participates in a
redox-dependent catalysis of group transfer (8). Though the enzyme is
classified as a methyltransferase, it is in reality a glutaryltransferase, catalyzing the reversible transfer of a 2-glutaryl group from linkage to an amino
group to linkage with a methylamino group.

$$\text{L-Glutamate} + \text{CH}_3\text{NH}_2 \longleftrightarrow \textit{N}\text{-methylglutamate} + \text{NH}_3$$

The transfer occurs via a glutaryl-enzyme intermediate, which can be isolated
in labeled form after incubating the enzyme with [$^{14}$C]glutamate or *N*-
[*glutaryl*-$^{14}$C]methylglutamate. Such incubations incorporate radioactivity
into the enzyme, whereas none is incorporated from *N*-[*methyl*-$^{14}$C]methyl-
glutamate. The labeled enzyme—whose rate of formation equals the rate of
the net reaction—yields glutamate or *N*-methylglutamate upon treatment
with ammonia or methylamine, respectively. When glutaryl enzyme is
denatured with acid or by heating, the radioactivity is released as α-ketoglutarate, proving that the substrate is oxidized when it is bound to the
enzyme. Concurrent with the formation of the glutaryl enzyme is the reduction of a stoichiometric amount of enzyme-bound FMN; and, conversely,
the removal of the glutaryl group from the enzyme by ammonia or methylamine causes the stoichiometric reoxidation of the reduced flavin. The rates
of flavin reduction and reoxidation are entirely compatible with that of the
overall reaction. Consistent with the role of glutaryl enzyme as an intermediate of the reaction is the catalysis by the enzyme of a methylamine—
*N*-methylglutamate exchange, but not of a glutamate—*N*-methylglutamate
exchange. The enzyme also exhibits ping-pong kinetics in both the forward
and reverse directions.

The foregoing facts are accommodated by a sequence of chemical events
such as are depicted in Fig. 1. These events include the oxidation of the α-
carbon of glutamate (or *N*-methylglutamate) by the FMN of the enzyme
followed by two "trans-Schiffizations" involving an amino function of the
enzyme. Completing the catalytic cycle is the reduction of the α-carbon to its
original state of oxidation.

At no time during the reaction does the proton on the α-carbon of the
substrate exchange with protons of the medium. Thus α-tritiated glutamate,
reacting with the enzyme, yields glutaryl enzyme in which the tritium is
retained; and methylamine, reacting with the tritiated intermediate, returns
tritiated *N*-methylglutamate. Closer study reveals that the hydrogen on the
α-carbon of substrate is transferred reversibly to the N-5 position of FMN
during the redox phase of the catalytic cycle, the N-5 of FMN being sheltered

(8) R. J. Pollock and L. B. Hersh, *JBC* 248, 6724–6733 (1973).

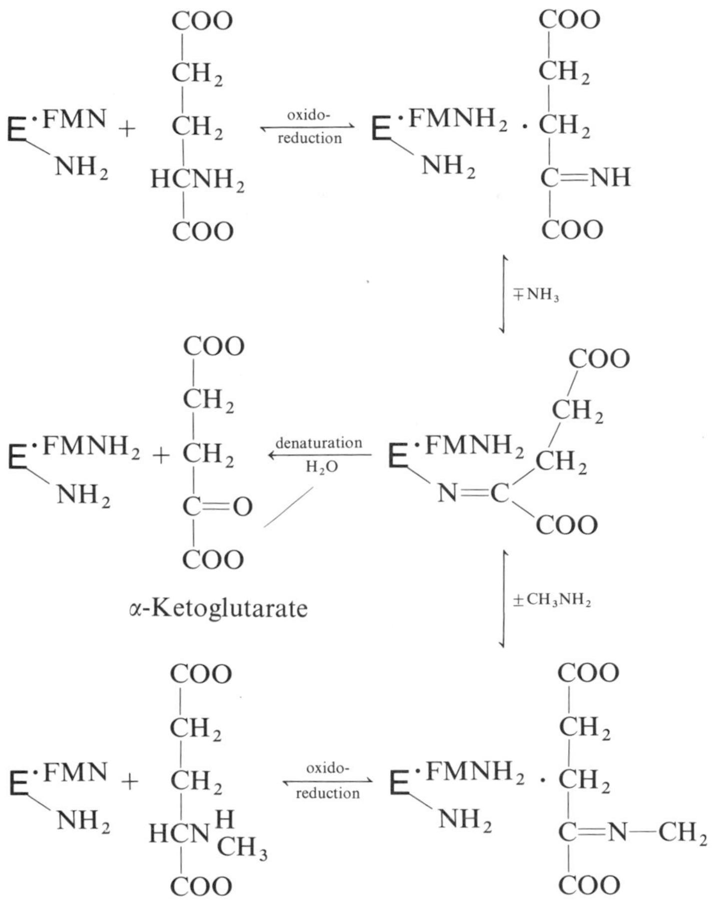

$$E\diagdown{}^{\cdot FMN}_{NH_2} + \begin{matrix} COO \\ | \\ CH_2 \\ | \\ CH_2 \\ | \\ HCNH_2 \\ | \\ COO \end{matrix} \xrightarrow[\text{reduction}]{\text{oxido-}} E\diagdown{}^{\cdot FMNH_2}_{NH_2} \cdot \begin{matrix} COO \\ | \\ CH_2 \\ | \\ CH_2 \\ | \\ C=NH \\ | \\ COO \end{matrix}$$

$\mp NH_3$

$$E\diagdown{}^{\cdot FMNH_2}_{NH_2} + \begin{matrix} COO \\ | \\ CH_2 \\ | \\ CH_2 \\ | \\ C=O \\ | \\ COO \end{matrix} \xleftarrow[\text{H}_2\text{O}]{\text{denaturation}} E\diagdown{}^{\cdot FMNH_2}_{N=C} \begin{matrix} COO \\ | \\ CH_2 \\ CH_2 \\ \diagdown COO \end{matrix}$$

α-Ketoglutarate

$\pm CH_3NH_2$

$$E\diagdown{}^{\cdot FMN}_{NH_2} + \begin{matrix} COO \\ | \\ CH_2 \\ | \\ CH_2 \\ | \\ HCN{}^{H}_{CH_3} \\ | \\ COO \end{matrix} \xrightarrow[\text{reduction}]{\text{oxido-}} E\diagdown{}^{\cdot FMNH_2}_{NH_2} \cdot \begin{matrix} COO \\ | \\ CH_2 \\ | \\ CH_2 \\ | \\ C=N-CH_2 \\ | \\ COO \end{matrix}$$

**Fig. 1.** A proposed mechanism for the N-methylglutamate synthase reaction (8).

from the medium (9). This last recalls, in some of its aspects, the proton retention observed in the action of D-amino-acid oxidase.

# Serine Hydroxymethyltransferase [EC 2.1.2.1]

With pyridoxal 5-phosphate as prosthetic group, serine hydroxymethyl-transferase catalyzes the reversible conversion of serine into glycine, and tetrahydrofolate into 5,10-methylenetetrahydrofolate.

L-Serine + tetrahydrofolate ⟶ glycine + 5,10-methylenetetrahydrofolate    (1)

(9) M. S. Jorns and L. B. Hersh, *JACS* 96, 4012–4014 (1974); *idem.*, *JBC* 250, 3620–3628 (1975).

It is established that serine and glycine form Schiff bases with the holoenzyme during the course of the reaction (10, 11). The glycyl-enzyme intermediate has a characteristic absorption spectrum (11). Also catalyzed by serine hydroxy-methyltransferase is the conversion of serine into glycine and formaldehyde in the absence of tetrahydrofolate (12, 13).

$$\text{L-Serine} \longleftrightarrow \text{glycine} + \text{formaldehyde}$$

This reaction is much slower than reaction 1. After the $\alpha,\beta$-bond of serine is cleaved, as evidenced by the appearance of the glycyl-enzyme spectrum, the formaldehyde dissociates very slowly from the enzyme. This slow dissociation contrasts sharply with the rapid dissociation of acetaldehyde and benzalde-hyde after the cleavage, respectively, of the $\beta$-hydroxyamino acids allo-threonine and $\beta$-phenylserine.

$$\text{Allothreonine} \longleftrightarrow \text{glycine} + \text{acetaldehyde}$$

$$\beta\text{-Phenylserine} \longleftrightarrow \text{glycine} + \text{benzaldehyde}$$

The enzyme catalyzes these latter reactions at rates which are fully com-parable to that of reaction 1. Inferred from these facts is that formaldehyde, unlike the bulkier acetaldehyde and benzaldehyde, has a specific subsite within the active center to which it joins when in transit between glycyl enzyme and tetrahydrofolate. Schematically, reaction 1 can be pictured as in Fig. 2 (12). Serine and glycine are of course fixed in the active center of the enzyme as Schiff bases, while formaldehyde is thought to form first a hydroxylmethylamine and then an imine with an amino group which is within reaction range of tetrahydrofolate and the glycyl-Schiff base (13):

(10) L. Schirch and W. T. Jenkins, *JBC* 239, 3801–3807 (1964).
(11) L. Schirch and A. Diller, *JBC* 246, 3961–3966 (1971).
(12) M. S. Chen and L. Schirch, *JBC* 248, 3631–3635 (1973).
(13) M. S. Chen and L. Schirch, *JBC* 248, 7979–7984 (1973).

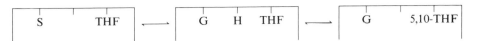

**Fig. 2.** Schematic representation of the serine hydroxymethyltransferase reaction (12). (S, serine; G, glycine; H, hydroxymethyl; THF, tetrahydrofolate; 5,10-THF, 5,10-methylenetetrahydrofolate.)

## Transcarboxylase [EC 2.1.3.1]

Transcarboxylase fulfills a pivotal role in propionic acid fermentation by catalyzing the reversible transfer of carbon dioxide from $S$-methylmalonyl-CoA to pyruvate to yield oxaloacetate and propionyl-CoA.

$$CH_3CH(COO)COSCoA + CH_3COCOO \longleftarrow$$
$$CH_3CH_2COSCoA + OOCCH_2COCOO \tag{2}$$

Covalently bound biotin is a prosthetic group of the enzyme, and is fixed to the smallest of its three dissimilar subunits, which have been isolated from each other and purified. The two larger subunits are carboxyltransferases. The larger of these catalyzes the reversible carboxylation of the biotinyl subunit by methylmalonyl-CoA,

$$\overset{14COO}{\underset{|}{CH_3CHCOSCoA}} + \text{biotinyl subunit} \xrightarrow[\text{subunit}]{\text{larger}}$$
$$OO^{14}C\text{-biotinyl subunit} + CH_3CH_2COSCoA$$

while the smaller catalyzes the reversible carboxylation of the biotinyl subunit by oxaloacetate (14).

$$OO^{14}C\text{-biotinyl subunit} + CH_3COCOO \xrightarrow[\text{subunit}]{\text{smaller}}$$
$$\text{biotinyl subunit} + OO^{14}CCH_2COCOO \tag{3}$$

These two partial reactions add up to the net reaction (Eq. 2). Isolation of the carboxylated holoenzyme (15) and of the carboxylated biotinyl subunit (14) are easily accomplished. In these intermediates the carbon dioxide is found linked to the 1'-nitrogen of biotin.

(14) M. Chuang, F. Ahmad, B. Jacobson, and H. G. Wood, *B* 14, 1611–1619 (1975).

(15) H. G. Wood, H. Lochmüller, C. Riepertinger, and F. Lynen, *Biochem. Z.* 337, 247–266 (1963).

$$\text{structure: } \begin{array}{c} C \stackrel{O}{\diagup} \\ \diagdown CHCH_2CH_2CH_2CH_2-NH-COCH_2CH_2CH_2CH_2 \\ NH \end{array} \quad \begin{array}{c} O \\ \| \\ HN \stackrel{C}{\diagup} N-C \stackrel{O}{\diagdown} \\ H-\!\!-\!\!-H \quad O \\ H \\ S \quad H \end{array}$$

Transcarboxylase possesses a second prosthetic group—a Co(II) or Zn(II) ion—which is firmly fixed to the smaller of the two carboxyltransferase subunits. How cobalt might enter catalytically into the carboxylation of pyruvate (reaction 3) is shown in Fig. 3 (16). Coordination of pyruvate to the transition metal ought to loosen one of the methyl protons, the departure of which allows the carboxylation of pyruvate to proceed. An altogether similar transition metal intermediate figures in the nonenzymic decarboxylation of oxaloacetate (17).

The proton which leaves pyruvate to make room for carbon dioxide is not immediately lost to the medium. Instead it is transferred to propionyl-CoA, via biotin, when the latter is next carboxylated by methylmalonyl-CoA. As carbon dioxide leaves the 2-position of propionyl-CoA upon transfer to biotin, its place is filled by the former pyruvate proton, till now stored on the biotin. Once transferred to propionyl-CoA, the proton exchanges rapidly with water in the presence of transcarboxylase (18).

Transcarboxylase also has unusual kinetic properties. For both the forward and reverse reactions it exhibits initial velocity behavior typical of standard ping-pong kinetics, but the inhibition patterns diverge notably from those predicted for such kinetics. The term "hybrid ping-pong" has been introduced to describe the particular kinetics of transcarboxylase. They point to a model of transcarboxylase action which is depicted in Fig. 4 (19). The active center of the enzyme is shown equipped with two independent subsites (corresponding to the two larger subunits of the enzyme), one adapted to the two keto acids and the other to the two coenzyme A esters. Between them is poised the carboxyl-carrying biotinyl subunit, the ureido function of which stands at the end of a long chain of atoms extending out from the surface of the enzyme. The biotin structure and its attached chain constitute a "swinging arm," which carries a molecule of carbon dioxide

(16) D. B. Northrop and H. G. Wood, *JBC* 244, 5801–5807 (1969).
(17) R. Steinberger and F. H. Westheimer, *JACS* 73, 429–435 (1951).
(18) I. A. Rose, E. L. O'Connell, and F. Solomon, *JBC* 251, 902–904 (1976).
(19) D. B. Northrop, *JBC* 244, 5808–5819 (1969).

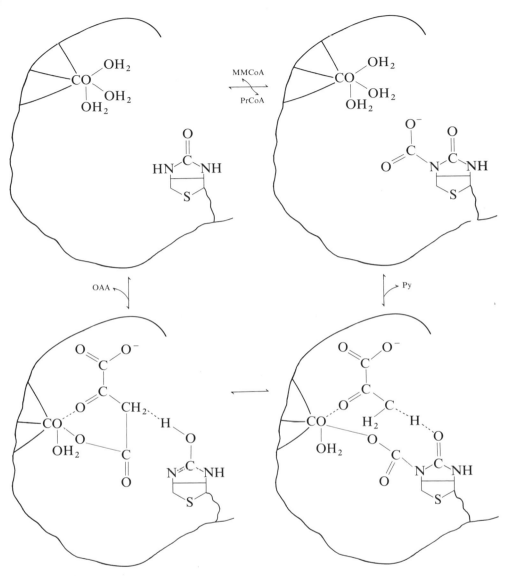

**Fig. 3.** Proposed role of cobalt in the transcarboxylase reaction (16). After biotin is carboxylated by methylmalonyl-CoA, pyruvate coordinates to the cobalt, aiding thus the departure of a methyl proton and subsequent transfer of carbon dioxide from carboxylbiotin to the enolized pyruvate. (MMCoA, methylmalonyl-CoA; PrCoA, propionyl-CoA; Py, pyruvate; OAA, oxaloacetate.)

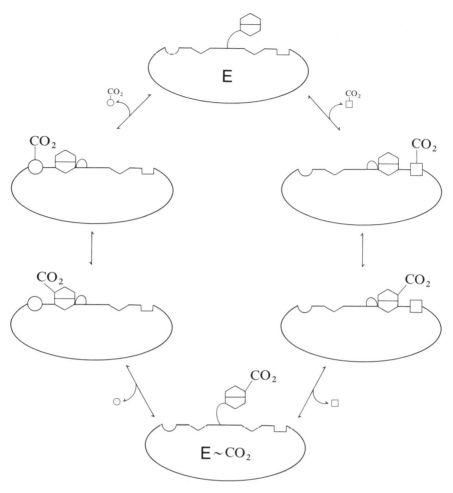

**Fig. 4.** Proposed model of the transcarboxylase reaction showing the "swinging arm" (19). (Meaning of symbols: free circle, pyruvate; carboxylated circle, oxaloacetate; free square, propionyl-CoA; carboxylated square, methylmalonyl-CoA; hexagonal structure, biotin; carboxylated hexagonal structure, carboxylbiotin.)

between the two flanking subsites, each gripping their respective donor or acceptor substrates. The "swinging arm" is thus revealed as a handy device, which is widely used in covalent catalysis by enzymes.

## Glycine Amidinotransferase [EC 2.1.4.1]

Glycine amidinotransferase participates in the first phase of creatine biosynthesis by catalyzing the transfer of an amidino group from arginine to glycine.

$$
\begin{array}{c}
\overset{+}{N}H_2 \\
\parallel \\
NH-C-NH_2 \\
\mid \\
(CH_2)_3 \\
\mid \\
H_3\overset{+}{N}-CH-COO^-
\end{array}
\quad + \; H_3\overset{+}{N}CH_2COO^- \;\longrightarrow
$$

L-Arginine                Glycine

$$
\begin{array}{c}
\overset{+}{N}H_3 \\
\mid \\
(CH_2)_3 \\
\mid \\
H_3\overset{+}{N}-CH-COO^-
\end{array}
\quad + \quad
\begin{array}{c}
\overset{+}{N}H_2 \\
\parallel \\
C-NH_2 \\
\mid \\
HN-CH_2-COO^-
\end{array}
$$

L-Ornithine          Guanidinoacetate

In the process, the enzyme itself is amidinated at an intermediary stage (20), with a cysteine sulfhydryl in the active center as the site of amidination (21).

$$
\text{L-Arginine} + \text{enzyme-SH} \;\longleftrightarrow\; \text{L-ornithine} + \text{enzyme-S}-\overset{\overset{\displaystyle NH_2^+}{\parallel}}{C}-NH_2
$$

$$
\text{Enzyme-S}-\overset{\overset{\displaystyle NH_2^+}{\parallel}}{C}-NH_2 + \text{glycine} \;\longleftrightarrow\; \text{enzyme-SH} + \text{guanidinoacetate}
$$

Incubation of the enzyme with L-[*guanidino*-$^{14}$C]arginine or L-[U-$^{14}$C]-arginine, followed by gel filtration, affords labeled enzyme in which one amidino group has combined with one molecule of enzyme (21). Upon reaction of the amidinated enzyme with glycine or ornithine, the amidino group is transferred to these acceptors, forming guanidinoacetate and L-arginine, respectively (20). Consistent with its role in the double-displacement process, glycine amidinotransferase catalyzes a glycine–guanidinoacetate exchange (22, 23) and an L-ornithine—L-arginine exchange (23, 24). The rates of these partial reactions are comparable to the rate of the net transamidination (23). The enzyme also exhibits ping-pong kinetics (25).

## Transketolase [EC 2.2.1.1]

Prominent in the pentose phosphate pathway of glucose metabolism is the enzyme transketolase, which catalyzes the reversible transfer of a ketol group from a keto sugar to an aldehyde acceptor.

(20) E. Grazi, F. Conconi, and V. Vigi, *JBC* 240, 2465–2467 (1965).
(21) E. Grazi and N. Rossi, *JBC* 243, 538–542 (1968).
(22) J. B. Walker, *JBC* 224, 57–66 (1957).
(23) S. Ratner and O. Rochovansky, *Arch. Biochem. Biophys.* 63, 296–315 (1956).
(24) J. B. Walker, *JBC* 221, 771–776 (1956).
(25) G. Ronca, V. Vigi, and E. Grazi, *JBC* 241, 2589–2595 (1966).

$$
\begin{array}{ccc}
\boxed{\begin{array}{c} CH_2OH \\ | \\ C{=}O \end{array}} & & HC{\nearrow}^{O} \\
| & & | \\
HOCH & + & HCOH \\
| & & | \\
HCOH & & HCOH \\
| & & | \\
CH_2OP & & HCOH \\
 & & | \\
 & & CH_2OP
\end{array}
$$

D-Xylulose-5-P        D-Ribose-5-P

(4)

$$
\begin{array}{ccc}
\boxed{\begin{array}{c} CH_2OH \\ | \\ C{=}O \end{array}} & & \\
| & & \\
HOCH & & HC{\nearrow}^{O} \\
| & + & | \\
HCOH & & HCOH \\
| & & | \\
HCOH & & CH_2OP \\
| & & \\
HCOH & & \\
| & & \\
CH_2OP & &
\end{array}
$$

D-Sedoheptulose-7-P        D-Glyceraldehyde-3-P

For prosthetic group, transketolase has thiamine pyrophosphate (TPP). There is also a requirement for magnesium ion. The ketol group which undergoes transfer may be regarded formally as the hypothetical ionized form of glycolaldehyde.

$$
\underset{}{\overset{O}{\underset{\|}{HOCH_2CH}}} \quad \xrightarrow{\;\;\times\;\;} \quad H^+ + [\overset{O}{\underset{\|}{HOCH_2C\colon}}]^-
$$

The enzymic equivalent of "ionized glycolaldehyde" is derived from enzyme-bound $\alpha,\beta$-dihydroxyethyl-TPP (**5** in Fig. 5), which can be prepared chemically (26) and is isolable from enzyme incubation mixtures (27).

The first enzymic event in transketolase action is thought to be adduct (**1**) formation between xylulose-5-P and the ylid form of TPP-enzyme. Fragmentation of the sugar portion of the adduct between C-2 and C-3 releases glyceraldehyde-3-P as one of the products of the net reaction. At the same time **2**—the enzymic equivalent of "ionized glycolaldehyde"—is formed. It is stabilized by resonance (**2** and **3**). It can reversibly acquire a proton from the medium to give $\alpha,\beta$-dihydroxyethyl-TPP-enzyme ("active glycolaldehyde," **5**); or in the presence of an acceptor (for instance, ribose-5-P

(26) L. O. Krampitz, I. Suzuki, and G. Greull, *Fed. Proc.*, 20, 971–977 (1961).
(27) A. G. Datta and E. Racker, *JBC* 236, 624–628 (1961).

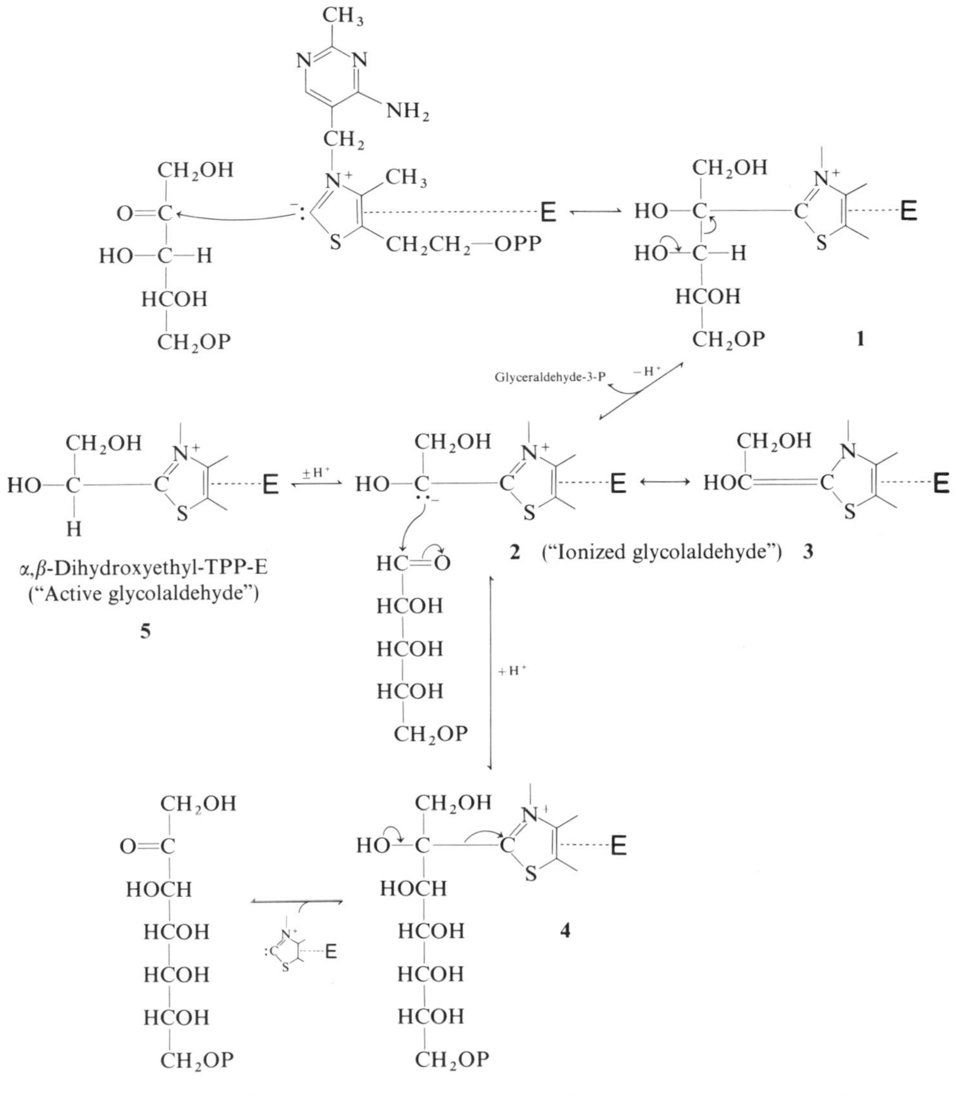

**Fig. 5.** The mechanism of the transketolase reaction showing the roles of "active glycolaldehyde" and "ionized glycolaldehyde." Thiamine pyrophosphate is firmly bound to the enzyme throughout the catalytic cycle.

as shown in Fig. 5), a new sugar adduct with TPP-enzyme, **4**, is formed, which is longer by two carbon atoms than the first adduct, **1**. Collapse of the new adduct to sedoheptulose-7-P and free holoenzyme completes the catalytic cycle. α,β-Dihydroxyethyl-TPP can be synthesized chemically by reacting TPP with two equivalents of formaldehyde at pH 8.8 (26). It combines readily with apotransketolase to give "active glycolaldehyde." A substrate quantity of the latter acts upon ribose-5-P to yield sedoheptulose-7-P.

"Active glycolaldehyde," **5**, can also be isolated from an incubation mixture in which transketolase acts upon uniformly labeled D-fructose-6-P—the enzyme being also active on this keto sugar—in the absence of a ketol acceptor. The intermediate, after isolation by ion exchange chromatography, can transfer its ketol group to ribose-5-P (27). In conformity with all of the foregoing, transketolase also catalyzes a rapid exchange between radioactive erythrose-4-P and fructose-6-P (28).

## Transaldolase [EC 2.2.1.2]

Equally prominent with transketolase in the pentose phosphate pathway is the enzyme transaldolase, which works in tandem with the former enzyme. It acts upon the two products of transketolase action (Eq. 4) to catalyze the following reversible reaction:

D-Sedoheptulose-7-P        D-Glyceraldehyde-3-P

D-Erythrose-4-P        D-Fructose-6-P

While transketolase transfers a 2-carbon segment of the keto sugar between acceptor aldehydes (Eq. 4), transaldolase transfers a 3-carbon segment—the dihydroxyacetonyl group. Transaldolase has no prosthetic group. Transfer

---

(28) M. G. Clark, J. F. Williams, and P. F. Blackmore, *Biochem. J.* 125, 381–384 (1971).

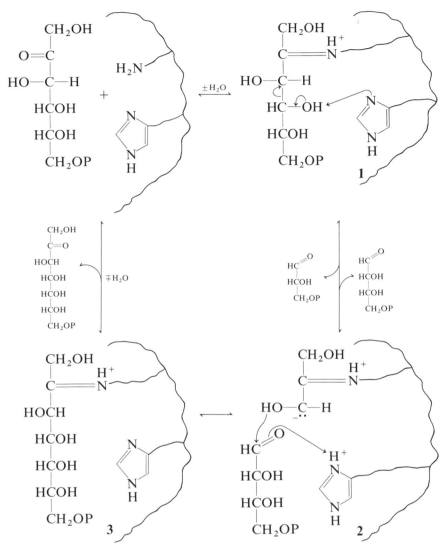

**Fig. 6.** Mechanism of transaldolase action, showing the role of the dihydroxyacetonyl-enzyme intermediate. The formation and hydrolysis of the two Schiff bases take place via the appropriate carbinolamines which, for brevity, are omitted.

is mediated by the ε-amino function of a lysyl residue in the active center which binds the dihydroxyacetonyl group in Schiff base linkage (29), and carries it reversibly between the acceptor aldehydes (Fig. 6). By incubating

(29) E. Grazi, P. T. Rowley, T. Cheng, O. Tchola, and B. L. Horecker, *Biochem. Biophys. Res. Commun.* 9, 38–43 (1962); E. Kuhn and K. Brand, *B* 12, 5217–5223 (1973); O. Tsolas and B. L. Horecker, *Arch. Biochem. Biophys.* 173, 577–585 (1976); E. Grazi, G. Balboni, K. Brand, and O. Tsolas, *Arch. Biochem. Biophys.* 179, 131–135 (1977).

transaldolase with labeled fructose-6-P in the absence of acceptor, labeled dihydroxyacetonyl-enzyme (**2**, Fig. 6) can be prepared and isolated (30, 31). This intermediate contains nearly one mole of dihydroxyacetone per mole of enzyme. The covalently bound dihydroxyacetone of the isolated intermediate is easily transferred to glyceraldehyde-3-P or erythrose-4-P (30, 32).

The catalytic cycle of transaldolase (Fig. 6) is believed to begin with the formation of a Schiff base, **1**, between substrate (fructose-6-P) and the catalytic amino group in the active center. Dealdolization follows. This process requires the removal of the hydroxyl proton on C-4 of the sugar. The proton, once removed from the substrate, is so held by the enzyme as to be nonexchangeable with the medium (33). The proton is transferred quantitatively to C-3 of the dihydroxyacetonyl group when the complex, **2**, is denatured or reduced with borohydride There are good grounds for believing that the enzymic base which detaches the proton from the substrate and holds it until released to product is the imidazole ring of a histidine residue, which is spatially close to the functional ε-amino group of the active center (33). Upon reaction of **2** with erythrose-4-P, a new Schiff base, **3**, forms, and is later hydrolyzed to sedoheptulose-7-P and regenerated enzyme.

## Arylamine Acetyltransferase [EC 2.3.1.5]

Among the many acetyl-transferring enzymes that are known one of the best studied is arylamine acetyltransferase, which catalyzes the acetylation of a variety of arylamines by acetyl-CoA.

$$\text{Acetyl-CoA} + \text{arylamine} \longleftrightarrow \text{CoA} + N\text{-acetylarylamine}$$

In place of acetyl-CoA the enzyme also acts on *p*-nitrophenyl acetate and *p*-nitroacetanilide as acetyl donors (34). The catalysis by the enzyme of pertinent exchange reactions—for instance, the 4-aminoazobenzene-4'-sulfonate–4-acetylaminoazobenzene-4'-sulfonate (35), CoA—acetyl-CoA, and aniline–acetanilide (36) exchanges—intimates the participation of an acetyl-enzyme intermediate in its action. In accord with this is the observation that the enzyme from diverse sources consistently follows ping-pong kinetics (34, 37). When the enzyme is incubated with [1-*acetyl*-$^{14}$C]acetyl-

(30) R. Venkataraman and E. Racker, *JBC* 236, 1883–1886 (1961).

(31) B. L. Horecker, S. Pontremoli, C. Ricci, and T. Cheng, *PNAS* 47, 1949–1955 (1961).

(32) B. L. Horecker, T. Cheng, and S. Pontremoli, *JBC* 238, 3428–3431 (1963).

(33) K. Brand, O. Tsolas, and B. L. Horecker, *Arch. Biochem. Biophys.* 130, 521–529 (1969).

(34) B. Riddle and W. P. Jencks, *JBC* 246, 3250–3258 (1971).

(35) S. P. Bessman and F. Lipmann, *Arch. Biochem. Biophys.* 46, 252–254 (1953).

(36) M. S. Steinberg, S. N. Cohen, and W. W. Weber, *BBA* 235, 89–98 (1971).

(37) J. W. Jenne and P. D. Boyer *BBA* 65, 121–127 (1962); W. W. Weber and S. N. Cohen, *Mol. Pharmacol.* 3, 266–273 (1967); W. W. Weber, S. N. Cohen, and M. S. Steinberg, *Ann. N.Y. Acad. Sci.* 151, 734–741 (1968).

CoA and gel filtered, the labeled acetyl enzyme can be isolated; and the acetyl group bound thus to the enzyme is easily transferable to acceptor (36).

Using p-nitrophenyl acetate as acetyl donor and anilines of varying basicity as acceptors, it was demonstrated that the acetylation of the enzyme is the rate-determining step in the presence of strongly basic anilines; and, conversely, the deacetylation of the enzyme is rate-determining when weakly basic anilines are the acceptors (34, 38). It also emerges from this work that the acetyl enzyme is kinetically significant, and that the acetyl group is almost certainly fixed to the enzyme in thiol ester linkage.

# Acetyl-CoA Acetyltransferase [EC 2.3.1.9]

The acetylation of a carbon atom takes place when two molecules of acetyl-CoA combine under the influence of acetyl-CoA acetyltransferase (thiolase) to form acetoacetyl-CoA and CoA.

$$CH_3\overset{O}{\overset{\|}{C}}-SCoA + CH_3\overset{O}{\overset{\|}{C}}-SCoA \longrightarrow CH_3\overset{O}{\overset{\|}{C}}CH_2\overset{O}{\overset{\|}{C}}-SCoA + HSCoA$$

As with arylamine acetyltransferase, the reaction catalyzed by thiolase is mediated by an acetyl-enzyme intermediate. Incubation of the enzyme with [acetyl-$^{14}$C]acetyl-CoA and subsequent gel filtration allows isolation of the acetyl enzyme (39, 40). Thiolase also exhibits ping-pong kinetics (41–43), and catalyzes the CoA—acetyl-CoA and acetyl-CoA—acetoacetyl-CoA exchanges (40). The acetyl group is known to be linked to a cysteine sulfhydryl of thiolase (39). But a second catalytic group in the active center appears also to hold the substrate covalently at some stage of the reaction (42, 44). This second group is thought to be an amino group because the enzyme is inactivated by sodium borohydride in the presence of substrate (acetoacetyl-CoA). How an amino group and a sulfhydryl might coordinate their activities in the active center to consummate the thiolase reaction is depicted in Fig. 7.

(38) W. P. Jencks, M. Gresser, M. S. Valenzuela, and F. C. Huneens, JBC 247, 3756–3760 (1972).
(39) U. Gehring and J. I. Harris, FEBS lett. 1, 150–152 (1968).
(40) U. Gehring, C. Riepertinger, and F. Lynen, EJB 6, 264–280 (1968); G. R. Duncombe and F. E. Frerman, Arch. Biochem. Biophys. 176, 159–170 (1976).
(41) D. S. Goldman, JBC 208, 345–357 (1954).
(42) J. Kornblatt and H. Rudney, JBC 246, 4417–4423 (1971).
(43) B. Middleton, Biochem. J. 139, 109–121 (1974); W. Huth, R. Jonas, I. Wunderlich, and W. Seubert, EJB 59, 475–489 (1975).
(44) P. C. Holland, M. G. Clark, and D. P. Bloxham, B 12, 3309–3315 (1973).

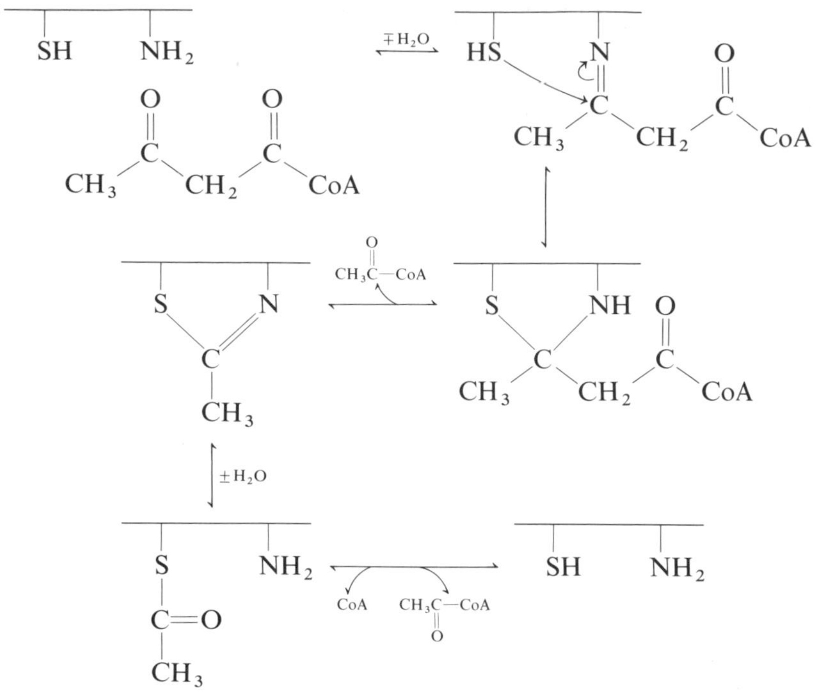

**Fig. 7.** The proposed roles of the sulfhydryl and amino groups in the active center of thiolase. The details of proton transfer are omitted. Adapted from (44).

## [Acyl-Carrier-Protein]Malonyltransferase [EC 2.3.1.39]

The acylation of a sulfur atom by an acyl-CoA is exemplified by the malonylation of the functional sulfur of the acyl-carrier-protein (ACP).

$$\text{Malonyl-CoA} + \text{HS—ACP} \longleftrightarrow \text{Malonyl—S—ACP} + \text{CoA}$$

ACP malonyltransferase is a component of the fatty-acid-synthesizing complex of enzymes which has been studied in a number of tissues. The enzyme follows ping-pong kinetics (45). Like the acetyltransferase above described, ACP malonyltransferase acts via an acyl enzyme. A malonyl enzyme can be isolated by gel filtration after incubation of the enzyme with malonyl-CoA or with malonyl-ACP (45, 46). The malonyl enzyme so prepared is chemically competent, transferring its malonyl group either to coenzyme A or to ACP. Unexpectedly, the point of covalent attachment of the malonyl group is the hydroxyl of a seryl residue in the active center

(45) V. C. Joshi and S. J. Wakil, *Arch. Biochem. Biophys.* 143, 493–505 (1971).
(46) F. E. Ruch and P. R. Vagelos, *JBC* 248, 8095–8106 (1973).

(46, 47). It is a notable fact that the malonyl-enzyme bond is an energy-rich one, judging from the capacity of the malonyl group to transfer reversibly between the thiolester condition in malonyl-CoA and malonyl-ACP. The native malonyl enzyme is readily disrupted by hydroxylamine and is easily hydrolyzed at alkaline pH. But when malonyl enzyme is denatured with urea it is rendered stable to hydroxylamine and to alkaline pH, and the malonyl group is no longer transferable. Removal of the denaturant, however, restores the sensitivity to these reagents along with the transferability to coenzyme A and ACP (46). Since a simple oxygen-ester, such as malonyl-$O$-serine, is not an energy-rich compound, it is clear that the tertiary structure of the native malonyl enzyme creates a microenvironment in the active center which somehow endows an ordinary oxygen-ester function with the quality of energy-richness. In activating the catalytic serine and its ester in the active center, the intact enzyme is seen here in its role as energy reservoir (p. 20).

# Transglutaminase [EC 2.3.2.13]

Yet another category of acyltransferases is formed by the aminoacyl-transferring enzymes, of which transglutaminase is a fully studied representative (48). In a reaction requiring $Ca^{2+}$ ion, transglutaminase catalyzes the replacement by a primary amine of the amino group in the $\gamma$-carboxamide function of peptide- or protein-bound glutamine residues.

$$
\begin{array}{ccc}
\underset{\displaystyle \text{C}-\text{NH}_2}{\overset{\displaystyle \text{O}}{\|}} & & \underset{\displaystyle \text{C}-\text{NHR}}{\overset{\displaystyle \text{O}}{\|}} \\
| & & | \\
\text{CH}_2 & + \text{RNH}_2 \xrightarrow{\ Ca^{2+}\ } & \text{CH}_2 \quad + \text{NH}_3 \\
| & & | \\
\overset{\displaystyle \text{O}}{\underset{}{\|}}\ \text{CH}_2 & & \overset{\displaystyle \text{O}}{\underset{}{\|}}\ \text{CH}_2 \\
\text{C}-\text{CH}-\text{NH} & & \text{C}-\text{CH}-\text{NH}
\end{array}
$$

When the primary amine is the $\varepsilon$-amino group of a lysyl residue of a second molecule of protein, a crosslink is established between the two peptide chains. Such crosslinking capacity is probably the basis of the transglutaminase participation in the blood clotting process. Transglutaminase also catalyzes the hydrolysis of the $\gamma$-carboxamide group, and, conveniently for mechanistic studies, the esterolysis of $p$-nitrophenyl esters, like $p$-nitrophenyl trimethyl-acetate. Reaction of the enzyme with the latter substrate proceeds through a trimethylacetyl-transglutaminase intermediate (49). This is revealed by the "burst" liberation of $p$-nitrophenolate ion as the enzyme reacts with $p$-nitrophenyl trimethylacetate, and by the concurrent incorporation of one

(47) G. T. Phillips, J. E. Nixon, A. S. Abramovitz, and J. W. Porter, *Arch. Biochem. Biophys.* 138, 357–371 (1970); E. Schweizer, F. Piccinini, C. Duba, S. Günther, E. Ritter, and F. Lynen, *EJB* 15, 483–499 (1970); V. C. Joshi, C. A. Plate, and S. J. Wakil, *JBC* 245, 2857–2867 (1970).
(48) J. E. Folk, *Ann. N.Y. Acad. Sci.* 202, 59–76 (1972).
(49) J. E. Folk, P. W. Cole, and J. P. Mullooly, *JBC* 242, 4329–4333 (1967).

mole of trimethylacetyl per mole of enzyme. An essential cysteine residue provides the catalytic sulfhydryl for binding to the acyl group of the substrate (49). In accord with the acyl-enzyme mechanism, transglutaminase catalyzes the predicted exchange reactions and displays ping-pong kinetics with some substrates (50). But with other substrates sequential kinetics are followed (51). In the latter case the acyl enzyme is formed from the ternary complex of the enzyme and its two substrates, and has no free existence.

## Sucrose Phosphorylase [EC 2.4.1.7]

Sucrose phosphorylase is one of that large category of enzymes known as the hexosyl transferases. It has the historical distinction of being the first enzyme for which a covalent enzyme–substrate complex was claimed to be an intermediate (52). It catalyzes the reversible transfer of an α-D-glucosyl group between D-fructose and orthophosphate.

Sucrose

α-D-Glucose-1-P                    β-D-Fructose

It was early discovered that the enzyme catalyzes a phosphate—glucose-1-P exchange (52), as well as a fructose–sucrose exchange (53), in the absence of

(50) J. E. Folk, *JBC* 244, 3707–3713 (1969); S. I. Chung and J. E. Folk, *JBC* 247, 2798–2807 (1972).
(51) M. Gross and J. E. Folk, *JBC* 248, 1301–1306 (1973).
(52) M. Doudoroff, H. A. Barker, and W. Z. Hassid, *JBC* 168, 725–732 (1947).
(53) H. Wolochow, E. W. Putnam, M. Doudoroff, W. Z. Hassid, and H. A. Barker, *JBC* 180, 1237–1242 (1949).

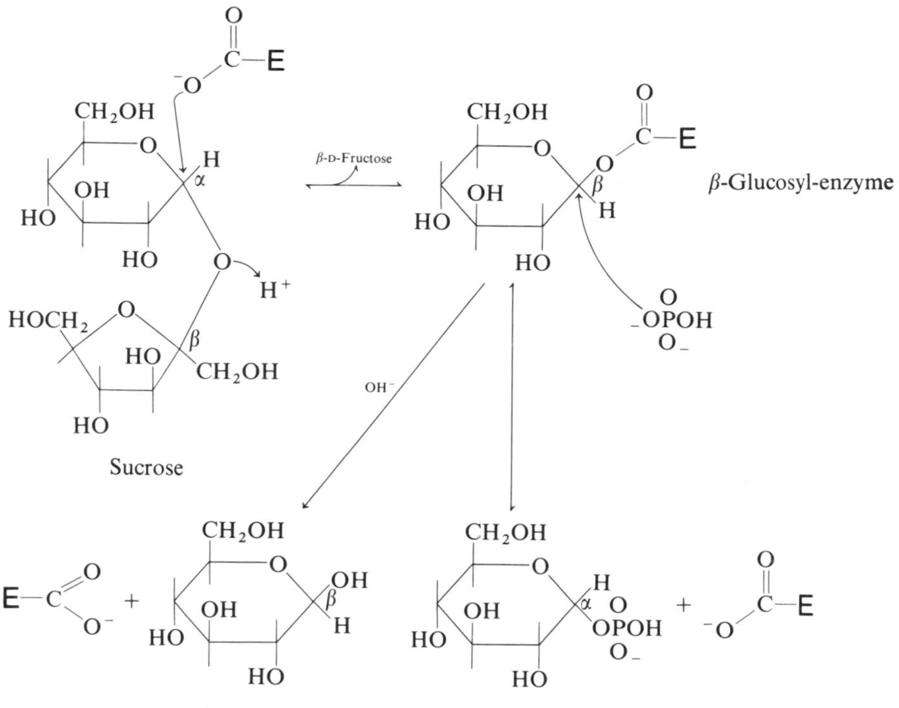

**Fig. 8.** Mechanism of the sucrose phosphorylase reaction showing the two steric inversions which account for the net retention of configuration at C-1 of glucose.

the respective cosubstrates. These observations prompted the notion that the transfer reaction is a composite of two partial reactions.

$$\alpha\text{-Glucosyl-fructoside (sucrose)} + E \longleftrightarrow \beta\text{-glucosyl—E} + \text{fructose} \quad (6)$$

$$\beta\text{-Glucosyl—E} + P_i \longleftrightarrow E + \alpha\text{-glucose-1-P} \quad (7)$$

Fortifying the glucosyl-enzyme concept are the stereochemical manifestations of the sucrose phosphorylase reaction. The α configuration of C-1 of the glucosyl portion of the sucrose molecule is retained in the product, α-glucose-1-P. It could be argued that such retention results from the front-side approach of orthophosphate to C-1 in a single-displacement reaction. But this is justly regarded as being most improbable (54). Much more reasonable is that the retention of configuration is a consequence of a pair of successive Walden inversions; the first being a backside displacement of the fructose from C-1 of glucose by a catalytic group in the active center of the enzyme, followed by a second backside displacement, this time of the catalytic group by the second substrate (Fig. 8). The catalytic group of the enzyme is known to be a carboxyl (55).

(54) D. E. Koshland, in *Mechanism of Enzyme Action,* W. D. McElroy and B. Glass, eds., Johns Hopkins Press, Baltimore, 1954, p. 608.
(55) F. De Toma and R. H. Abeles, *Fed. Proc.* 29, 461 (1970).

That the reaction between sucrose and enzyme (Eq. 6) does indeed take place with a steric inversion can be shown by rapidly denaturing a mixture of enzyme and [U$^{14}$C]sucrose. The denatured protein contains no fructose (56), but does bind a mole of glucose per mole of enzyme (57). The glucosyl-enzyme bond is alkali-labile; and the hydroxide ion, acting at the carbonyl carbon of the ester function of the intermediate, releases the sugar as $\beta$-D-glucose (Fig. 8). (58). Thus the formation of $\beta$-glucosyl enzyme in reaction 6 proceeds with inversion of configuration at C-1 of the glucosyl portion of sucrose. Since the net reaction (Eq. 5) results in overall retention of configuration, it follows that the reaction of $\beta$-glucosyl enzyme with orthophosphate (Eq. 7) must also proceed with inversion.

## Adenine Phosphoribosyltransferase
## [EC 2.4.2.7]

Pentosyl transferases, like the hexosyl transferases, catalyze the transfer of a sugar residue between a donor and an acceptor. A case in point is adenine phosphoribosyltransferase which moves a phosphoribosyl group between adenine and pyrophosphate.

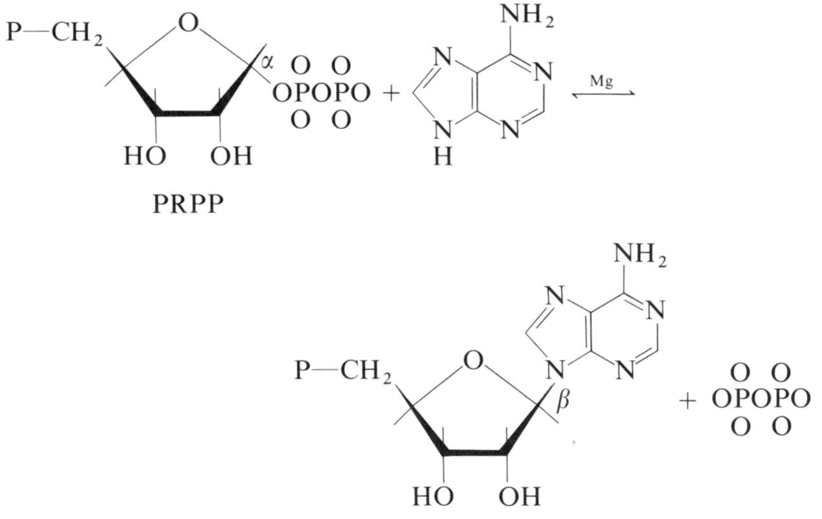

PRPP

(8)

AMP

Unlike the sucrose phosphorylase reaction in which the configuration of C-1 of the hexosyl group is retained in the product, the stereochemical course of phosphoribosyl transfer results in the inversion of C-1 of the pentosyl group.

---

(56) J. Voet and R. H. Abeles, *JBC* 241, 2731–2732 (1966).
(57) R. Silverstein, J. Voet, D. Reed, and R. H. Abeles, *JBC* 242, 1338–1346 (1967).
(58) J. Voet and R. H. Abeles, *JBC* 245, 1020–1031 (1970).

It might be supposed that the simplest way to account for the inversion is for adenine and PRPP to react directly with each other by a single-displacement mechanism. But such seems not to be the case. The enzyme, as isolated from rat liver or human erythrocytes, holds a phosphoribosyl group already fixed to it (59). This is revealed in the initial "burst" of AMP synthesis upon incubation of the isolated phosphoribosyl enzyme with $[^{14}C]$adenine. The "burst" synthesis has no requirement for magnesium ion, whereas the net synthesis of AMP from adenine and PRPP (Eq. 8) does. During the "burst," about one mole of $[^{14}C]$AMP is synthesized per mole of enzyme. The "burst" phenomenon requires the presence of PRPP which, however, cannot react with adenine according to Eq. 8 in the presence of EDTA (that is, in the absence of magnesium) or at 0°C, under which conditions the "burst" is nevertheless fully manifest. The PRPP is thought to facilitate the "burst" by putting the enzyme into its active conformation. It is noteworthy that though a phosphoribosyl enzyme mediates the reaction of Eq. 8, the enzyme follows sequential kinetics (60), and does *not* catalyze the predicted exchange reactions in the absence of cosubstrates (60, 61).

Somewhat in the same vein as the above are the findings made upon the enzyme orotate phosphoribosyltransferase [EC 2.4.2.10] which catalyzes the following reaction:

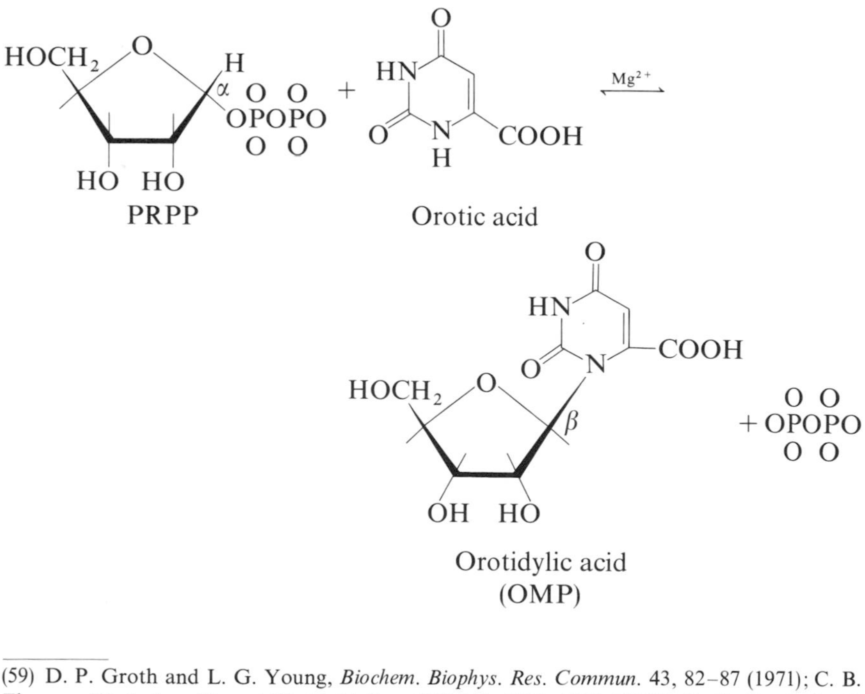

PRPP                    Orotic acid

Orotidylic acid
(OMP)

(59) D. P. Groth and L. G. Young, *Biochem. Biophys. Res. Commun.* 43, 82–87 (1971); C. B. Thomas, W. J. Arnold, and W. N. Kelley, *JBC* 248, 2529–2535 (1973); D. P. Groth, L. G. Young, and J. G. Kenimer, *Methods Enzymol.* 5, 574–580 (1978).
(60) R. E. A. Gadd and J. F. Henderson, *BBA* 191, 735–737 (1969).
(61) J. G. Kenimer, L. G. Young, and D. P. Groth, *BBA* 384, 87–101 (1975).

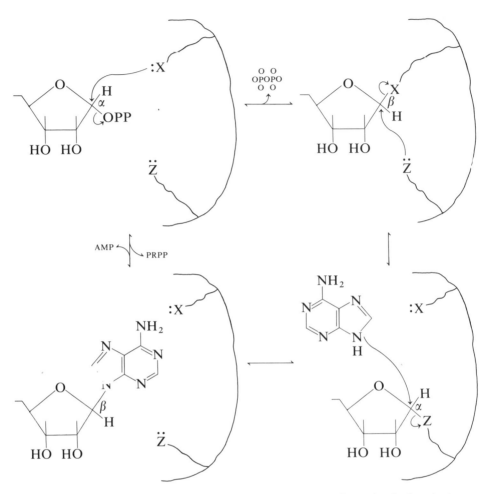

**Fig. 9.** A proposed mechanism to account for inversion of configuration in the adenine phosphoribosyltransferase reaction. The main feature is a triple displacement involving three steric inversions and two phosphoribosyl-enzyme intermediates.

In its chemical and stereochemical form this reaction is obviously identical with that of reaction 8, and so its chemical mechanism ought likewise to be identical. Though no phosphoribosyl enzyme has so far been reported, the reversible synthesis of OMP does follow authentic ping-pong kinetics in both the forward and reverse directions. The appropriate inhibition patterns for ping-pong kinetics are also observed. And further in accord with mediation by a phosphoribosyl enzyme is the catalysis by the enzyme of a pyrophosphate–PRPP and an orotate—OMP exchange in the absence of cosubstrates (62). Since a secondary isotope effect is observed when the proton on C-1 of

(62) J. Victor, L. B. Greenberg, and D. L. Sloan, *JBC* 254, 2647–2655 (1979).

PRPP is replaced by tritium (63), it is proposed that the enzyme-bound phosphoribosyl group exists as a carbonium ion (62), presumably stabilized as an ion-pair with a (carboxylate?) group of the enzyme. From the standpoint of covalent catalysis, the difference between the suggested ion-pair and a bonafide covalent intermediate is largely a semantic one.

There remains the problem of reconciling the steric course of these reactions with their mediation by a phosphoribosyl enzyme. If, as seems likely, all displacements on C-1 of the transferring pentosyl group give steric (Walden) inversions (54), then only an *odd* number of them can result in net inversion. We are forced to entertain the possibility that as the reaction unfolds the phosphoribosyl group is bonded covalently in succession to two catalytic groups of the enzyme, each bonding being the consequence of an $S_N2$ reaction with steric inversion (Fig. 9). Such a triple displacement gives, of course, a net inversion of configuration. Thus, the work on adenine and orotate phosphoribosyltransferase shows that, contrary to former belief, *a net inversion of configuration can be realized through covalent catalysis.*

## Thiaminase I [EC 2.5.1.2]

Thiaminase I is an enzyme capable of transferring the aralkyl group of thiamine to a variety of acceptor molecules, one of which is quinoline.

Thiamine                                    Quinoline

PM-quinoline                          5-(2-Hydroxyethyl)-
                                              4-methylthiazole

The aralkyl group undergoing transfer is the (4-amino-2-methyl-5-pyrimidinyl)methyl, abbreviated for convenience as the PM group. It happens that thiaminase I is active on other PM-donors besides thiamine; for instance, PM-3-chloropyridinium and pyrithiamine.

(63) R. K. Goitein, D. Chelsky, and S. M. Parsons, *JBC* 253, 2963–2971 (1978).

PM-3-chloropyridinium                    Pyrithiamine

It is observed that quinoline reacts with all of these PM-donors at the same maximum velocity. From this is deduced the mediation of a PM-enzyme intermediate, the reaction of which with quinoline must be the rate-limiting step of the net reaction (64). Were quinoline to react directly with each of the PM-donors in a single-displacement reaction, the rates ought to differ noticeably from each other, depending on the basicities of the respective leaving groups.

In like manner, when PM-quinoline serves as PM-donor to differing PM-acceptors (pyridine, benzenethiol, and $p$-nitrobenzenethiol) it is found that the $V_{max}$ values are virtually identical for the three reactions. Again, the most reasonable basis for this phenomenon is the participation in the reaction of an intermediate composed of the enzyme and the PM group. But this time the rate-determining step is the formation of the intermediate and not its further reaction.

## Aspartate Aminotransferase [EC 2.6.1.1]

Of the many transaminases we know of, aspartate aminotransferase has doubtless been the most searchingly scrutinized. It catalyzes the following amino group transfer:

L-Aspartate + 2-oxoglutarate ⟷ oxaloacetate + L-glutamate

The chemical mechanism of aspartate aminotransferase—of which the coenzyme is pyridoxal-5-P (PLP)—probably holds, in its essentials, for all the other PLP-requiring transaminases.

It was learned at an early date that aspartate aminotransferase exhibits authentic ping-pong kinetics (65). When aspartic acid gives up its amino group to the holoenzyme the latter is transformed into the intermediary phosphopyridoxamine form of the enzyme (E—PMP) (Fig. 10), which can be isolated free of substrates (66). E—PMP in turn, acting on its keto-acid substrates, is restored to the phosphopyridoxal form (E—PLP), to complete

(64) G. E. Lienhard, *B* 9, 3011–3020 (1970).

(65) S. F. Velick and J. Vavra, *JBC* 237, 2109–2122 (1962); C. P. Henson and W. W. Cleland, *B* 3, 338–345 (1964).

(66) W. T. Jenkins and F. W. Sizer, *JBC* 235, 620–624 (1960); H. Wada and E. E. Snell, *JBC* 237, 127–132 (1962); W. T. Jenkins and L. D'Ari, *JBC* 241, 2845–2854 (1966).

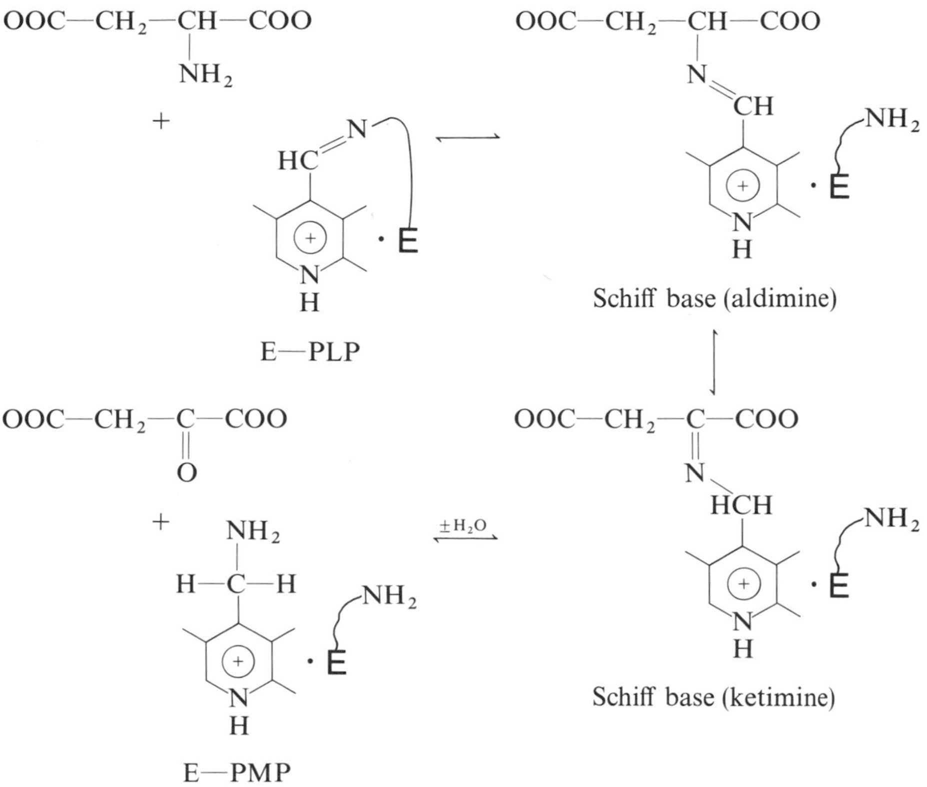

$$OOC-CH_2-CH-COO$$
$$\underset{NH_2}{|}$$

+

E—PLP

⟶

$$OOC-CH_2-CH-COO$$

Schiff base (aldimine)

$$OOC-CH_2-\underset{O}{\overset{||}{C}}-COO$$

+

$$NH_2$$
$$H-\overset{|}{C}-H$$

E—PMP

$\pm H_2O$

$$OOC-CH_2-\overset{||}{C}-COO$$

Schiff base (ketimine)

**Fig. 10.** A simple scheme for the first half-reaction catalyzed by aspartate aminotransferase, in which three covalent enzyme–substrate intermediates (two Schiff bases and E—PMP) have a share. The catalytic cycle is completed when E—PMP reacts with 2-oxoglutarate, in reversal of the above reactions, to yield glutamate.

the catalytic cycle. The chemical route to and from E—PMP includes two more covalent enzyme–substrate intermediates—the Schiff base forms of the enzyme. Starting with amino acid, the first Schiff base to appear is the aldimine. This isomerizes to the second Schiff base—the ketimine—as a result of the tautomeric migration of the proton on the α-carbon of the amino acid. Appearance of the Schiff bases can be followed spectrophotometrically (67, 68). A chemical verification of Schiff base formation is also possible through use of the substrate analogue α-methylaspartate. Lacking hydrogen on the

(67) P. Fasella, A. Giartosio, and G. G. Hammes, *B* 5, 197–202 (1966).
(68) P. A. Briley, R. Eisenthal, R. Harrison, and G. D. Smith, *Biochem. J.* 167, 193–200 (1977); G. Eichele, D. Karabelnik, R. Halonbrenner, J. N. Jansonius, and P. Christen, *JBC* 253, 5239–5242 (1978): C. M. Metzler, D. E. Metzler, D. S. Martin, R. Newman, A. Arnone, and P. Rogers, *JBC* 253, 5251–5254 (1978).

α-carbon, the analogue can undergo only the first part of the catalytic cycle with the enzyme to form the aldimine intermediate (67).

α-Methyl-L-aspartate                    E—PLP

Schiff base (aldimine)

The aldimine, upon reduction with borohydride and separation of the protein, yields the phosphopyridoxyl-α-methylaspartate derivative (69).

A key step in the transamination cycle is the prototropic shift alluded to above, which attends the aldimine–ketimine transformation. Intensive study of this process with a variety of PLP-transaminases, including aspartate aminotransferase, reveals that the proton transfer is a highly stereospecific one. It is consistent only with a cis ("one-side") removal and addition of the proton via a single basic group of the enzyme; that is, the loss and addition of the proton occur from the same side of the imine π system (70).

(69) A. Braunstein, *Vitam. Horm.* (*NY*) 22, 451–484 (1964).
(70) J. E. Ayling, H. C. Dunathan, and E. E. Snell, *B* 7, 4537–4542 (1968); H. C. Dunathan, *Adv. Enzymol.* 35, 79–134 (1971).

# Hexokinase [EC 2.7.1.1]

Hexokinase, which catalyzes the phosphorylation of the 6-hydroxyl group of glucose,

$$\text{D-Glucose} + \text{ATP} \longleftrightarrow \text{D-glucose-6-P} + \text{ADP} \qquad (9)$$

has long been an object of investigation as to its kinetic and chemical mechanism. We know now that the $\gamma$-phosphoryl group of ATP, upon transfer to glucose, undergoes a steric inversion on the phosphorus atom (71). At first view, it might be inferred from this fact that hexokinase catalyzes a single-displacement reaction. But steric inversion is also compatible with triple-displacement catalysis (72). And since there is some evidence for a phosphoenzyme intermediate in hexokinase action, triple-displacement catalysis by this enzyme seems the more likely prospect.

Eminently consistent with this view is the discovery that a phosphorylated hexokinase [with one phosphoserine group per monomeric unit (73)] is formed when the enzyme is incubated with ATP in the presence of the non-phosphorylatable sugar, D-xylose (74, 75, 76).

$$\text{Hexokinase (active)} + \text{ATP} \xrightarrow{\text{D-xylose}} \text{P-hexokinase (inactive)} + \text{ADP} \quad (10)$$

This results in the inactivation of the enzyme for the phosphorylation of glucose and for the hydrolysis of ATP (77). For this reason it is said to be a "lethal" phosphorylation. But reaction 10, we note, is reversible. In the presence of thiols or D-xylose (for stabilization), the phosphoenzyme can react with ADP, in reversal of reaction 10, to synthesize ATP (76, 78). The phosphoserine bond of the phosphoenzyme must therefore be energy-rich. On this view, it seems doubtful that the phosphoenzyme is a dead-end product of an aberrant chemical reaction. The very existence of phospho-hexokinase as an energy-rich entity, which can phosphorylate ADP, hints rather at a role for it in the normal catalytic process.

In light of the possibility that hexokinase is a triple-displacement enzyme, reaction 10 assumes the aspect of an interruption by D-xylose of progress along the normal catalytic pathway. The five-carbon sugar, D-xylose, is a competitive inhibitor of the six-carbon D-glucose, and sits in the glucose subsite during the hydrolysis of ATP, which it stimulates. One can imagine

---

(71) W. A. Blättler and J. R. Knowles, *JACS* 101, 510–512 (1979).
(72) See the discussions of adenine phosphoribosyltransferase (p. 80) and acetate kinase (p. 92).
(73) L. C. Menezes and J. Pudles, *EJB* 65, 41–47 (1976).
(74) S. P. Colowick, in *The Enzymes*, 3rd ed., P. D. Boyer, ed., Academic Press, New York, 1973, Vol. 9, pp. 1–48.
(75) L. Y. Cheng, T. Inagami, and S. P. Colowick, *Fed. Proc.* 32, 667 (1973).
(76) L. C. Menezes and J. Pudles, *Arch. Biochem. Biophys.* 178, 34–42 (1977).
(77) Hexokinase has a weak intrinsic ATPase activity proceeding at a rate of about 0.02 $\mu$mole/min/mg of protein compared with a value of 800 for the phosphorylation of glucose (74).
(78) L. Y. Cheng, T. Inagami, and S. P. Colowick, *Int. Cong. Biochem. Abstr. 9th* p. 90 (1973).

that as the protein closes around D-xylose in the glucose subsite (79), a molecule of water is enclosed with it, filling the space normally occupied by the 6-hydroxyl of glucose. The water is thus phosphorylated in place of the 6-hydroxyl. But once in about 300 turns of the ATPase cycle (80), a D-xylose molecule is gripped by the enzyme in the glucose subsite *unattended by the usual water molecule.* This of course leaves the 6-hydroxyl space vacant, and the phosphoryl group of the phosphoenzyme has no place to go—except back to an ADP molecule (Eq. 10). Given the restricted exit from and entry into the sugar subsite after the sugar-induced structural change (81), the phosphoenzyme finds itself locked into a condition in which it can phosphorylate neither water nor glucose.

The hexokinase dimer in solution, though made up of two identical subunits, is nonetheless asymmetrically associated, resulting in nonequivalent environments for the two subunits. D-Glucose and D-xylose are known, moreover, to have different preferences as to which subunit they bind to in the asymmetric dimer. And while each sugar can induce a conformational change in the protein upon binding, these changes are somewhat different for each sugar (82). Thus, D-xylose, we may suppose, induces an inactivating structural change in phosphohexokinase which only it (D-xylose) can reverse. From this perspective, reaction 10 is perceived as a phosphorylation of hexokinase by ATP, which, though beginning in catalytically normal fashion, is shunted by D-xylose (once in 300 cycles) into an inactivating direction, which is however reversible.

Also in harmony with the phosphoenzyme concept is the catalysis by hexokinase of a pair of pertinent exchange reactions. It is now generally agreed that the steady-state kinetics of hexokinase are not ping-pong (74). An intermediary phosphoenzyme can exist therefore only in the presence of bound substrate(s). In the forward direction the phosphoenzyme must form out of the ternary complex, E·glucose·ATP; and out of E·glucose-6-P·ADP in the reverse direction. Despite these constraints, hexokinase can catalyze an ADP–ATP exchange in the absence of "any trace of glucose or glucose-6-P" (83, 84, 85). The exchange is an intrinsic activity of the enzyme, and "cannot be explained trivially in terms of reversible equilibrium of the overall reaction" (86). Hexokinase also catalyzes the related glucose—glucose-6-P exchange in the assured absence of ADP and ATP (85). These exchange reactions, being independent of cosubstrates, speak strongly for their mediation

(79) W. S. Bennett, Jr., and T. A. Steitz, *PNAS* 75, 4848–4852 (1978); R. C. McDonald, T. A. Steitz, and D. M. Engelman, *B* 18, 338–342 (1979).
(80) Ref. (74), p. 25.
(81) W. S. Bennett, Jr., and T. A. Steitz, *JMB* 140, 211–230 (1980).
(82) W. F. Anderson and T. A. Steitz, *JMB* 92, 279–287 (1975).
(83) S. Kaufman, *JBC* 216, 153–164 (1955).
(84) A. Kagi and S. P. Colowick, *JBC* 240, 4454–4462 (1965).
(85) C. T. Walsh and L. B. Spector, *Arch. Biochem. Biophys.* 145, 1–5 (1971).
(86) Ref. (74), p. 24.

by a *free* phosphoenzyme. But since the normal kinetic habit of hexokinase is to act through ternary complexes, it is only natural that the independent exchanges should proceed at rates slower than that of the net reaction (Eq. 9). But the net reaction, we hasten to add, is a rapid one; and the exchanges, though slower in a relative sense (0.01%), have a substantial inherent velocity. They are, in fact, four times faster than the intrinsic ATPase activity of hexokinase (86), which is easily measured. The relative slowness of the exchanges is imputable therefore to the unfavorable conformation of the enzyme in the absence of cosubstrate. Hexokinase, as noted above, is subject to extensive conformational change when glucose binds to the protein (79, 87). If an enzyme is at its conformational "best" (and thus maximally active) only in the presence of *all* its reactant substrates, it follows that the absence of any one of them will have, in some degree, a "crippling" effect on its activity. In the presence therefore of *only* ADP plus ATP (i.e., in the absence of glucose and glucose-6-P) the enzyme must be in such a "crippled" state. It is all the more remarkable that it catalyzes any exchange at all. In any case, the relative speed of an exchange reaction is but a minor consideration in the question of its chemical mechanism. Succinic thiokinase (p. 209) also catalyzes a relatively slow ADP–ATP exchange, but the participation of a phosphoenzyme in its action is beyond doubt. All that matters about an exchange, apart from its being measurable, is that it be genuinely independent of cosubstrates. The ADP–ATP and glucose—glucose-6-P exchanges catalyzed by hexokinase are measurably fast, and are indeed independent of cosubstrates. They therefore carry weight in the question of phosphoenzyme mediation of hexokinase catalysis (88).

Of peculiar relevance to hexokinase and its chemical mechanism is its inhibition by *N*-acetylglucosamine (84). This molecule binds in the glucose subsite of hexokinase (79), and prevents the phosphorylation of glucose without being phosphorylated itself. It also inhibits the ATPase activity of the enzyme, but *does not inhibit the ADP–ATP exchange* (89). These facts were regarded as contradictory, since, if a phosphoenzyme mediates all of the

---

(87) G. De la Fuente, R. Lagunas, and A. Sols, *EJB* 16, 226–233 (1970); S. P. Colowick, F. C. Womack, and J. Neilsen, in *The Role of Nucleotides for the Function and Conformation of Enzymes*, H. M. Kalckar, ed., Munksgaard, Copenhagen, 1969, pp. 15–37.

(88) In this connection, it is noteworthy that an enzyme can operate by covalent catalysis and yet fail altogether to catalyze predicted exchange reactions. A case in point is adenine phosphoribosyltransferase (p. 80). This enzyme is actually isolated from rat liver and human erythrocytes as the phosphoribosylenzyme (59), acting undeniably through a covalent enzyme–substrate intermediate. But it does not catalyze the independent exchange reactions predicted for it by Eq. 8 (p. 80); that is, the exchanges are so slow as to be undetectable. For this, the quirks of enzyme conformation and "substrate synergism" are probably to blame. In less extreme cases (e.g., succinic thiokinase and hexokinase), the predicted exchanges are slow, but clearly detectable. An enzyme is evidently under no imperative to catalyze an exchange just because it engages in covalent catalysis. It follows that the failure of an enzyme to catalyze a predicted exchange is no argument against covalent catalysis by that enzyme.

(89) F. Solomon and I. A. Rose, *Arch. Biochem. Biophys.* 147, 349–350 (1971).

chemical activities of hexokinase, $N$-acetylglucosamine ought to inhibit all of them if it inhibits any one of them. It was reasoned, therefore, that the ADP–ATP exchange is a "side reaction" of the enzyme (89, 90). Yet two very careful investigations of the ADP–ATP exchange have shown it to be an intrinsic activity of hexokinase (84, 85).

The failure of $N$-acetylglucosamine to inhibit the ADP–ATP exchange is indeed difficult to explain if hexokinase is a single-displacement enzyme, but is readily accounted for by covalent catalysis. By excluding water from the glucose subsite, $N$-acetylglucosamine will of course block the ATPase activity at the same time as it blocks the phosphorylation of glucose by the phosphorylated form of the enzyme. But the phosphorylation of ADP, and its reversal (that is, the ADP–ATP exchange), will not be blocked *if the glucose subsite and the nucleotide subsite are sufficiently distant from each other.* Were the two subsites contiguous—and thus close enough for direct phosphoryl transfer by single-displacement catalysis—then the presence of an inhibitor in the glucose subsite ought to exert a perturbing effect on events in the contiguous nucleotide subsite. The total absence of such perturbance argues strongly for a substantial space between the two subsites, a space bridgeable only by a catalytic group (or groups) which carries the phosphoryl between the donor and acceptor subsites (91).

Some credence is given to this picture of the active center by crystallographic studies recently made on hexokinase and its complex with glucose (92). When glucose settles into its subsite, there follows a major structural change of the protein, resulting in the engulfment of glucose by the protein. When ATP is placed in its calculated position on a model of the hexokinase–glucose complex, it is found that the $\gamma$-phosphorus of ATP is too far (6 Å) from the 6-hydroxyl of glucose for a direct phosphoryl transfer to be possible (92, 93). A further structural change, it is said, would bring the reacting groups within direct reaction range. But how can one know that the additional change—if it occurs—would not put the reacting groups even further apart? And even if the groups could be brought together, covalent catalysis by the enzyme remains no less a possibility, as is argued in Chapter 1. But be that as it may, the X-ray data available until April 1981 are compatible only with covalent catalysis by hexokinase.

The X-ray study of crystalline hexokinase reveals, further, the presence of a serine hydroxyl and a carboxyl group in the active center of the enzyme. These findings are mirrored in a set of chemical studies, showing the stoichiometric phosphorylation of a serine in the inactivated phosphohexokinase (see above), and in the modification of a single essential carboxyl group of

(90) M. J. Wimmer and I. A. Rose, *Ann. Rev. Biochem.* 47, 1040 and 1071–1072 (1978).

(91) The effect of $N$-acetylglucosamine on the activities of hexokinase finds a remarkable parallel in the effect of mercuric ion on the activities of acetate kinase (p. 94).

(92) C. M. Anderson, F. H. Zucker, and T. A. Steitz, *Science* 204, 375–380 (1979).

(93) M. Shoham and T. A. Steitz, *JMB* 140, 1–14 (1980).

hexokinase with concurrent total inactivation (94). The serine hydroxyl and the carboxyl group form, thus, a pair of credible phosphoryl carriers in a triple-displacement catalysis by hexokinase.

## Pyruvate Kinase [EC 2.7.1.40]

One of the two "energy-producing" enzymes of glycolysis is pyruvate kinase, which provides an ATP molecule when phosphoenolpyruvate donates a phosphoryl group to ADP.

$$\text{Phosphoenolpyruvate} + \text{ADP} \xrightleftharpoons{\text{Mg}^{2+}, \text{K}^+} \text{ATP} + \text{pyruvate}$$
(PEP)

There is no good direct evidence at this time that a phosphoenzyme mediates the reaction, and for this reason pyruvate kinase is not included as a covalent enzyme in Table 3.1. It is nonetheless fitting to record here that five enzymic reactions are known in which PEP, acting as phosphoryl donor, phosphorylates the enzyme at one stage of the catalytic cycle. These enzymes are:

PEP-histidinoprotein phosphotransferase [EC 2.7.3.9] (95)
Pyruvate, phosphate dikinase [EC 2.7.9.1] (96) (p. 102)
PEP synthase [EC 2.7.9.2] (97)
Alkaline phosphatase [EC 3.1.3.1] (p. 117) (catalyzes the phosphorylation of Tris and water by PEP) (98)
Phosphoglycerate phosphomutase [EC 5.4.2.1] (catalyzes the phosphorylation of D-glycerate by PEP) (99)

It could be argued that there is no reason in logic which dictates that pyruvate kinase, like these five enzymes, must act upon PEP via a phosphoenzyme intermediate. Yet there is the feeling, verging upon conviction, that when nature happens upon a good way to do a chemical job she tends to use that way over and over again. This is the old argument from chemical analogy, which was and still is inseparable from the growth of chemical science. If in five enzymic reactions PEP delivers its phosphoryl group to an acceptor via

(94) D. B. Pho, C. Roustan, G. Desvages, L.-A. Pradel, and N. v. Thoai, *FEBS lett.* 45, 114–117 (1974); D. B. Pho, C. Roustan, A. N. T. Tot, and L.-A. Pradel, *B* 16, 4533–4537 (1977).
(95) R. D. Simoni, J. B. Hays, T. Nakazawa, and S. Roseman, *JBC* 248, 957–965 (1973); R. Stein, O. Schrecker, H. F. Lauppe, and H. Hengstenberg, *FEBS lett.* 42, 98–100 (1974); H. Hoving, J. S. Lolkema, and G. T. Robillard, *B* 20, 87–93 (1981).
(96) H. J. Evans and H. G. Wood, *PNAS* 61, 1448–1453 (1968); Y. Milner and H. G. Wood, *PNAS* 69, 2463–2468 (1972).
(97) R. A. Cooper and H. L. Kornberg, *Biochem. J.* 105, 49C (1967); *idem.*, *BBA* 141, 211–213 (1967); K. M. Berman and M. Cohn, *JBC* 245, 5309–5325 (1970); S. Narindrasorasak and W. A. Bridger *JBC* 252, 3121–3127 (1977).
(98) H. Barrett, R. Butler, and I. B. Wilson, *B* 8, 1042–1047 (1969).
(99) R. Breathnach and J. R. Knowles, *B* 16, 3054–3060 (1977).

a phosphoenzyme, it seems a safe bet that in a sixth reaction PEP will do the same. What is more, since the pyruvate kinase reaction results in a net steric inversion on the transferred phosphorus atom (71), a triple-displacement mechanism is in prospect for this enzyme, with two phosphoenzyme intermediates sharing in the catalysis (100).

## Acetate Kinase [EC 2.7.2.1]

Of the enzymes officially designated as carboxylate kinases (101), acetate kinase was the first for which a phosphoenzyme was invoked as a reaction intermediate (102). Acetate kinase catalyzes the reversible synthesis of acetyl phosphate from ATP and acetate,

$$\text{ATP} + \text{acetate} \xrightarrow{\text{Mg}^{2+}, \text{K}^+} \text{acetyl phosphate} + \text{ADP} \qquad (11)$$

though the thermodynamics of the reaction actually favor the synthesis of ATP from acetyl phosphate and ADP. The $\gamma$-phosphoryl group of ATP, during transit to acetate, undergoes a net steric inversion on the phosphorus atom (103). Catalysis of steric inversion via a phosphoenzyme intermediate thus implies for acetate kinase a triple-displacement mechanism (104).

When the enzyme (from *E. coli*) is incubated with ATP or acetyl phosphate in the absence of cosubstrates, a phosphoenzyme is formed which can be isolated by gel filtration (102, 105, 106). The phosphoenzyme prepared with ATP and the one prepared with acetyl phosphate are identical with respect to pH stability, reactivity with hydroxylamine, and the ability to phosphorylate ADP and acetate (107). The phosphoryl group of the phospho—

---

(100) It has been reported that when ATP is incubated with rabbit muscle pyruvate kinase—in the absence of added pyruvate—a randomization of the oxygen atoms on the $\beta$-phosphorus with the $\beta,\gamma$-bridge oxygen of recovered ATP takes place [G. Lowe and B. S. Sproat, *J. Chem. Soc., Perkin Trans. 1* 1622–1630 (1978)]. This randomization was regarded as evidence for the occurrence of a "tightly bound metaphosphate ion intermediate." But it may (more reasonably) be seen as the first *direct* evidence for the phosphoenzyme intermediate in pyruvate kinase catalysis—assuming of course that the randomization is not due to a trace of pyruvate in the experimental system.

(101) An enzyme can have carboxylate kinase activity as a component of an altogether different net activity. Thus, succinyl-CoA synthetase (p. 209) and ATP citrate lyase (p. 168) are not designated officially as carboxylate kinases; yet the former synthesizes succinyl phosphate and the latter citryl phosphate—both as enzyme-bound intermediates—through the action of ATP upon the respective carboxylates. And, as with acetate kinase, a phosphoenzyme mediates the synthesis of both acyl phosphates.

(102) R. S. Anthony and L. B. Spector, *JBC* 245, 6739–6741 (1970).

(103) W. A. Blättler and J. R. Knowles, *B* 18, 3927–3933 (1979).

(104) L. B. Spector, *PNAS* 77, 2626–2630 (1980).

(105) R. S. Anthony and L. B. Spector, *JBC* 247, 2120–2125 (1972).

(106) B. C. Webb, J. A. Todhunter, and D. L. Purich, *Arch. Biochem. Biophys.* 173, 282–292 (1976).

(107) R. S. Anthony, Doctoral Dissertation, The Rockefeller University, New York, 1971, pp. 47–50.

enzyme is present as an acyl phosphate (105), joined to the $\gamma$-carboxyl of a glutamyl residue (108).

Phosphorylated acetate kinase can transfer its phosphoryl group quantitatively to ADP (within 1 min or less) in the absence of cosubstrates (105). The rapid, quantitative synthesis of ATP is, we note, in the thermodynamically "downhill" direction of reaction 11. The same transfer can be made to GDP, in conformity with the intrinsic purine nucleosidediphosphate kinase activity of acetate kinase (102). Phosphorylated acetate kinase also phosphorylates acetate, in the absence of added nucleotide, in a reaction which is slower than the above and less extensive (70% yield of acetyl-P in 15 min) (105). The phosphorylation of acetate is of course in the thermodynamically "uphill" direction of reaction 11.

Acetate kinase also catalyzes an ADP–ATP exchange which is independent of its cosubstrates (acetate and acetyl-P). The exchange activity is intrinsic to acetate kinase; for upon gel electrophoresis of the enzyme, the exchange activity coincides precisely, in its migration, with the kinase activity (109). Moreover, the ADP–ATP exchange and the kinase reaction proceed at the same rate (109). It would be an amazing coincidence if the exchange, were it catalyzed by an impurity, had the same rate as the kinase reaction catalyzed by acetate kinase (110). These observations essentially exclude the possibility that the exchange is catalyzed by an enzyme other than acetate kinase.

Further to the same point is the finding that as phosphorylated acetate kinase reacts with hydroxylamine—thus transforming the acyl phosphate function into the corresponding hydroxamate—the exchange and kinase activities are lost at precisely the same rate (106). A close tie is thereby established between these two activities of acetate kinase, centering on a catalytic carboxyl group in the active site (111).

If in reaction 11 the adenine nucleotides are replaced by the guanine or inosine nucleotides (but not by the pyrimidine nucleotides), acetate kinase still catalyzes the reversible phosphorylation of acetate (102, 112). And just as the enzyme catalyzes phosphoryl transfer from ATP to ADP in the absence of cosubstrates, so can it catalyze the phosphoryl transfer from

(108) J. A. Todhunter and D. L. Purich, *Biochem. Biophys. Res. Commun.* 60, 273–280 (1974).

(109) R. S. Anthony and L. B. Spector, *JBC* 246, 6129–6135 (1971).

(110) Since the rates of ATP synthesis are the same in the exchange and in the kinase reactions, it follows that the phosphoenzyme, which mediates both reactions, is kinetically (as well as chemically) competent in the overall acetate kinase reaction.

(111) Acetate kinase also catalyzes an acetate–acetyl-P exchange in the absence of added nucleotide (109). The rate is much slower than that of the ADP–ATP exchange, because nucleotide is necessary to optimize the enzyme's conformation (109, 112). But unlike the ADP–ATP exchange, the acetate–acetyl-P exchange may not be a truly independent reaction, because the enzyme's endogenous content of nucleotide is unknown. Apart from its mere mention here, no weight is given to the acetate–acetyl-P exchange as an argument for covalent catalysis by acetate kinase.

(112) J. A. Todhunter, K. B. Reichel, and D. L. Purich, *Arch. Biochem. Biophys.* 174, 120–128 (1976)

ATP to GDP, also in the absence of cosubstrates. The latter reaction is recognizable as the classical nucleosidediphosphate (NDP) kinase reaction (p. 98). The purine NDP kinase activity is intrinsic to acetate kinase, since, upon reaction of phosphorylated acetate kinase with hydroxylamine, the purine NDP kinase and acetate kinase activities are lost at the same rate (112). It is further of interest that the purine NDP kinase activity of acetate kinase has a ping-pong kinetic mechanism, pointing to a free phosphoenzyme as mediator of the reaction (112). Since reaction 11 tends, on the other hand, to follow sequential kinetics (106), it is clear that covalent catalysis by acetate kinase is compatible with either ping-pong or sequential kinetics, depending upon circumstances.

The aggregate catalytic properties of acetate kinase enumerated above accord only with an intermediary phosphoenzyme, and exclude the possibility of single-displacement catalysis. This preeminent fact has then to be reconciled with the net steric inversion which takes place on phosphorus during the acetate kinase reaction (Eq. 11). Since enzyme-catalyzed displacements on phosphorus do not involve pseudorotation (113), it follows that a net steric inversion on phosphorus results from an *odd number* of in-line ($S_N2$ type) reactions. From all of the foregoing, it is clear that the odd number in question cannot be 1, but could be 3 (104). How a triple-displacement catalysis might be realized in the active center of acetate kinase is shown schematically in Fig. 11.

Clearly indicated in Fig. 11 is the noncontiguous disposition in space of the respective binding regions for the nucleotides and for acetate (and acetyl-P). The reality of a substantial distance between the two binding regions follows from the remarkable effects of $Hg^{2+}$ on acetate kinase and its reactions (104, 109). At a low concentration of $Hg^{2+}$ (10 $\mu M$), the overall kinase reaction and the acetate—acetyl-P exchange (111) are both 90% or more inhibited. But even at a 33-fold higher concentration of $Hg^{2+}$, the ADP–ATP exchange retains its normal rate. And in nice accord with the above is the effect of $Hg^{2+}$ on phosphoryl transfer from phosphorylated acetate kinase to acetate and ADP. Phosphoryl transfer to acetate is altogether stopped by $Hg^{2+}$; but the transfer to ADP, in synthesis of ATP, is utterly indifferent to the presence of $Hg^{2+}$. The mercurated enzyme thus blocks all chemical activity in the acetate (and acetyl-P) subregion of the active center, while leaving undisturbed the activity normal to the nucleotide subregion.

These facts speak for a substantial separation in space of the nucleotide and acetate (and acetyl-P) subregions of the active center. The distance between subregions must be sufficiently large to permit the disruption by $Hg^{2+}$ of chemical events in the acetate (and acetyl-P) subregion, while

(113) Ref. (103); B. M. Dunn, C. DiBello, and C. B. Anfinsen, *JBC* 248, 4769–4774 (1973); K.-F. R. Sheu, J. P. Richard, and P. A. Frey, *B* 18, 5548–5556 (1979); F. H. Westheimer, in *Molecular Rearrangements*, P. deMayo, ed., Interscience, New York, in press.

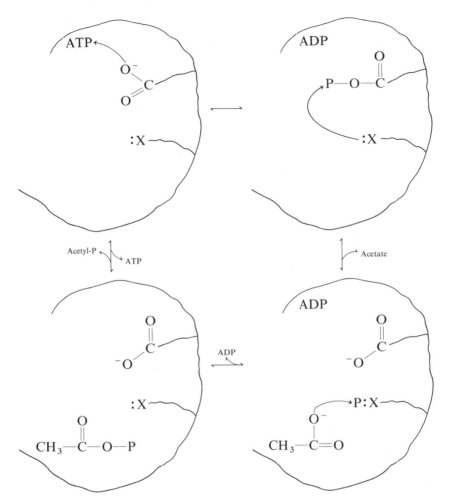

**Fig. 11.** A schematic representation of the acetate kinase reaction. The main feature is the triple displacement on the transferring phosphorus atom, involving three steric inversions and two phosphoenzyme intermediates. The nucleotides and acetate (and acetyl-P) are pictured as binding in noncontiguous regions of the active center. Adjacent to the nucleotide subregion is the catalytic carboxyl group, which has ATP as its immediate phosphoryl donor. And adjacent to the acetate (and acetyl-P) subregion is an unidentified catalytic group (X), which has acetyl-P as its immediate phosphoryl donor. All transfers of the phosphoryl group among the substrates and catalytic groups occur with steric inversion on the phosphorus. The intermediate E—X—P is deemed to be at a higher energy level than E—COO—P because only the latter is isolated when the enzyme is phosphorylated by ATP or acetyl-P.

leaving unperturbed the events normal to the nucleotide subregion. For if the two subregions were contiguous, a perturbance in one of them could hardly escape some degree of transmission to the other. But not the slightest perturbance is in fact transmitted. There is a total disruption of activity in one subregion and a total failure to disrupt in the other. Separated in space as these subregions seem thus to be, *a direct phosphoryl transfer from donor to acceptor, in a single displacement, becomes an improbable event.* Much more likely is a bridging of the spatial gap by a catalytic group (or two), which can swing the phosphoryl group from donor to acceptor across the intervening space.

The noncontiguous arrangement of the nucleotide and acetate (and acetyl-P) subregions enables acetate kinase to avoid the "three-body" problem which is inseparable from single-displacement catalysis (114). By binding its reacting substrates noncontiguously, acetate kinase transforms the (hypothetical) "three-body" problem into the more easily managed "two-body" problem, in which the catalytic group of the enzyme is one of the two bodies, while the donor and acceptor molecules take turns at being the other (115). Such division of the acetate kinase reaction (Eq. 11) into several partial reactions (Fig. 11) provides a pathway which, though seemingly more complex than the hypothetical single displacement, is energetically easier (116).

Why acetate kinase should need two phosphoenzymes in order to consummate its reaction is not immediately obvious. Maybe the space to be bridged is too broad to be negotiated by a single catalytic group. Or the phosphoryl acceptor, in its binding site, is perhaps so oriented as to need a second phosphoenzyme to bring the phosphoryl group in from the best direction. In the latter case, the phosphoryl group engages in the kind of "surface walk" we spoke of in Chapter 1.

## Phosphoglycerate Kinase [EC 2.7.2.3]

In its overall chemical form, the reaction catalyzed by phosphoglycerate kinase

$$\text{ATP} + \text{3-P-glycerate} \longleftrightarrow \text{3-P-glyceryl-P} + \text{ADP} \qquad (12)$$

is entirely analogous to the reaction catalyzed by acetate kinase (Eq. 11). It is reasonable, therefore, to suppose that phosphoglycerate kinase—like acetate kinase, succinyl-CoA synthetase, and ATP citrate lyase (101)— would have a phosphoenzyme to mediate its catalysis. Such *positive* evidence as bears upon this controversial subject argues for the fulfillment of this

---

(114) See the discussion in Chapter 1 (p. 13) (section on 3-body and 2-body interactions).
(115) In Fig. 11, when the phosphoryl group is being transferred between the two catalytic groups of acetate kinase, this too is reckoned a "two-body" reaction.
(116) See the discussion in Chapter 1 (p. 10) (section on inversion).

expectation. Because the enzyme exhibits sequential kinetics only (117), its phosphoenzyme develops from a ternary complex, and is therefore present throughout its lifetime in a complex with cosubstrate(s). In this respect, 3-P-glycerate kinase may be likened to acetate kinase when the latter operates under conditions of sequential kinetics.

A phosphoenzyme complex is prepared and isolated by incubation of yeast 3-P-glycerate kinase with ATP—in the presumed presence of some enzyme-bound 3-P-glycerate (118)—followed by gel filtration (119). Hydroxylamine acts on the phosphoenzyme complex to release orthophosphate. The resulting enzyme-acylhydroxamate, subjected to the Lossen rearrangement and subsequent hydrolysis, yields 2,4-diaminobutyric acid in amounts proportional to the degree of phosphorylation of the enzyme (119). The finding of 2,4-diaminobutyric acid signals the existence in the phosphoenzyme of a covalent link between the phosphoryl group and the $\gamma$-carboxyl of a glutamyl residue of the enzyme, just as was found in the case of acetate kinase. And, again like acetate kinase, phosphoglycerate kinase catalyzes a net steric inversion on the transferring phosphorus atom (120), with the inferred triple displacement.

It is further of interest that 3-P-glycerate kinase is one of three enzymes that use 3-P-glyceryl phosphate as phosphoryl donor in a phosphotransferase reaction. The others are glucose-1,6-diP synthase [EC 2.7.1.–] and bisphosphoglycerate synthase [EC 2.7.5.4]. The first of these catalyzes the phosphorylation of glucose-1-P as follows (121):

$$3\text{-P-Glyceryl-P} + \text{glucose-1-P} \longrightarrow 3\text{-P-glycerate} + \text{glucose-1,6-diP}$$

The second enzyme catalyzes—by sequential or by ping-pong kinetics, depending upon the enzyme source—the following transformation (122):

$$3\text{-P-Glyceryl-P} + 3\text{-P-glycerate (or 2-P-glycerate)} \longleftrightarrow$$

$$2,3\text{-diP-glycerate} + 3\text{-P-glycerate}$$

What is important here is that a phosphoenzyme is on the pathway of both reactions. The individual phosphoenzymes can be isolated and shown to be chemically and kinetically competent. Again, on grounds of chemical analogy, 3-P-glycerate kinase, acting on 3-P-glyceryl-P as phosphoryl

(117) M. Larsson-Raznikiewicz and L. Arvidsson, *EJB* 22, 506–512 (1971); C. A. Janson, and W. W. Cleland, *JBC* 249, 2567–2571 (1974).

(118) P. E. Johnson, S. J. Abbott, G. A. Orr, M. Sémériva, and J. R. Knowles, *B* 15, 2893–2899 (1976).

(119) A. Brevet, C. Roustan, G. Desvages, L.-A. Pradel, and N. van Thoai, *EJB* 39, 141–147 (1973).

(120) M. R. Webb and D. R. Trentham, *JBC* 255, 1775–1779 (1980).

(121) L.-J. Wong and I. A. Rose, *JBC* 251, 5431–5439 (1976); I. A. Rose, J. V. B. Warms, and L.-J. Wong, *JBC* 252, 4262–4268 (1977).

(122) Z. B. Rose and R. G. Whalen, *JBC* 248, 1513–1519 (1973); Z. B. Rose and S. Dube, *JBC* 251, 4817–4822 (1976); *idem.*, *Arch. Biochem. Biophys.* 177, 284–292 (1976).

donor—in reversal of reaction 12—ought to hold to the same pattern and use the same mode of phosphoryl transfer. What positive evidence is available suggests that it does indeed do so.

## Nucleosidediphosphate Kinase [EC 2.7.4.6]

The enzyme that catalyzes the reversible transfer of a phosphoryl group from ATP to GDP

$$\text{ATP} + \text{GDP} \longleftrightarrow \text{ADP} + \text{GTP}$$

happens also to be the very model of a ping-pong enzyme. Isolated from at least eight different sources, the enzyme has never been found to follow any but ping-pong kinetics (123, 124). And in conformity with this, the enzyme from all sources has been isolated in phosphorylated condition after incubation with $[^{32}P]$ATP in the absence of cosubstrates (123–129). Thus prepared, the phosphoenzyme is capable of transferring its phosphoryl group to any of a number of nucleosidediphosphates that are substrates of the enzyme. The rates of enzyme phosphorylation and dephosphorylation

$$\text{E} + [^{32}P]\text{ATP} \longleftrightarrow \text{E—}^{32}P + \text{ADP}$$

$$\text{E—}^{32}P + \text{GDP} \longleftrightarrow \text{E} + [^{32}P]\text{GTP}$$

are somewhat greater than the rate of the net reaction in the forward direction (128). The phosphoryl group in the phosphoenzyme is joined to a histidyl residue in the active center, linking to the N-1 of the imidazole ring (124, 126, 129, 130). In keeping with the ping-pong kinetics, nucleosidediphosphate kinase also catalyzes a rapid ADP–ATP exchange, as well as a variety of other exchanges among substrate nucleoside di- and triphosphates (131). As a straightforward double-displacement enzyme, nucleosidediphosphate kinase also catalyzes, as expected, a net steric retention of configuration on the transferring phosphorus atom (132).

(123) N. Mourad and R. E. Parks, Jr., *JBC* 241, 271–278 (1966); E. Garces and W. W. Cleland, *B* 8, 633–640 (1969).

(124) J. Sedmark and R. Ramaley, *JBC* 246, 5365–5372 (1971).

(125) N. Mourad and R. E. Parks, Jr., *Biochem. Biophys. Res. Commun.* 19, 312–316 (1965); M. G. Colomb, J. G. Laturaze, and P. V. Vignais, *Biochem. Biophys. Res. Commun.* 24, 909–915 (1966).

(126) A. W. Norman, R. T. Wedding, and M. K. Black, *Biochem. Biophys. Res. Commun.* 20, 703–709 (1965).

(127) P. L. Pedersen, *JBC* 243, 4305–4311 (1968); M. G. Colomb, A. Chéruy, and P. V. Vignais, *B* 8, 1926–1939 (1969).

(128) O. Wålinder, Ö. Zetterqvist, and L. Engström, *JBC* 244, 1060–1064 (1969); B. Edlund and O. Wålinder, *FEBS lett.* 38, 225–228 (1974).

(129) M. G. Colomb, A. Chéruy, and P. V. Vignais, *B* 11, 3378–3386 (1972).

(130) B. Edlund, L. Rask, P. Olsson, O. Wålinder, Ö. Zetterqvist, and L. Engström, *EJB* 9, 451–455 (1969).

(131) M. G. Colomb, A. Chéruy, and P. V. Vignais, *B* 11, 3370–3378 (1972).

(132) K. F. R. Sheu, J. P. Richard, and P. A. Frey, *B* 18, 5548–5556 (1979).

# Phosphoglucomutase [EC 2.7.5.1]

Among the first enzymes to be recognized as a covalent catalyst is phosphoglucomutase, which catalyzes the reversible transfer of a phosphoryl
group between C-1 and C-6 of glucose (133, 134).

$$\alpha\text{-D-Glucose-1-P} \xrightleftharpoons{\text{Mg}^{2+}} \alpha\text{-D-Glucose-6-P} \tag{13}$$

The reaction is in effect an isomerization. The enzyme is often isolated in the
phosphorylated state (133). Phosphoglucomutase isolated from rabbit
muscle and yeast contains an equivalent of covalently bound phosphorus
(133, 134), while other sources yield enzyme with lesser degrees of phosphorylation (136). Enzyme from *E. coli* (137) and *M. lysodeikticus* (138) are
altogether devoid of phosphorus when isolated, but they can be phosphorylated upon incubation with glucose 1,6-diphosphate. A serine hydroxyl in
the active center of the enzyme is the point of attachment of the phosphoryl
group (139). For sources as diverse as rabbit muscle (140) and *M. lysodeikticus* (141) the amino acid sequence surrounding the active serine is the
same. The phosphoenzyme from rabbit muscle can transfer its phosphoryl
group reversibly to glucose-1-P with a turnover rate compatible in all
respects with the rate of the net reaction (142). Though linked to the primary
hydroxyl of a serine residue, the phosphoryl group in the phosphoenzyme
has the properties of an energy-rich phosphate. Thus, fluoride ion, acting
upon the native phosphoenzyme, releases the phosphoryl as phosphorofluoridate, whereas the denatured phosphoenzyme is unreactive (143). In this
respect the phosphorylated phosphoglucomutase is like the acylated ACP-
malonyltransferase (p. 76), in which the acyl bond to the active-site serine is
also an energy-rich bond, but only in the native acyl enzyme.

As to the mechanism of the phosphoglucomutase reaction, it was formerly
thought that glucose 1,6-diphosphate and dephosphoenzyme were *free*
intermediates of the reaction. It seems certain, however, that these entities,

(133) V. Jagannathan and J. M. Luck, *JBC* 179, 569–575 (1949).

(134) V. A. Najjar and M. E. Pullman, *Science* 119, 631–634 (1954).

(135) J. B. Sudbury, Jr., and V. A. Najjar, *JBC* 227, 517–522 (1957); J. A. Yankeelov, Jr., H. R.
Horton, and D. E. Koshland, Jr., *B* 3, 349–355 (1964); J. P. Daugherty, W. F. Kraemer, and
J. G. Joshi, *EJB* 57, 115–126 (1975).

(136) W. J. Ray, Jr., and E. J. Peck, in *The Enzymes*, 3rd ed., P. D. Boyer, ed., Academic Press,
New York, 1972, Vol. 6, pp. 407–457.

(137) J. G. Joshi and P. Handler, *JBC* 239, 2741–2751 (1964).

(138) J. B. Clarke, M. Birch, and H. G. Britton, *Biochem. J.* 137, 463–467 (1974).

(139) L. Anderson and G. R. Jolles, *Arch. Biochem. Biophys.* 70, 121–128 (1957); E. P. Kennedy
and D. E. Koshland, Jr., *JBC* 228, 419–431 (1957).

(140) C. Milstein and F. Sanger, *Biochem. J.* 79, 456–469 (1961).

(141) C. P. Milstein, J. B. Clarke, and H. G. Britton, *Biochem. J.* 135, 551–553 (1973).

(142) O. Wålinder and J. G. Joshi, *JBC* 249, 3166–3169 (1974).

(143) P. P. Layne and V. A. Najjar, *JBC* 250, 966–972 (1975).

if they share in the reaction at all, do so only as part of an undissociated complex (142, 144, 145).

$$
\text{E—P + glucose-1-P} \longleftrightarrow
\begin{bmatrix}
\text{E—P} \text{ --- glucose-1-P} \\
\text{E} \text{ ------ glucose-1,6-diP} \\
\text{E—P} \text{ --- glucose-6-P}
\end{bmatrix}
$$

$$
\longleftrightarrow \text{E—P + glucose-6-P}
$$

Glucose 1,6-diphosphate is reported to interchange its phosphoryl groups with the enzyme one hundred times faster than it dissociates from the enzyme (146). Dissociation of the central complex is thus an infrequent event which, by setting free the dephosphoenzyme, eliminates that molecule of enzyme from the catalytic cycle. It can, however, be restored to activity by rephosphorylation with free glucose 1,6-diphosphate (144).

A long-standing puzzle has been the mode of phosphoryl transfer between the phosphoenzyme, on the one hand, and the 1- and 6-positions of glucose, on the other. If all phosphoryl transfer takes place at the active-site serine, then the glucose molecule must change its position on the enzyme in order to accommodate phosphoryl transfer to or from its C-1 and C-6 extremities. Alternatively, the glucose might stay fixed in its position, while the phosphoryl group moves on the surface of the enzyme. In this case a second phosphoryl transfer site would exist, to which the phosphoryl group could migrate reversibly from the serine before eventual transfer to glucose. Thus, one enzyme site would correspond to C-1 of glucose, and the other to C-6. A suggested candidate for the second phosphoryl-bonding site is tyrosine (147). It would be helpful to the solution of this puzzle to know if reaction 13 takes place with retention or inversion of configuration on phosphorus.

## Galactose-1-Phosphate Uridylyltransferase [EC 2.7.7.10]

To effect the transfer of a uridylyl group between glucose-1-P and galactose-1-P

UDPglucose + α-D-galactose-1-P $\longleftrightarrow$ α-D-glucose-1-P + UDPgalactose

this nucleotidyltransferase [from *E. coli* (148) and human red cells (149)] acts first upon UDPglucose to bind the uridylyl group covalently to itself, displacing α-D-glucose-1-P in the process.

(144)  W. J. Ray, Jr., and G. A. Roscelli, *JBC* 239, 1228–1236 (1964).
(145)  T. Hashimoto and P. Handler, *JBC* 241, 3940–3948 (1966); A. D. Gounaris, H. R. Horton, and D. E. Koshland, Jr., *BBA* 132, 41–55 (1967); J. B. Clarke and H. G. Britton, *Biochem. J.* 137, 453–461 (1974).
(146)  A. S. Mildvan, *Ann. Rev. Biochem.* 43, 371 (1974).
(147)  P. P. Layne and V. A. Najjar, *PNAS* 76, 5010–5013 (1979).
(148)  L. J. Wong and P. A. Frey, *JBC* 249, 2322–2324 (1974).
(149)  J. W. Wu, T. A. Tedesco, R. G. Kallen, and W. J. Mellman, *JBC* 249, 7038–7039 (1974).

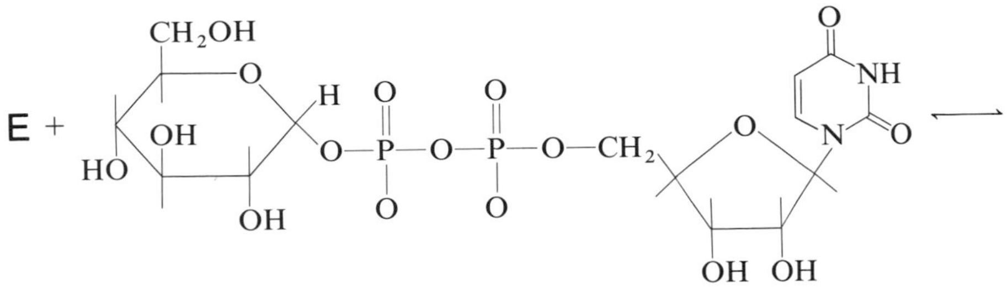

UDPglucose

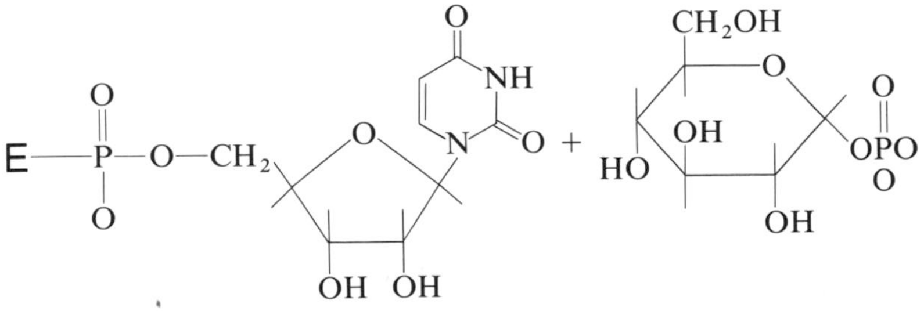

Uridylyl enzyme                              α-D-Glucose-1-P

Thereafter, the uridylyl enzyme reacts with α-D-galactose-1-P to deliver up the uridylyl group as UDPgalactose.

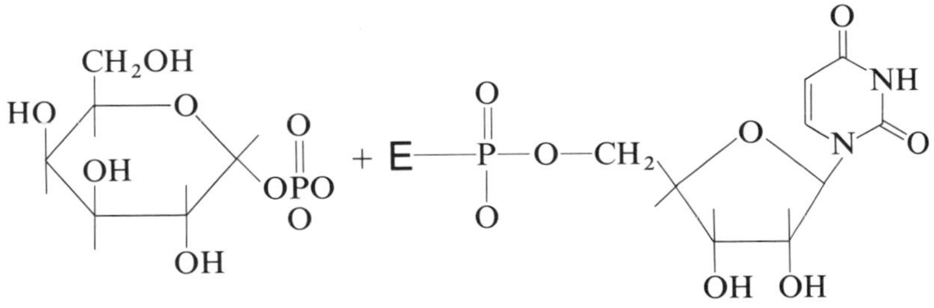

α-D-Galactose-1-P                    Uridylyl enzyme

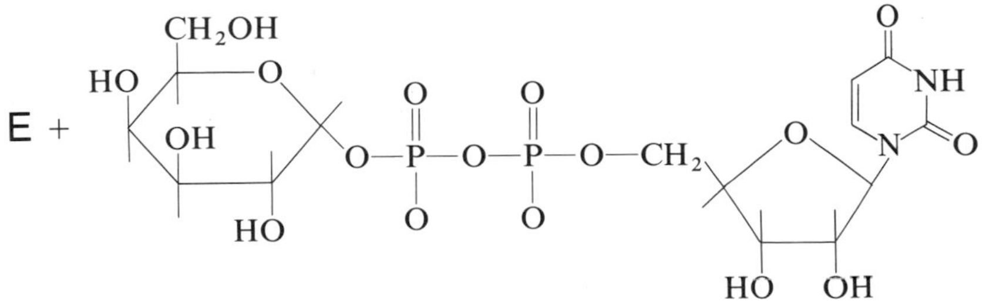

UDPgalactose

By incubating the enzyme with UDPglucose labeled with tritium in the uracil portion of the molecule and with $^{14}C$ in the glucose, in the absence of galactose-1-P, only tritium adheres to the enzyme after gel filtration. Consonant with this finding, the enzyme exhibits authentic ping-pong kinetics (150, 151). It also catalyzes a glucose-1-P—UDPglucose exchange (in the absence of cosubstrates) at a rate which exceeds that of the net reaction (150). The uridylyl enzyme is thus on the enzymic pathway, being both chemically and kinetically competent. The uridylyl group is linked covalently to the enzyme via nitrogen-3 of a histidine residue (152). As expected for a straight-forward double-displacement enzyme, galactose-1-P uridylyltransferase catalyzes a retention of configuration on the uridylyl phosphorus atom (132, 153).

## Pyruvate, Orthophosphate Dikinase [EC 2.7.9.1]

This unusual enzyme (from *P. shermanii* and *B. symbiosus*) effects the synthesis of phosphoenolpyruvate by transferring the central phosphorus of ATP to pyruvate and the terminal phosphorus to orthophosphate to make inorganic pyrophosphate (154).

$$\text{ATP} + \text{pyruvate} + \text{P}_i \xrightarrow{\text{Mg}^{2+}} \text{phosphoenolpyruvate} + \text{AMP} + \text{PP}_i$$
$$\text{(PEP)}$$

This complex process begins with the transfer of the pyrophosphoryl group of ATP to the enzyme, forming a pyrophosphoenzyme intermediate.

$$\text{E} + \text{Ad}-\overset{\alpha}{\text{P}}-\overset{\beta}{\text{P}}-\overset{\gamma}{\text{P}} \xrightarrow{\text{Mg}^{2+}} \text{E} \sim \overset{\beta\gamma}{\text{PP}} + \overset{\alpha}{\text{AMP}}$$

The pyrophosphoenzyme then yields up its outer phosphoryl to ortho-phosphate, resulting in inorganic pyrophosphate and a new intermediate—the phosphoenzyme—in which the former central phosphorus of ATP is still covalently fixed to the enzyme.

$$\text{E} \sim \overset{\beta\gamma}{\text{PP}} + \text{P}_i \xrightarrow{\text{Mg}^{2+}} \text{E} \sim \overset{\beta}{\text{P}} + \overset{\gamma}{\text{PP}}_i \qquad (14)$$

(150) L. J. Wong and P. A. Frey, *B* 13, 3889–3894 (1974).

(151) H. B. Markus, J. W. Wu, F. S. Boches, T. A. Tedesco, W. J. Mellman, and R. G. Kallen, *JBC* 252, 5363–5369 (1977).

(152) S. L. Yang and P. A. Frey, *B* 18, 2980–2984 (1979).

(153) Several other nucleotidyltransferases have been examined as to their stereochemical course and found to proceed with *inversion* of configuration at the nucleotidyl phosphorus [K. F. R. Sheu and P. A. Frey, *JBC* 253, 3378–3380 (1978); P. M. J. Burgers and F. Eckstein, *PNAS* 75, 4798–4800 (1978); *idem.*, *B* 18, 450–454 (1979); *idem.*, *JBC* 254, 6889–6893 (1979); S. P. Langdon and G. Lowe, *Nature (London)* 281, 320–321 (1979)]. Alone on steric grounds, these enzymes were declared to be single-displacement enzymes. But experience with acetate kinase (p. 92) suggests that steric inversion on phosphorus results from triple-displacement catalysis.

(154) H. G. Evans and H. G. Wood, *PNAS* 61, 1448–1453 (1968); Y. Milner and H. G. Wood, *PNAS* 69, 2463–2468 (1972); *idem.*, *JBC* 251, 7920–7928 (1976); Y. Milner, G. Michaels, and H. G. Wood, *JBC* 253, 878–883 (1978).

Lastly, pyruvate appropriates the phosphoryl group from the phospho-enzyme to complete the catalytic cycle.

$$\overset{\beta}{E \sim P} + \text{pyruvate} \xrightarrow{\text{Mg}^{2+}} E + \overset{\beta}{\text{PEP}} \tag{15}$$

The phosphoenzyme can be prepared by incubating the enzyme with PEP, in reversal of reaction 15; and in accord with the existence of the phospho-enzyme as intermediate is the $[^{14}\text{C}]$pyruvate-PEP exchange, which occurs in the absence of all other substrates at a rate which is more than twice the rate of the overall reaction.

Pyrophosphoenzyme is accessible from phosphoenzyme by incubating the latter with inorganic pyrophosphate in reversal of reaction 14. One mole of $E \sim PP$ can be formed per mole of enzyme; and consistent with the existence of the pyrophosphoenzyme as intermediate is the $[^{14}\text{C}]$AMP–ATP exchange which occurs in the absence of cosubstrates at 30% of the rate of the overall forward reaction. The pyrophosphoenzyme is chemically competent, acting on orthophosphate to yield inorganic pyrophosphate and on AMP to yield ATP. When $E \sim PP$ is labeled in the outer phosphoryl, reaction with AMP results in ATP labeled solely in the terminal position, which accords with the transfer of an integral pyrophosphoryl group. Gentle acid hydrolysis of $E \sim PP$ produces mainly inorganic pyrophosphate, thus excluding the possibility that the two phosphoryl groups in $E \sim PP$ are separately bound to the enzyme. The pyrophosphoryl group is covalently joined to a histidine residue in the active center of the enzyme (155).

## Rhodanese [EC 2.8.1.1]

In the course of catalyzing the transfer of the sulfenyl sulfur atom of thio-sulfate to cyanide,

$$S\overset{O}{\underset{O}{\overset{\|}{\underset{\|}{S}}}}O^{2-} + CN^- \rightleftarrows :S\overset{O}{\underset{O}{\overset{\|}{\underset{\|}{S}}}}O^{2-} + SCN^-$$

rhodanese is itself transiently bound to the transferring sulfur. In fact, crystalline rhodanese is usually isolated from beef liver with active, trans-ferable sulfur atoms already attached (156–158). After incubation with $^{35}\text{SSO}_3$—but not with $S^{35}\text{SO}_3$—the crystalline enzyme can be isolated as the enzyme–sulfur intermediate containing about two atoms of labeled sulfur per dimeric molecule of enzyme (157). The labeled enzyme reacts

(155) A. D. Spronk, H. Yoshida, and H. G. Wood, *PNAS* 73, 4415–4419 (1976).
(156) J. R. Green and J. Westley, *JBC* 236, 3047–3050 (1961).
(157) J. Westley and T. Nakamoto, *JBC* 237, 547–549 (1962).
(158) C. Cannella, L. Pecci, B. Pensa, M. Costa, and D. Cavallini, *FEBS lett.* 49, 22–24 (1974).

rapidly with sulfite or cyanide to yield labeled thiosulfate and thiocyanate, respectively,

$$E + 2\,SSO_3^{2-} \longrightarrow E{-}S_2 + 2\,SO_3^2 \qquad (16)$$

$$E{-}S_2 + 2\,CN^- \longrightarrow E + 2\,SCN^-$$

supporting thus the double-displacement mechanism for the rhodanese reaction. In conformity with Eq. 16 rhodanese catalyzes a rapid $SO_3^{2-}$–$SSO_3^{2-}$ exchange (159). Kinetic analysis has also confirmed the double-displacement character of rhodanese-catalyzed reactions (160). Within the active center of rhodanese are found a cationic site (161)—possibly a metal ion—and an essential cysteinyl sulfhydryl group (162). The transferring sulfur atom is lined to the cysteine (Cys 247) during catalysis as an enzyme–persulfide intermediate (158, 163), a fact confirmed by X-ray analysis of crystalline rhodanese (164).

## 3-Ketoacid CoA-Transferase [EC 2.8.3.5]

In the reaction between succinyl-CoA and acetoacetate,

$$\underset{\quad\;\; \|}{\overset{\quad\;\; O}{OOCCH_2CH_2CSCoA}} + \underset{\quad\; \|}{\overset{\quad\; O}{CH_3CCH_2COO}} \longrightarrow$$

$$OOCCH_2CH_2COO + \underset{\qquad\quad\;\; \|\;\;\;\;\, \|}{\overset{\qquad\quad\;\; O\;\;\;\;\, O}{CH_3CCH_2CSCoA}}$$

the reversible transfer of coenzyme A to the carboxyl of acetoacetate occurs concurrently with the transfer of a carboxyl oxygen of acetoacetate to the carboxyl of succinate (165); that is, coenzyme A and the oxygen atom move in "opposite" directions. During the catalytic cycle the transferring oxygen atom (166) and coenzyme A (167) are separately joined to the enzyme in covalent linkage (Figs. 12 and 13). The enzyme-CoA intermediate can be isolated by gel filtration after incubation of the enzyme with acetoacetyl-CoA

(159) B. Sörbo, *Acta Chem. Scand.* 16, 243–245 (1962).

(160) M. Volini and J. Westley, *JBC* 241, 5168–5176 (1966); J. Westley and D. Heyse, *JBC* 246, 1468–1474 (1971); S. Oi, *J. Biochem. (Tokyo)* 78, 825–834 (1975).

(161) R. Mintel and J. Westley, *JBC* 241, 3386–3389 (1966); M. Volini, B. Van Sweringen, and F. S. Chen, *Arch. Biochem. Biophys.* 191, 205–215 (1978).

(162) S. F. Wang and M. Volini, *JBC* 243, 5465–5470 (1968).

(163) K. R. Leininger and J. Westley, *JBC* 243, 1892–1899 (1968); P. Schlesinger and J. Westley, *JBC* 249, 780–788 (1974).

(164) J. H. Ploegman, G. Drent, K. H. Kalk, W. G. J. Hol, R. L. Heinrikson, P. Keim, L. Weng, and J. Russell, *Nature (London)* 273, 124–129 (1978).

(165) A. B. Falcone and P. D. Boyer, *Arch. Biochem. Biophys.* 83, 337–344 (1959).

(166) R. W. Benson and P. D. Boyer, *JBC* 244, 2366–2371 (1969).

(167) L. B. Hersh and W. P. Jencks, *JBC* 242, 3481–3486 (1967).

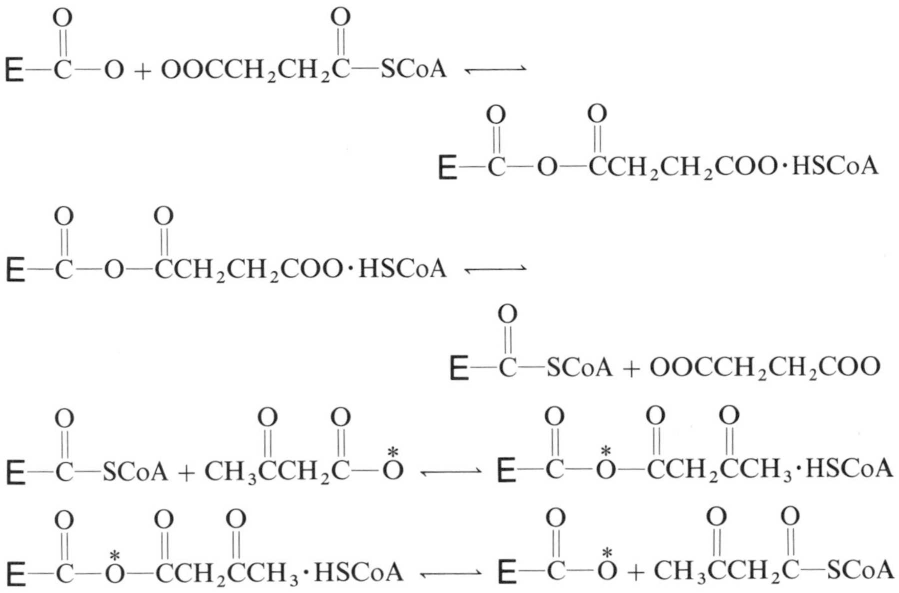

$$E-\overset{O}{\overset{\|}{C}}-O + OOCCH_2CH_2\overset{O}{\overset{\|}{C}}-SCoA \longrightarrow$$

$$E-\overset{O}{\overset{\|}{C}}-O-\overset{O}{\overset{\|}{C}}CH_2CH_2COO\cdot HSCoA$$

$$E-\overset{O}{\overset{\|}{C}}-O-\overset{O}{\overset{\|}{C}}CH_2CH_2COO\cdot HSCoA \longrightarrow$$

$$E-\overset{O}{\overset{\|}{C}}-SCoA + OOCCH_2CH_2COO$$

$$E-\overset{O}{\overset{\|}{C}}-SCoA + CH_3\overset{O}{\overset{\|}{C}}CH_2\overset{O}{\overset{\|}{C}}-\overset{*}{O} \longrightarrow E-\overset{O}{\overset{\|}{C}}-\overset{*}{O}-\overset{O}{\overset{\|}{C}}CH_2\overset{O}{\overset{\|}{C}}CH_3\cdot HSCoA$$

$$E-\overset{O}{\overset{\|}{C}}-\overset{*}{O}-\overset{O}{\overset{\|}{C}}CH_2\overset{O}{\overset{\|}{C}}CH_3\cdot HSCoA \longrightarrow E-\overset{O}{\overset{\|}{C}}-\overset{*}{O} + CH_3\overset{O}{\overset{\|}{C}}CH_2\overset{O}{\overset{\|}{C}}-SCoA$$

**Fig. 12.** A proposed mechanism of the CoA-transferase reaction, showing the transfer of coenzyme A between donor and acceptor via a catalytic carboxyl group of the enzyme. Sharing in the catalysis are two different carboxylic anhydrides involving the enzyme as a covalent partner. The progress of an oxygen atom (starred) from one substrate to the other via the enzyme is illustrated in part. A second turnover of the enzyme is required to complete the oxygen transfer between substrates. Coenzyme A does not dissociate from the enzyme during the catalytic cycle.

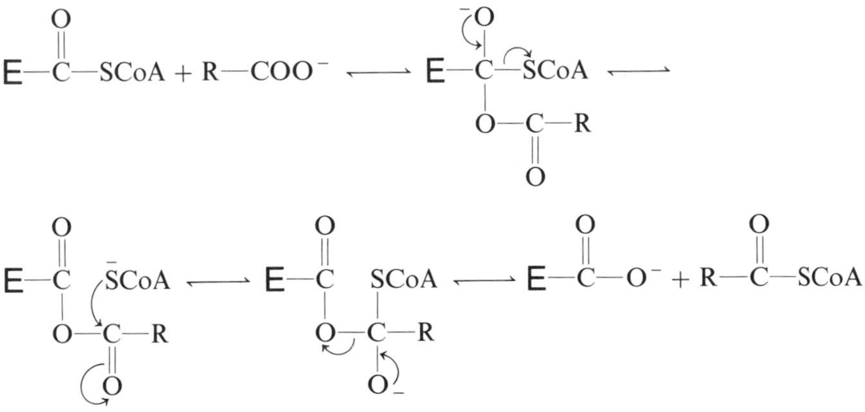

$$E-\overset{O}{\overset{\|}{C}}-SCoA + R-COO^- \longrightarrow E-\overset{O}{\underset{O-C-R}{\overset{\|}{\underset{\|}{C}}}}-SCoA \longrightarrow$$

$$E-\overset{O}{\underset{O-C-R}{\overset{\|}{\underset{\|}{\underset{O}{C}}}}}SCoA \longrightarrow E-\overset{O}{\underset{O-C-R}{\overset{\|}{\underset{\|}{\underset{O}{C}}}}}SCoA \longrightarrow E-\overset{O}{\overset{\|}{C}}-O^- + R-\overset{O}{\overset{\|}{C}}-SCoA$$

**Fig. 13.** The detailed mechanism of coenzyme A transfer between the carboxyl group of the enzyme and the carboxyl of substrate, showing mediation by a carboxylic anhydride in which the enzyme is a partner (169).

or succinyl-CoA (167). A thiolester bond to the $\gamma$-carboxyl group of a glutamyl residue is the means by which the coenzyme A is bound to the enzyme (168). The transferring oxygen atom is believed to pass in and out of the same glutamyl carboxyl group as the latter forms successive acid anhydrides with succinate and acetoacetate (165, 166, 169, 170). The same anhydrides afford the chemical means of coenzyme A transfer between substrates and enzyme (Fig. 13). Obeying ping-pong kinetics, the enzyme also catalyzes succinate—succinyl-CoA and acetoacetate—acetoacetyl-CoA exchanges, in the absence of cosubstrates, at rates as rapid as the net reaction (171). We have in this CoA-transferase a very carefully studied enzyme which catalyzes a seemingly "simple" transferase reaction by segmenting it into a succession of four partial reactions involving three different covalent intermediates (Fig. 12).

**Table 3.1.** Transferases Known to Act by Covalent Catalysis

| EC no. | Familiar name of enzyme[a] | Criteria[b] | References |
|---|---|---|---|
| 2.1.1.13 | Tetrahydropteroylglutamate methyltransferase [cobalamin] | I[c], E, K | (172) |
| 2.1.1.21 | N-Methylglutamate synthase [FMN] | I, E, K, M, R | (173) |
| 2.1.1.45 | Thymidylate synthase [H$_4$-folate] | M | (174) |
| 2.1.2.1 | Serine hydroxymethyltransferase [PLP] | I[d], M | (175) |
| 2.1.2.10 | Glycine synthase [PLP, lipoate] | I, E | (176) |
| 2.1.2 | Deoxyuridylate hydroxymethylase [H$_4$-folate] | M | (177) |
| 2.1.3.1 | Methylmalonyl-CoA carboxyltransferase [biotin] | I, K | (178) |
| 2.1.4.1 | Glycine amidinotransferase | I, E, K | (179) |
| 2.2.1.1 | Transketolase [TPP] | I, E | (180) |
| 2.2.1.2 | Transaldolase | I, E, K, B, S | (181) |
| 2.3.1.5 | Arylamine acetyltransferase | I, E, K, $V_{max}$ | (182) |
| 2.3.1.9 | Acetyl-CoA acetyltransferase | I, E, K, M | (183) |
| 2.3.1.12 | Lipoate acetyltransferase [lipoate] | I, K | (184) |
| 2.3.1.31 | Homoserine acetyltransferase | E, K | (185) |
| 2.3.1.37 | $\delta$-Aminolevulinate synthase [PLP] | I[d], M | (186) |
| 2.3.1.38 | [Acyl-carrier-protein]acetyltransferase | I | (187) |
| 2.3.1.39 | [Acyl-carrier-protein]malonyltransferase | I, K | (188) |
| 2.3.1.41 | 3-Oxoacyl-[acyl-carrier-protein]synthase | I | (189) |
| 2.3.1.54 | Formate acetyltransferase | I[f], E, K | (190) |
| 2.3.1.61 | Lipoate succinyltransferase [lipoate] | I | (191) |
| 2.3.2.2 | $\gamma$-Glutamyltransferase | I, E, K, G | (192) |
| 2.3.2 | Transglutaminase | I[e], E, K, B | (193) |
| 2.4.1.1 | Phosphorylase | M, C, I[f] | (194) |
| 2.4.1.2 | Dextrin dextranase | C | (195) |
| 2.4.1.4 | Amylosucrase | E, C | (196) |
| 2.4.1.5 | Dextransucrase | E, M, C | (197) |
| 2.4.1.7 | Sucrose phosphorylase | I, E, K, C | (198) |
| 2.4.1.9 | Inulosucrase | C | (199) |
| 2.4.1.10 | Levansucrase | I[f], E, K, M, C | (200) |

**Table 3.1.** Transferases Known to Act by Covalent Catalysis  (*Continued*)

| EC no. | Familiar name of enzyme[a] | Criteria[b] | References |
|---|---|---|---|
| 2.4.1.11 | Glycogen synthase | C | (201) |
| 2.4.1.13 | Sucrose synthase | C | (202) |
| 2.4.1.14 | Sucrose-phosphate synthase | C | (203) |
| 2.4.1.15 | α,α-Trehalose-phosphate synthase | C | (204) |
| 2.4.1.18 | 1,4-α-Glucan branching enzyme | C | (205) |
| 2.4.1.19 | Cyclomaltodextrin glucanotransferase | C | (206) |
| 2.4.1.21 | Starch (bacterial glycogen)synthase | C | (207) |
| 2.4.1.24 | 1,4-α-D-glucan 6-α-D-glucosyltransferase | C | (208) |
| 2.4.1.25 | 4-α-D-Glucanotransferase | C | (209) |
| 2.4.1.26 | UDPglucose-DNA α-D-glucosyltransferase | C | (210) |
| 2.4.1.36 | α,α-Trehalose-phosphate synthase (GDP-forming) | C | (211) |
| 2.4.1.37 | Blood-group-substance α-D-galactosyltransferase | C | (212) |
| 2.4.1.40 | Fucosyl-galactose acetylgalactosaminyltransferase | C | (213) |
| 2.4.1.41 | UDPacetylgalactosamine-protein acetylgalactos-aminyltransferase | C | (214) |
| 2.4.1.43 | UDPgalacturonate-polygalacturonate α-D-galacturonosyltransferase | C | (215) |
| 2.4.1.44 | UDPgalactose-lipopolysaccharide galactosyltransferase | C | (216) |
| 2.4.1.48 | GDPmannose α-D-mannosyltransferase | C | (217) |
| 2.4.1.52 | UDPglucose-poly(glycerol phosphate) α-D-glucosyl-transferase | C | (218) |
| 2.4.1.57 | GDPmannose-phosphatidyl-*myo*-inositol α-D-mannosyl-transferase | C | (219) |
| 2.4.1.58 | UDPglucose-lipopolysaccharide glucosyltransferase I | C | (220) |
| 2.4.1.67 | Galactinol-raffinose galactosyltransferase | E, C | (221) |
| 2.4.1.73 | UDPglucose-lipopolysaccharide glucosyltransferase II | C | (222) |
| 2.4.1.82 | Galactinol-sucrose galactosyltransferase | E, C | (223) |
| 2.4.1.87 | UDPgalactose-tetraglycosylceramide α-D-galactosyltransferase | C | (224) |
| 2.4.1.88 | UDPacetylgalactosamine-globoside α-N-acetyl-D-galactosaminyltransferase | C | (225) |
| 2.4.1.95 | Bilirubin-glucuronoside glucuronosyltransferase | C | (226) |
| 2.4.1.96 | UDPgalactose-*sn*-glycerol-3-phosphate galactosyl-transferase | C | (227) |
| 2.4.2.6 | Nucleoside deoxyribosyltransferase | E, K, C | (228) |
| 2.4.2.7 | Adenine phosphoribosyltransferase | I[c] | (229) |
| 2.4.2.8 | Hypoxanthine phosphoribosyltransferase | E, K | (230) |
| 2.4.2.10 | Orotate phosphoribosyltransferase | E, K | (231) |
| 2.4.2.14 | Amidophosphoribosyltransferase [Fe—S] | G | (232) |
| 2.4.2 | tRNA transglycosylase | C | (233) |
| 2.5.1.2 | Thiaminase I | $V_{max}$ | (234) |
| 2.5.1.7 | Enoylpyruvate transferase | I, E | (235) |
| 2.5.1.16 | Spermidine synthase | K | (236) |
| 2.6.1.1 | Aspartate aminotransferase [PLP] | I, K, M | (237) |
| 2.6.1.2 | Alanine aminotransferase [PLP] | I[d], K, M | (238) |

**Table 3.1.** Transferases Known to Act by Covalent Catalysis   (*Continued*)

| EC no. | Familiar name of enzyme[a] | Criteria[b] | References |
|--------|---------------------------|-------------|------------|
| 2.6.1.5 | Tyrosine aminotransferase [PLP] | I[d], M | (239) |
| 2.6.1.9 | Histidinol-phosphate aminotransferase [PLP] | K, M | (240) |
| 2.6.1.13 | Ornithine—oxo-acid aminotransferase [PLP] | I, K | (241) |
| 2.6.1.15 | Glutamine—oxo-acid aminotransferase [PLP] | K | (242) |
| 2.6.1.19 | Aminobutyrate aminotransferase [PLP] | I, K, M | (243) |
| 2.6.1.21 | D-Alanine aminotransferase [PLP] | E, K | (244) |
| 2.6.1.30 | Pyridoxamine-pyruvate transaminase | I[c], M | (245) |
| 2.6.1.31 | Pyridoxamine-oxaloacetate transaminase | M | (246) |
| 2.6.1.36 | L-Lysine 6-aminotransferase [PLP] | I, K, M | (247) |
| 2.6.1.42 | Branched-chain-amino-acid aminotransferase [PLP] | M | (248) |
| 2.7.1.1 | Hexokinase | E, M | (249) |
| 2.7.1.23 | NAD kinase | E, K | (250) |
| 2.7.1.37 | Protein kinase | I, K | (251) |
| 2.7.1.62 | Phosphoramidate-hexose phosphotransferase | I[f], E | (252) |
| 2.7.1.77 | Nucleoside phosphotransferase | C | (253) |
| 2.7.1.98 | PEP—fructose 1-phosphotransferase | K, M | (254) |
| 2.7.1 | Glucose-1,6-diphosphate synthase | I, K | (255) |
| 2.7.2.1 | Acetate kinase | I, E, K | (256) |
| 2.7.2.3 | Phosphoglycerate kinase | E, M | (257) |
| 2.7.2.9 | Carbamoyl-phosphate synthase (glutamine) | M, G | (258) |
| 2.7.3.9 | Phosphoenolpyruvate–protein phosphotransferase | I, K | (259) |
| 2.7.4.6 | Nucleosidediphosphate kinase | I, E, K, C | (260) |
| 2.7.5.1 | Phosphoglucomutase | I[c], K | (261) |
| 2.7.5.2 | Acetylglucosamine phosphomutase | I, K | (262) |
| 2.7.5.3 | Phosphoglyceromutase | I, K, M, C | (263) |
| 2.7.5.4 | Bisphosphoglyceromutase | I | (264) |
| 2.7.7.5 | Sulfate adenylyltransferase | E, M, K | (265) |
| 2.7.7.7 | DNA polymerase [Zn, Mg or Mn] | Ma | (266) |
| 2.7.7.10 | Galactose-1-P uridylyltransferase | I, E, K, C | (267) |
| 2.7.7.35 | Ribose 5-phosphate adenylyltransferase | E, K | (268) |
| 2.7.7 | Capping enzyme | E, I | (269) |
| 2.7.9.1 | Pyruvate, orthophosphate dikinase | I, E, K | (270) |
| 2.7.9.2 | Phosphoenolpyruvate synthase | I, E, K, M | (271) |
| 2.8.1.1 | Rhodanese | I[c], E, K | (272) |
| 2.8.1.2 | 3-Mercaptopyruvate sulfurtransferase | I | (273) |
| 2.8.3.5 | 3-Ketoacid CoA-transferase | I, E, K, M | (274) |
| 2.8.3.8 | Acetate CoA-transferase | I, E, K | (275) |
| 2.8.3 | Acetyl-CoA acetoacetate CoA-transferase | I, E, K, M | (276) |

[a] Any parenthetical expression is part of the official name of the enzyme. Prosthetic groups are indicated in brackets. Abbreviations: TPP, thiamine pyrophosphate; PLP, pyridoxal-5'-P; Fe—S, a nonheme iron-labile–sulfur center.

[b] The symbols mean the following:

   I, the holoenzyme links covalently with the substrate or a fragment of it to form a chemically competent intermediate;

E, the enzyme catalyzes one or more exchange reactions consistent with the participation of a covalent enzyme–substrate intermediate; where E is given as the sole criterion it will be understood that the presence of cosubstrate in the exchange reaction is definitely excluded;

K, the enzyme exhibits kinetic properties (exclusive of "burst" and $V_{max}$ kinetics) consistent with the participation of a covalent enzyme–substrate intermediate;

B, the enzyme displays "burst" kinetics; $V_{max}$, the enzyme acts on diverse substrates in conformity with the maximum velocity principle (W. P. Jencks, *Catalysis in Chemistry and Enzymology*, McGraw-Hill, New York, 1969, pp. 50–53);

M, miscellaneous data and derivative arguments which are peculiar to the enzyme in question;

G, the enzyme is irreversibly inactivated for glutamine utilization by stoichiometric alkylation with a glutamine analogue; inferred from this is the participation of a glutamyl-enzyme intermediate;

C, the enzyme catalyzes a reaction in which steric configuration is retained at the carbon or phosphorus atom at which reaction occurs, thereby excluding the possibility of a single-displacement mechanism;

S, the enzyme forms a Schiff base with substrate or fragment thereof;

Ma, magnetic resonance or other physical studies show that this enzyme binds a substrate molecule by coordinate linkage to its prosthetic group—a metal ion;

R, the oxidized holoenzyme accepts electrons directly from reduced substrate.

$^c$ This enzyme is actually isolated from tissue as the covalent enzyme–substrate intermediate; that is, the transferring portion of the substrate is already fixed to the enzyme.

$^d$ Isolated and identified spectroscopically.

$^e$ Transglutaminase does not form a stable intermediate with its natural substrate. But, like chymotrypsin and other enzymes, transglutaminase reacts with unnatural substrates, such as p-nitrophenyl acetate and p-nitrophenyl trimethylacetate, to form, respectively, an acetyl- and a trimethylacetyl-enzyme. The latter is quite stable to hydrolysis.

$^f$ The covalent enzyme–substrate intermediate was isolated in denatured condition.

(168) F. Solomon and W. P. Jencks, *JBC* 244, 1079–1081 (1969).

(169) H. White and W. P. Jencks, *JBC* 251, 1688–1699 (1976).

(170) C. M. Pickart and W. P. Jencks, *JBC* 254, 9120–9129 (1979).

(171) L. B. Hersh and W. P. Jencks, *JBC* 242, 3468–3480 (1967).

(172) Refs. (3, 4, 6); H. Rüdiger and L. Jaenicke, *EJB* 16, 92–95 (1970); R. T. Taylor and H. Weissbach, *Arch. Biochem. Biophys.* 123, 109–126 (1968); R. T. Taylor and H. Weissbach, *Arch. Biochem. Biophys.* 129, 745–766 (1969).

(173) Refs. (8, 9).

(174) R. J. Langenbach, P. V. Danenberg, and C. Heidelberger, *Biochem. Biophys. Res. Commun.* 48, 1565–1571 (1972); D. V. Santi and C. S. McHenry, *PNAS* 69, 1855–1857 (1972); P. V. Danenberg, R. J. Langenbach, and C. Heidelberger, *B* 13, 926–933 (1974); A. L. Pogolotti, C. Weill, and D. V. Santi, *B* 18, 2794–2798 (1979); C. Garrett, Y. Wataya, and D. V. Santi, *B* 18, 2798–2804 (1979).

(175) Refs. (10, 11, 12, 13); R. J. Ulevitch and R. G. Kallen, *B* 16, 5350–5363 (1977); J. Hansen and L. Davis, *BBA* 568, 321–330 (1979).

(176) S. M. Klein and R. D. Sagers, *JBC* 241, 197–205 (1966); H. Kochi and G. Kikuchi, *J. Biochem. (Tokyo)* 75, 1113–1127 (1974); Y. Motokawa and G. Kikuchi, *Arch. Biochem. Biophys.* 164, 634–640 (1974); H. Kochi and G. Kikuchi, *Arch. Biochem. Biophys.* 173, 71–81 (1976); K. Fujiwara, K. Okamura, and Y. Motokawa, *Arch. Biochem. Biophys.* 197, 454–462 (1979).

(177) M. G. Kunitani and D. V. Santi, *B* 19, 1271–1275 (1980).

(178) Ref. (15); D. B. Northrop, *JBC* 244, 5808–5819 (1969); Ref. (14).

(179) Refs. (20–25).

(180) Refs. (26–28).

(181) Refs. (30–32); A. G. Datta and E. Racker, *Fed. Proc.* 19, 82 (1960); L. Ljungdahl, H. G. Wood, E. Racker, and D. Couri, *JBC* 236, 1622–1625 (1961); E. Kuhn and K. Brand, *B* 12, 5217–5223 (1973); O. Tsolas and B. L. Horecker, *Arch. Biochem. Biophys.* 173, 577–585 (1976).

(182) Refs. (34–38).

(183) Refs. (39–44).

(184) C. S. Tsai and L. J. Reed, *Fed. Proc.* 32, 627 (1973); P. J. Butterworth, C. S. Tsai, M. H. Eley, T. E. Roche, and L. J. Reed, *JBC* 250, 1921–1925 (1975); D. C. Speckhard, B. H. Ikeda, S. S. Wong, and P. A. Frey, *Biochem. Biophys. Res. Commun.* 77, 708–713 (1977); J. H. Collins and L. J. Reed, *PNAS* 74, 4223–4227 (1977); M. C. Ambrose–Griffin, M. J. Danson, W. G. Griffin, G. Hale, and R. N. Perham, *Biochem. J.* 187, 393–401 (1980).

(185) R. Miyajima and I. Shiio, *J. Biochem. (Tokyo)* 73, 1061–1068 (1973); A. Wyman and H. Paulus, *JBC* 250, 3897–3903 (1975).

(186) G. Kikuchi, A. Kumar, P. Talmadge, and D. Shemin, *JBC* 233, 1214–1219 (1958); Z. Zasman, P. M. Jordan, and M. Akhtar, *Biochem. J.* 135, 257–263 (1973); D. L. Nandi, *JBC* 253, 8872–8877 (1978).

(187) I. P. Williamson and S. J. Wakil, *JBC* 241, 2326–2332 (1966); J. Ziegenhorn, R. Niedermeier, C. Nüssler, and F. Lynen, *EJB* 30, 285–300 (1972).

(188) Refs. (45–47).

(189) M. D. Greenspan, A. W. Alberts, and P. R. Vagelos, *JBC* 244, 6477–6485 (1969); G. D'Agnolo, I. S. Rosenfeld, and P. R. Vagelos, *JBC* 250, 5283–5288 (1975).

(190) M. F. Utter, F. Lipmann, and C. H. Werkman, *JBC* 158, 521–531 (1945); D. G. Lindmark, P. Paolella, and N. P. Wood, *JBC* 244, 3605–3612 (1969); R. K. Thauer, F. H. Kirchniawy, and K. A. Jungermann, *EJB* 27, 282–290 (1972); J. Knappe, H. P. Blaschkowski, P. Gröbner, and T. Schmitt, *EJB* 50, 253–263 (1974).

(191) F. H. Pettet, L. Hamilton, P. Munk, G. Namihira, M. H. Eley, C. R. Willms, and L. J. Reed, *JBC* 248, 5282–5290 (1973); J. H. Collins and L. J. Reed, *PNAS* 74, 4223–4227 (1977).

(192) S. S. Tate and A. Meister, *JBC* 249, 7593–7602 (1974); J. S. Elce and B. Broxmeyer, *Biochem. J.* 153, 223–232 (1976); J. H. Strømme and L. Theodersen, *Clin. Chem.* 22, 417–421 (1976); J. W. London, L. M. Shaw, D. Fetterolf, and D. Garfinkel, *Fed. Proc.* 35, 1497 (1976); A. M. Karkowsky, M. V. W. Bergamini, and M. Orlowski, *JBC* 251, 4736–4743 (1976); S. S. Tate and A. Meister, *PNAS* 74, 931–935 (1977); S. S. Tate and M. E. Ross, *JBC* 252, 6042–6045 (1977); J. S. Elce, *Biochem. J.* 185, 473–481 (1980).

(193) Refs. (48–50).

(194) J. I. Tu, G. R. Jacobson, and D. J. Graves, *B* 10, 1229–1236 (1971); L. M. Firsov, T. I. Bogacheva, and S. E. Bresler, *EJB* 42, 605–609 (1974); F. C. Kokesh and Y. Kakuda, *B* 16, 2467–2473 (1977); L. N. Johnson, J. A. Jenkins, K. S. Wilson, E. A. Stura, and G. Zanotti, *JMB* 140, 565–580 (1980); E. J. M. Helmreich and H. W. Klein, *Angew. Chem. Int. Ed.* 19, 441–455 (1980).

(195) E. J. Hehre, *JBC* 192, 161–174 (1951).

(196) G. Okada and E. J. Hehre, *JBC* 249, 126–135 (1974).

(197) E. J. Bourne, J. Peters, and H. Weigel, *J. Chem. Soc.* 4605–4607 (1964); E. J. Hehre and H. Suzuki, *Arch. Biochem. Biophys.* 113, 675–683 (1966); J. F. Robyt, B. K. Kimble, and T. F. Walseth, *Arch. Biochem. Biophys.* 165, 634–640 (1974); J. F. Robyt and H. Taniguchi, *Arch. Biochem. Biophys.* 174, 129–135 (1976); J. F. Robyt and T. F. Walseth, *Carbohydr. Res.* 61, 433–445 (1978).

(198) Refs. (52, 56–58).

(199) J. Edelman and J. S. D. Bacon, *Biochem. J.* 49, 529–540 (1951).

(200) G. Avigad, D. S. Feingold, and S. Hestrin, *BBA* 20, 129–134 (1956); G. Rapoport, R. Dionne, E. Toulouse, and R. Dedonder, *Bull. Soc. Chim. Biol.* 48, 1323–1348 (1966); R. Dedonder, in *Biochemistry of the Glycosidic Linkage*, R. Piras and H. G. Pontis, eds., Academic Press, New York, 1972, pp. 49–52; R. Chambert, G. Treboul, and R. Dedonder, *EJB* 41, 285–300 (1974); R. Chambert and G. Gonzy–Treboul, *EJB* 62, 55–64 (1976); R. Chambert and G. Gonzy–Treboul, *EJB* 71, 493–508 (1976).

(201) L. F. Leloir and C. E. Cardini, *JACS* 79, 6340–6341 (1957).

(202) C. E. Cardini, L. F. Leloir, and J. Chiriboga, *JBC* 214, 149–155 (1955).

(203) L. F. Leloir and C. E. Cardini, *JBC* 214, 157–165 (1955).

(204) E. Cabib and L. F. Leloir, *JBC* 231, 259–275 (1958).

(205) E. J. Hehre, *Adv. Enzymol.* 11, 324–326 (1951).

(206) E. J. Hehre, *Adv. Enzymol.* 11, 322–324 (1951).

(207) R. B. Frydman and C. E. Cardini, *BBA* 96, 294–303 (1965).

(208) M. Abdullah and W. J. Whelan, *Biochem. J.* 75, 12P (1960).

(209) J. H. Pazur and S. Okada, *JBC* 243, 4732–4738 (1968).

(210) S. R. Kornberg, S. B. Zimmerman, and A. Kornberg, *JBC* 236, 1487–1493 (1961).

(211) A. D. Elbein, *JBC* 242, 403–406 (1967).

(212) C. Race, D. Ziderman, and W. M. Watkins, *Biochem. J.* 107, 733–735 (1968).

(213) V. M. Hearn, Z. G. Smith, and W. M. Watkins, *Biochem. J.* 109, 315–317 (1968).

(214) R. D. Marshall, *Ann. Rev. Biochem.* 41, 681 (1972).

(215) C. L. Villemez, A. L. Swanson, and W. Z. Hassid, *Arch. Biochem. Biophys.* 116, 446–452 (1966).

(216) A. Endo and L. Rothfield, *B* 8, 3500–3507 (1969).

(217) H. Ankel, E. Ankel, J. S. Schutzbach, and J. C. Garancis, *JBC* 245, 3945–3955 (1970).

(218) L. Glaser and M. M. Burger, *JBC* 239, 3187–3191 (1964).

(219) Y. C. Lee and C. E. Ballou, *B* 4, 1395–1404 (1965); P. Brennan and C. E. Ballou, *Biochem. Biophys. Res. Commun.* 30, 69–75 (1968).

(220) L. Rothfield, M. J. Osborn, and B. L. Horecker, *JBC* 239, 2788–2795 (1964).

(221) W. Tanner and O. Kandler, *EJB* 4, 233–239 (1968).

(222) R. D. Edstrom and E. C. Heath, *JBC* 242, 3581–3588 (1967).

(223) L. Lehle and W. Tanner, *EJB* 38, 103–110 (1973).

(224) M. Basu and S. Basu, *JBC* 248, 1700–1706 (1973).

(225) S. Kijimoto, T. Ishibashi, and A. Makita, *Biochem. Biophys. Res. Commun.* 56, 177–184 (1974).

(226) B. H. Billing, P. G. Cole, and G. H. Lathe, *Biochem. J.* 65, 774–784 (1957).

(227) H. Kauss and B. Schobert, *FEBS lett.* 19, 131–135 (1971).

(228) W. S. Macnutt, *Biochem. J.* 50, 384–397 (1952); H. M. Kalckar, W. S. Macnutt, and E. Hoff-Jorgensen, *Biochem. J.* 50, 397–400 (1952); C. Danzin and R. Cardinaud, *EJB* 48, 255–262 (1974); C. Danzin and R. Cardinaud, *EJB* 62, 365–372 (1976).

(229) Ref. (59).

(230) T. A. Krenitsky and R. Papaioannou, *JBC* 244, 1271–1277 (1969).

(231) Ref. (62).

(232) S. Hartman, *JBC* 238, 3036–3047 (1963).

(233) N. Okada and S. Nishimura, *JBC* 254, 3061–3066 (1979).

(234) Ref. (64).

(235) P. J. Cassidy and F. M. Kahan, *B* 12, 1364–1374 (1973); R. I. Zemell and R. A. Anwar, *JBC* 250, 4959–4964 (1975).

(236) V. Zappia, G. Cacciapuoti, G. Pontoni, and A. Oliva, *JBC* 255, 7276–7280 (1980).

(237) Refs. (65–68); E. A. Malakhova and Yu. M. Torchinskii, *Doklady* 161, 100–102 (1965) (Engl. transl.).

(238) W. T. Jenkins, *Fed. Proc.* 20, 978–981 (1961); N. Katunuma, *Symp. Enz. Chem.* 16, 70 (1962); S. Hopper and H. L. Segal, *JBC* 237, 3189–3195 (1962); B. Bulos and P. Handler, *JBC* 240, 3283–3294 (1965); A. J. L. Cooper, *JBC* 251, 1088–1096 (1976); T. A. Alston, D. J. T. Porter, L. Mela, and H. J. Bright, *Biochem. Biophys. Res. Commun.* 92, 299–304 (1980).

(239) T. I. Diamondstone and G. Litwack, *JBC* 238, 3859–3868 (1963); G. Litwack and W. W. Cleland, *B* 7, 2072–2079 (1968).

(240) R. G. Martin, *Arch. Biochem. Biophys.* 138, 239–244 (1970); D. A. Weigent and E. W. Nester, *JBC* 251, 6974–6980 (1976).

(241) H. Tsai and W. T. Jenkins, *Fed. Proc.* 27, 792 (1968); C. Peraino, *BBA* 289, 117–127 (1972).

(242) A. J. L. Cooper and A. Meister, *JBC* 248, 8489–8498 (1973); F. van Leuven, *EJB* 65, 271–274 (1976).

(243) E. M. Scott and W. B. Jakoby, *JBC* 234, 932–936 (1959); Z. K. Nikolaeva and V. Yu. Vasiliev, *Biokhimiya* 37, 469–474 (1972) (Engl. transl.); M. Maître, L. Ciesielski, C. Cash, and P. Mandel, *EJB* 52, 157–169 (1975); M. J. Jung and B. W. Metcalf, *Biochem. Biophys. Res. Commun.* 67, 301–306 (1975); R. A. John and L. J. Fowler, *Biochem. J.* 155, 645–651 (1976);

R. R. Rando, *B* 16, 4604–4610 (1977); T. Beeler and J. E. Churchich, *EJB* 85, 365–372 (1978).

(244) M. Martinez-Carrion and W. T. Jenkins, *JBC* 240, 3547–3552 (1965); K. Soda, K. Yonaha, and H. Misono, *FEBS lett.* 46, 359–363 (1971).

(245) H. Wada and E. E. Snell, *JBC* 237, 133–137 (1962); W. B. Dempsey and E. E. Snell, *B* 2, 1414–1419 (1963); P. J. Gilmer and J. F. Kirsch, *B* 16, 5246–5253 (1977); J. Hodson, H. Kolb, E. E. Snell, and R. D. Cole, *Biochem. J.* 169, 429–432 (1978).

(246) H. Wada and E. E. Snell, *JBC* 237, 127–132 (1962).

(247) K. Soda and H. Misono, *B* 7, 4110–4119 (1968).

(248) R. T. Taylor and W. T. Jenkins, *JBC* 241, 4396–4405 (1966).

(249) Refs. (75, 84, 85); S. P. Colowick, in *The Enzymes*, 3rd ed., P. D. Boyer, ed., Academic Press, New York, 1972, Vol. 9, pp. 23–25.

(250) B. P. Orringer and A. E. Chung, *BBA* 250, 86–91 (1971).

(251) S. N. Kochetkov, T. V. Bulargina, L. P. Sashchenko, and E. S. Severin, *FEBS lett.* 71, 212–214 (1976); G. W. Moll, Jr., and E. T. Kaiser, *JBC* 251, 3993–4000 (1976); S. N. Kochetkov, T. V. Bulargina, L. P. Sashchenko, and E. S. Severin, *EJB* 81, 111–118 (1977); S. N. Kochetkov. L. L. Khachatryan, E. M. Bagirov, L. P. Sashchenko, and E. S. Severin, *Biokhimiya* 43, 126–130 (1978) (Engl. transl.).

(252) A. Fujimoto and R. A. Smith, *BBA* 56, 501–511 (1962); J. R. Stevens-Clark, K. A. Conklin, A. Fujimoto, and R. A. Smith, *JBC* 243, 4474–4478 (1968).

(253) J. P. Richard, D. C. Prasher, D. H. Ives, and P. A. Frey, *JBC* 254, 4339–4341 (1979).

(254) J. Perret and P. Gay, *EJB* 102, 237–246 (1979).

(255) L.-J. Wong and I. A. Rose, *JBC* 251, 5431–5439 (1976); I. A. Rose, J. V. B. Warms and L.-J. Wong, *JBC* 252, 4262–4268 (1977).

(256) Refs. (102, 104–106, 112).

(257) Ref. (119); (see discussion of this enzyme on p. 96).

(258) E. Khedouri, P. M. Anderson, and A. Meister, *B* 5, 3552–3557 (1966); V. P. Wellner, P. M. Anderson, and A. Meister, *B* 12, 2061–2066 (1973); L. M. Pinkus and A. Meister, *JBC* 247, 6119–6127 (1977).

(259) R. D. Simoni, J. B. Hays, T. Nakazawa, and S. Roseman, *JBC* 248, 957–965 (1973); R. Stein, O. Schrecker, H. F. Lauppe, and H. Hengstenberg, *FEBS lett.* 42, 98–100 (1974); E. B. Waygood and T. Steeves, *Can. J. Biochem.* 58, 40–48 (1980).

(260) Refs. (123–132).

(261) Refs. (133–136, 142).

(262) P.-W. Cheng and D. M. Carlson, *JBC* 254, 8353–8357 (1979).

(263) L. I. Pizer, *JBC* 235, 895–901 (1960); S. Grisolia and W. W. Cleland, *B* 7, 1115–1121 (1968); Z. B. Rose, *Arch. Biochem. Biophys.* 140, 508–513 (1970); Z. B. Rose, *Arch. Biochem. Biophys.* 146, 359–360 (1971); H. G. Britton, J. Carreros, and S. Grisolia, *EJB* 36, 495–503 (1973); Z. B. Rose and S. Dube, *JBC* 251, 4817–4822 (1976); W. A. Blättler and J. R. Knowles, *B* 19, 738–743 (1980).

(264) Z. B. Rose and R. G. Whalen, *JBC* 248, 1513–1519 (1973).

(265) M. Grunberg-Manago, A. Del Campillo-Campbell, L. Dondon, A. M. Michelson, *BBA* 123, 1–16 (1966); C. A. Adams and D. J. D. Nicholas, *Biochem. J.* 128, 647–654 (1972); R. G. Nicholls, *Biochem. J.* 165, 149–155 (1977).

(266) J. P. Slater, I. Tamir, L. A. Loeb, and A. S. Mildvan, *JBC* 247, 6784–6794 (1972); C. F. Springate, A. S. Mildvan, R. Abramson, J. L. Engle, and L. A. Loeb, *JBC* 248, 5987–5993 (1973); D. L. Sloan, L. A. Loeb, A. S. Mildvan, and R. J. Feldmann, *JBC* 250, 8913–8920 (1975); A. S. Mildvan, *Adv. Enzymol.* 49, 110–113 (1979).

(267) Refs. (132, 148–152).

(268) A. I. Stern and M. Avron, *BBA* 118, 577–591 (1966); W. R. Evans and A. San Pietro, *Arch. Biochem. Biophys.* 113, 236–244 (1966).

(269) K. Mizumoto and F. Lipmann, *PNAS* 76, 4961–4965 (1979); S. Shuman, M. Surks, H. Furneaux, and J. Hurwitz, *JBC* 255, 11588–11598 (1980); S. Shuman and J. Hurwitz, *PNAS* 78, 187–191 (1981).

(270) Refs. (154, 155).

(271) R. A. Cooper and H. L. Kornberg, *Biochem. J.* 105, 49C (1967); R. A. Cooper and H. L. Kornberg, *BBA* 141, 211–213 (1967); K. Berman, N. Itada, and M. Cohn, *BBA* 141, 214–216 (1967); K. M. Berman and M. Cohn, *JBC* 245, 5309–5318 (1970); *idem.*, *JBC* 245, 5319–5325 (1970).
(272) Refs. (156–164).
(273) J. W. Hylin and J. L. Wood, *JBC* 234, 2141–2144 (1959); H. Vachek and J. L. Wood, *BBA* 258, 133–146 (1972).
(274) Refs. (165–171).
(275) K. K. Tung and W. A. Wood, *J. Bacteriol.* 124, 1462–1474 (1975); M. Schulman and D. Valentino, *J. Bacteriol.* 128, 372–381 (1976).
(276) S. J. Sramek and F. E. Frerman, *Arch. Biochem. Biophys.* 171, 27–35 (1975); *idem.*, *Arch. Biochem. Biophys.* 181, 178–184 (1977).

# Chapter 4
# Hydrolases

Hydrolases are transferases that have water as a common acceptor. In general, we know little of the details of how water participates in hydrolytic action. But in at least a few cases it looks as if water must undergo an activation by the enzyme before it can accept the transferring group. The few instances of water activation we know of all involve coordination of water to a metal ion held by the enzyme in a prosthetic capacity. Whether enzymes can activate water other than by coordination to metal is not clear at this time.

Many hydrolases can use acceptors other than water, the common ones being alcohols and amines. This chapter gives some examples of such transferase activity. It depends upon the ability of certain nucleophiles to compete successfully with water in the interception of the transferring group while it is covalently fixed to the enzyme. How an "unnatural" nucleophile can sometimes jostle a water molecule—and especially an activated one—out of the way, during hydrolase activity, is also not clear at this time.

At the end of the present chapter Table 4.1 lists 100 hydrolases—16% of all the known hydrolases—whose activity includes a covalent intermediate formed from the enzyme and its substrate or some fragment of the substrate. There is no known instance of hydrolase action in which it can be said, on grounds of *positive* evidence, that water attacks a substrate molecule directly; that is, without covalent intervention by the enzyme. The 16% figure may be fairly extrapolated to over 90%, given the chemical analogy that holds good within each of the 37 subsubgroups of hydrolases (cf. Chapter 8).

To accompany the table we give a sampling of hydrolases whose mode of chemical action has been much studied. From a large literature only those facts are introduced which have a bearing on covalent catalysis. And, since in no case is our knowledge complete, some speculation seems justified.

# Carboxylesterase [EC 3.1.1.1]

As an enzyme of rather broad specificity, carboxylesterase can catalyze the
hydrolysis of esters with widely diverse molecular structures (1).

$$R-\overset{\displaystyle O}{\overset{\|}{C}}-O-R' + H_2O \longrightarrow R'OH + R-\overset{\displaystyle O}{\overset{\|}{C}}-OH$$

The enzyme is also active in the hydrolysis of thiolesters and aromatic
amides. But water is not the only acceptor of the ester acyl group, since the
latter can also be transferred by the enzyme to alcohols (2) and amines (3),
which compete with water as nucleophiles when they are present. Such
competition experiments provide evidence for an acyl-enzyme mechanism,
in which the acyl-enzyme intermediate is partitioned between water and the
competing nucleophile (2, 4).

$$R-\overset{\displaystyle O}{\overset{\|}{C}}-O-R' + E \longleftrightarrow R-\overset{\displaystyle O}{\overset{\|}{C}}-O-R'\cdot E \longrightarrow R-\overset{\displaystyle O}{\overset{\|}{C}}-E + R'OH$$

$$R-\overset{\displaystyle O}{\overset{\|}{C}}-E \overset{H_2O}{\underset{N}{\diagdown\diagup}} \begin{array}{l} R-\overset{\displaystyle O}{\overset{\|}{C}}-OH + E \qquad\qquad [N = \text{nucleophile}] \\[2ex] R-\overset{\displaystyle O}{\overset{\|}{C}}-E\cdot N \longrightarrow R-\overset{\displaystyle O}{\overset{\|}{C}}-N + E \end{array}$$

Both the hydrolase and transferase activities of carboxylesterase are inhibited
by diethyl *p*-nitrophenylphosphate and diisopropyl phosphorofluoridate
(DFP), reagents which block the catalytic serine residue of the enzyme.
[$^{32}$P]DFP reacts stoichiometrically with carboxylesterase, one atom of
isotopic phosphorus combining with each of the two subunits of the enzyme
(5). Carboxylesterase also reacts very rapidly with diethyl *p*-nitrophenyl-
phosphate, liberating a stoichiometric "burst" of *p*-nitrophenol (6),
amounting in effect to a titration of the two active centers of the enzyme.
The hydrolysis of *p*-nitrophenyl dimethylcarbamate by carboxylesterase
also proceeds with a "burst" release of *p*-nitrophenol equivalent to the

(1) K. Krisch, in *The Enzymes*, 3rd ed., P. D. Boyer, ed., Academic Press, New York, 1971,
Vol. 5, pp. 43–69.
(2) P. Greenzaid and W. P. Jencks, *B* 10, 1210–1222 (1971).
(3) W. Franz and K. Krisch, *Z. Physiol. Chem.* 349, 1413–1422 (1968); M. I. Goldberg and
J. S. Fruton, *B* 9, 3371–3378 (1970).
(4) W. P. Jencks, *Catalysis in Chemistry and Enzymology*, McGraw-Hill, New York, 1969,
pp. 53–56.
(5) J. H. Boursnell and E. C. Webb, *Nature (London)* 164, 875 (1949).
(6) K. Krisch, *BBA* 122, 265–280 (1966).

quantity of enzyme used (7). Thus, though the acylated form of the enzyme has not so far been isolated, its participation in the enzyme's mechanism of action seems beyond doubt.

In conformity with this is the inhibition of carboxylesterase by benzil, which forms a spectrally observable tetrahedral adduct with the active-site serine (8).

Acting on a monooxime of benzil, the enzyme can also cause the unusual clevage of a carbon–carbon bond as follows (9):

(7) D. J. Horgan, J. R. Dunstone, J. K. Stoops, E. C. Webb, and B. Zerner, *B* 8, 2006–2013 (1969).

(8) M. C. Berndt, J. de Jersey, and B. Zerner, *JACS* 99, 8332–8334 (1977).

(9) M. C. Berndt, J. de Jersey, and B. Zerner, *JACS* 99, 8334–8335 (1977).

# Alkaline Phosphatase [EC 3.1.3.1]

This much-studied enzyme catalyzes the hydrolysis of almost any phospho-monoester, yielding as products orthophosphate and the corresponding hydroxylic component of the substrate.

$$R-O-\overset{\overset{\displaystyle O}{\|}}{\underset{\underset{\displaystyle O}{|}}{P}}-O + H_2O \xrightarrow{Zn^{2+}} HO-\overset{\overset{\displaystyle O}{\|}}{\underset{\underset{\displaystyle O}{|}}{P}}-O + R-OH \qquad (1)$$

The enzyme also possesses a powerful phosphotransferase capability, with a wide variety of alcoholic compounds participating as acceptors (10, 11).

$$R-O-\overset{\overset{\displaystyle O}{\|}}{\underset{\underset{\displaystyle O}{|}}{P}}-O + HO-R' \xrightarrow{Zn^{2+}} R-OH + O-\overset{\overset{\displaystyle O}{\|}}{\underset{\underset{\displaystyle O}{|}}{P}}-O-R' \qquad (2)$$

In 1 $M$ Tris buffer, for instance, Tris $O$-phosphate is formed, the Tris competing with water for the phosphoryl group (11, 12). The ratio of Tris-P to orthophosphate formed in the reaction proved to be constant for a large number of phosphoryl donors as diverse in chemical structure as $p$-nitrophenylphosphate and phosphoenolpyruvate (11, 12). Such constancy can result only if a phosphoryl-enzyme participates as an intermediate of the reaction,

$$E + RO-P \longrightarrow E \cdot RO-P \longrightarrow E-P + ROH$$

$$E-P \underset{Tris}{\overset{H_2O}{<}} \begin{array}{l} E + P_i \\ \\ Tris\text{-}P + E \end{array} \qquad (3)$$

since only thus can the reaction with water and Tris be independent of the leaving group (RO) of the phosphoester. The hydrolysis of a wide assortment of phosphoesters with a nearly constant $V_{max}$ also points to mediation by a phosphoenzyme (13).

Consonant with this conclusion is the observation that alkaline phosphatase incorporates a phosphoryl group into covalent linkage upon brief incubation at pH 5 with $[^{32}P]$glucose-6-P (14). The same phosphoenzyme is also formed from orthophosphate at acid pH, in reversal of the hydrolytic

(10) I. B. Wilson, J. Dayan, and K. Cyr, *JBC* 239, 4182–4185 (1964).
(11) H. Neumann, *EJB* 8, 164–173 (1969).
(12) H. Barrett, R. Butler, and I. B. Wilson, *B* 8, 1042–1047 (1969).
(13) D. R. Harkness, *Arch. Biochem. Biophys.* 126, 513–523 (1968).
(14) L. Engström, *Arkiv Kemi* 19, 129–140 (1962).

branch of Eq. 3 (15). A serine hydroxyl in the active center of the enzyme is the site of phosphorylation. Phosphoenzyme prepared from $[^{32}P]P_i$ at pH 5 is hydrolyzed at pH 7 at a rate consistent with its role as intermediary in phosphoester hydrolysis (16). Phosphorylation of the enzyme on a serine hydroxyl also occurs during the hydrolysis of inorganic pyrophosphate (17).

$$PP_i + E \longleftrightarrow E \cdot PP_i \longleftrightarrow E{-}P + P_i$$

$$\downarrow H_2O$$

$$E + P_i$$

Also in harmony with the above, alkaline phosphatase exhibits "burst" kinetics with $p$-nitrophenylphosphate at acid pH, as well as at basic pH when the reaction temperature is reduced to 5°C. (18). With $O$-$p$-phenylazophenyl-phosphorothioate as substrate, "burst" kinetics are observed at both acid and basic pH, pointing to the mediation of a thiophosphorylenzyme in the hydrolysis (18, 19).

Alkaline phosphatase acts on orthophosphate to exchange the phosphate oxygen atoms with those of water (20), in keeping with the reversible phos-phorylation of the enzyme by orthophosphate.

(15) L. Engström, *BBA* 52, 49–59 (1961); *idem.*, *BBA* 56, 606–609 (1962); J. H. Schwartz and F. Lipmann, *PNAS* 47, 1996–2005 (1961); M. Caswell and M. Caplow, *B* 19, 2907–2911 (1980).
(16) W. N. Aldridge, T. E. Barman, and H. Gutfreund, *Biochem. J.* 92, 23C–25C (1964); T. E. Barman and H. Gutfreund, *Biochem. J.* 101, 460–466 (1966).
(17) H. N. Fernley and S. Bisaz, *Biochem. J.* 107, 279–283 (1968).
(18) J. F. Chlebowski and J. E. Coleman, *JBC* 247, 6007–6010 (1972).
(19) J. F. Chlebowski and J. E. Coleman, *JBC* 249, 7192–7202 (1974).
(20) J. H. Schwartz, *PNAS* 49, 871–878 (1963); J. L. Bock and M. Cohn, *JBC* 253, 4082–4085 (1978).

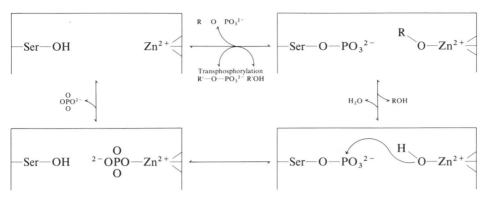

**Fig. 1.** A proposed mechanism for the action of alkaline phosphatase, showing the course of hydrolysis and transphosphorylation. The metal ion may aid the phosphorylation of serine by liganding the leaving alcoholate residue. Thereafter, rapid ligand exchange with water provides a nucleophilic hydroxyl—"activated water"—to remove the phosphoryl group from serine. If the liganded leaving alcohol is exchanged for an acceptor alcohol, transphosphorylation can occur.

As to the stereochemical course of alkaline phosphatase action, the transfer of a chiral phosphoryl group from phenyl phosphate to propane-1,2-diol occurs with net retention of configuration at phosphorus (21). This is of course the anticipated result for alkaline phosphatase if its mechanism is that of a simple double displacement.

Phosphorylation and dephosphorylation of the serine hydroxyl in the active center of alkaline phosphatase requires zinc ion (22). As the prosthetic group, zinc coordinates with the second substrate of the enzyme—a water molecule or a hydroxide ion (23)—providing what is in effect an "activated water." NMR studies reveal that the metal and the phosphoryl group of the phosphoenzyme must be adjoining in space, since they act upon each other to induce a mutual strain which is absent in the relevant nonenzymic models (24). And when inorganic phosphate is allowed to bind to the $Cd^{2+}$-substituted enzyme, some of the phosphate is found to be covalently fixed to serine, while *the remainder is coordinated to the metal* (25). Though the $Cd^{2+}$-enzyme is much less active than the native $Zn^{2+}$-enzyme, it is reasonable to suppose that both enzymes phosphorylate and dephosphorylate by the same pathway. Figure 1 proposes a mechanism to accommodate these

(21) S. R. Jones, L. A. Kindman, and J. R. Knowles, *Nature (London)* 275, 564–565 (1978).
(22) M. L. Applebury, B. P. Johnson, and J. E. Coleman, *JBC* 245, 4968–4976 (1970).
(23) R. S. Zukin and D. P. Hollis, *JBC* 250, 835–842 (1975).
(24) J. L. Bock and B. Sheard, *Biochem. Biophys. Res. Commun.* 66, 24–30 (1975); J. F. Chlebowski, I. M. Armitage, P. P. Tusa, and J. E. Coleman, *JBC* 251, 1207–1216 (1976); J. F. Chlebowski, I. M. Armitage, and J. E. Coleman, *JBC* 252, 7053–7061 (1977).
(25) J. D. Otvos, J. R. Alger, J. E. Coleman, and I. M. Armitage, *JBC* 254, 1778–1780 (1979).

facts (26). The idea is that after phosphorylation of the active-site serine by substrate, the phosphoryl group is displaced from serine by the zinc-bound hydroxyl. Subsequent loss of orthophosphate from zinc completes the hydrolytic cycle. The zinc ion may also have a role in phosphorylating the enzyme by liganding the alcoholic leaving group. Then by rapid ligand exchange at the metal ion, product alcohol can be removed in favor of water to give hydrolysis, or in favor of an acceptor alcohol to give phosphoryl transfer. A common mechanism is thus provided for the two activities of the enzyme.

## Acid Phosphatase [EC 3.1.3.2]

Like alkaline phosphatase, acid phosphatase is a nonspecific monophos-phoesterase (Eq. 1) which can also catalyze phosphoryl transfer to some alcohol acceptors (Eq. 2) (27). And, like alkaline phosphatase, acid phosphatase action proceeds over an intermediary phosphoenzyme, which undergoes partitioning between water and ethanol when the latter is a component of the reaction medium (28). During the enzymic hydrolysis of five different phosphoesters in an ethanol–water medium, the same ratio of hydrolysis to transphosphorylation was observed for each of the five substrates. A different, but still constant, ratio resulted when ethanolamine replaced ethanol as acceptor. These findings accord with the earlier report that a series of different phosphoesters were all hydrolyzed by the enzyme with the same $V_{max}$ (29). "Burst" kinetics have also been recorded for acid phosphatase (28, 30).

By incubating acid phosphatase (from wheat germ) with $[^{32}P]p$-nitro-phenylphosphate or $[^{32}P]$pyrophosphate, both of which are good substrates of the enzyme, a phosphoprotein results bearing an equivalent of covalently bound phosphorus (31). The enzyme from human prostate also reacts with labeled $p$-nitrophenylphosphate to yield a labeled enzyme (32), as does the rat liver enzyme incubated with labeled orthophosphate (30). All of the phosphoproteins, upon alkaline hydrolysis, yield 3'-phosphohistidine as

(26) J. E. Coleman and J. F. Chlebowski, *Adv. Inorg. Biochem.* 1, 1–66 (1979).

(27) At the time of writing, it is not known that acid phosphatase requires zinc ion for activity as is indicated in Eqs. 1 and 2.

(28) W. Ostrowski and E. A. Barnard, *B* 12, 3893–3898 (1973).

(29) G. S. Kilsheimer and B. Axelrod, *JBC* 227, 879–890 (1957); M. E. Hickey, P. P. Waymack, and R. L. Van Etten, *Fed. Proc.* 34, 599 (1975).

(30) M. Igarashi, H. Takahashi, and N. Tsuyama, *BBA* 220, 85–92 (1970).

(31) M. E. Hickey and R. L. Van Etten, *Arch. Biochem. Biophys.* 152, 423–425 (1972); R. L. Van Etten and M. E. Hickey, *Arch. Biochem. Biophys.* 183, 250–259 (1977).

(32) W. Ostrowski, *BBA* 526, 147–153 (1978).

the sole phosphorylated amino acid. It was further ascertained that the rat liver enzyme contains but one histidine residue per molecule of molecular weight 100,000, and that the loss of this histidine by photooxidation is directly correlated with the loss of enzymic activity (30). In alkaline phosphatase, by contrast, phosphorylation occurs on a serine hydroxyl group.

## Glucose-6-Phosphatase [EC 3.1.3.9]

This multifunctional catalyst is capable of both phosphohydrolase and phosphotransferase activities on a variety of substrates:

$$Glucose\text{-}6\text{-}P + H_2O \longrightarrow glucose + P_i$$

$$Glucose\text{-}6\text{-}P + sugar \longleftrightarrow glucose + sugar\text{-}P$$

$$R\text{-}P + glucose \longrightarrow R + glucose\text{-}6\text{-}P$$

$$R\text{-}P + H_2O \longrightarrow R + P_i$$

where R-P may be inorganic pyrophosphate, nucleosidetriphosphate, nucleosidediphosphate, mannose-6-P, fructose-6-P, carbamyl-P, phosphoenolpyruvate, and phosphoramide. Kinetic analysis and catalysis by the enzyme of a rapid $[^{14}C]$glucose—glucose-6-P exchange led early to the proposition that a phosphoenzyme is a component of the reaction pathway (33):

$$E + glucose\text{-}6\text{-}P \longleftrightarrow E \cdot glucose\text{-}6\text{-}P \longleftrightarrow E\text{—}P + glucose$$

$$\Big\downarrow H_2O$$

$$E + P_i$$

Compatible with this proposal is the observation that glucose-6-P, mannose-6-P, inorganic pyrophosphate, and carbamyl-P are all hydrolyzed with virtually the same $V_{max}$ by rat liver microsomal glucose 6-phosphatase (34). Moreover, when the enzyme is incubated with $[^{32}P]$glucose-6-P (35) or with $[^{32}P]$pyrophosphate (36), a labeled phosphoprotein results, which on alkaline hydrolysis, yields 3'-phosphohistidine.

(33) H. L. Segal, *JACS* 81, 4047–4050 (1959); L. F. Hass and W. L. Byrne, *JACS* 82, 947–954 (1960); W. J. Arion and R. C. Nordlie, *JBC* 239, 2752–2757 (1964).
(34) B. K. Wallin and W. J. Arion, *JBC* 248, 2380–2386 (1973).
(35) F. Feldman and L. G. Butler, *Biochem. Biophys. Res. Commun.* 36, 119–125 (1969); R. Parvin and R. A. Smith, *B* 8, 1748–1755 (1969).
(36) F. Feldman and L. G. Butler, *BBA* 268, 698–710 (1972).

## 5′-Nucleotide Phosphodiesterase [EC 3.1.4.1]

This enzyme, unlike alkaline and acid phosphatase and glucose-6-phosphatase, has little activity on phosphomonoesters, nor is it a phosphotransferase. It is active, however, in the hydrolysis of many phosphodiesters; as examples, the enzyme hydrolyzes TpT to 5′-TMP and thymidine, pApApApA to 5′-AMP, and 3′,5′-cyclic AMP to 5′-AMP (37). The enzyme has also the useful property of catalyzing the hydrolysis of phosphonic acid monoesters; e.g., 4-nitrophenyl phenylphosphonate:

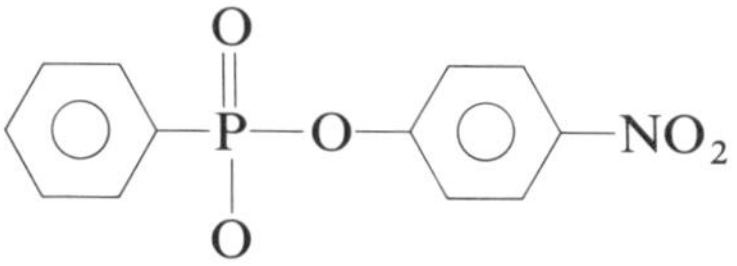

These compounds make convenient substrates for studying the chemical mechanism of 5′-nucleotide phosphodiesterase. Thus, 2-naphthyl phenylphosphonate,

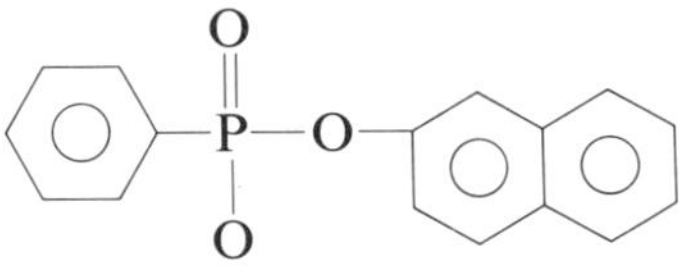

is hydrolyzed with the same $V_{max}$ as 4-nitrophenyl phenylphosphonate, a result which accords with the following simple reaction sequence (38):

$$E + R{-}O{-}\overset{\overset{O}{\|}}{\underset{\underset{O}{|}}{P}}{-}X \rightleftharpoons E{-}\overset{\overset{O}{\|}}{\underset{\underset{O}{|}}{P}}{-}X + R{-}OH \qquad (4)$$

$$E{-}\overset{\overset{O}{\|}}{\underset{\underset{O}{|}}{P}}{-}X + H_2O \rightleftharpoons E + HO{-}\overset{\overset{O}{\|}}{\underset{\underset{O}{|}}{P}}{-}X \qquad (5)$$

where X represents an alkyl or aryl group (phosphonate ester), or an alkoxy, aryloxy, or adenosyl group (phosphodiester). Other pairs of 4-nitrophenyl and 2-naphthyl esters of the same phosphonic acid (that is, different R but the same X in Eqs. 4 and 5) similarly adhere to the $V_{max}$ principle. These findings signal the intervention of a covalent enzyme–substrate intermediate containing X, whose hydrolysis (Eq. 5) is the rate-determining step. In accord with this view is the observation of a "burst" release of p-nitrophenol when 5′-nucleotide phosphodiesterase acts on the phosphodiester, bis(4-nitrophenyl) phosphate:

---

(37) S. J. Kelly, D. E. Dardinger, and L. G. Bulter, *B* 14, 4983–4988 (1975).
(38) S. J. Kelly and L. G. Butler, *B* 16, 1102–1104 (1977); M. Landt and L. G. Butler, *B* 17, 4130–4135 (1978).

$$O_2N-\langle\bigcirc\rangle-O-\overset{\overset{\displaystyle O}{\|}}{\underset{\underset{\displaystyle O}{|}}{P}}-O-\langle\bigcirc\rangle-NO_2$$

The amount of $p$-nitrophenol released in the burst is one mole per mole of dimeric enzyme, suggesting half-of-the-sites reactivity for this enzyme.

In conformity with all of the foregoing is the labeling of the enzyme with [$^3$H]3′,5′-cyclic AMP. When the enzyme (from bovine intestine) is incubated briefly with [$^3$H]cAMP and quickly quenched with phenol, labeled enzyme can be isolated after gel filtration under unfolding conditions. Longer incubations result in hydrolysis. The bond between the protein and the labeled adenylyl group is stable in acid but labile at pH 13. It appears to be a phosphodiester bond (to serine or threonine), because active phosphodiesterase acts on the labeled, denatured enzyme to release 5′-AMP (38).

In harmony with a double-displacement mechanism for 5′-nucleotide phosphodiesterase is the finding that the enzyme (from snake venom) catalyzes a net retention of configuration on phosphorus (39).

## Arylsulfatase [EC 3.1.6.1]

An enzyme that catalyzes the hydrolysis of arylsulfates

$$R-\langle\bigcirc\rangle-O-SO_3^- + H_2O \longrightarrow R-\langle\bigcirc\rangle-OH + SO_4^{2-} + H^+$$

has been long known but, until recently, little studied from the standpoint of chemical mechanism. Chemical analogy with the glucose-6-phosphatase and acid and alkaline phosphatase reactions (see above) prompts the view that a sulfonylenzyme ought to mediate the action of arylsulfatase.

$$R-\langle\bigcirc\rangle-O-SO_3^- + E \rightleftharpoons R-\langle\bigcirc\rangle-O-SO_3^-\cdot E \longrightarrow$$

$$E-SO_3^- + R-\langle\bigcirc\rangle-OH$$

$$\overset{H_2O}{\swarrow}$$

$$E + SO_4^{2-}$$

That such may indeed be the case is strongly indicated by the fact that seven different arylsulfates are hydrolyzed by the enzyme with the same $V_{max}$ (40), and that the hydrolysis of nitrocatechol sulfate (2-hydroxy-5-nitrophenylsulfate) occurs with a "burst" release of nitrocatechol (41). Moreover, the

---

(39) F. R. Bryant and S. J. Benkovic, *B* 18, 2825–2828 (1979); P.M.J. Burgers, F. Eckstein, and D. H. Hunneman, *JBC* 254, 7476–7478 (1979).
(40) S. J. Benkovic, E. V. Vergara, and R. C. Hevey, *JBC* 246, 4926–4933 (1971).
(41) A. B. Roy, *BBA* 227, 129–138 (1971).

titration of the active centers of the enzyme (two per molecule) with *o*-nitrophenyl oxalate has been described, and is also marked by a quantitative "burst" release of *o*-nitrophenol (42).

As with many other hydrolases, arylsulfatase possesses both hydrolytic and transferase activity. With tyramine as sulfonyl acceptor and nitrocatechol sulfate as donor, the transfer of the sulfonyl group follows ping-pong kinetics, from which is inferred a sulfonylenzyme as mediator of the reaction (43).

Of peculiar relevance to the chemical mechanism of arylsulfatase is the remarkable behavior of a specially prepared synthetic "enzyme" (a "synzyme"). A polyethyleneimine polymer, bearing dodecyl chains for binding sites and imidazole rings for catalytic groups, catalyzes the hydrolysis of nitrocatechol sulfate by a two-step mechanism which closely resembles that of the natural enzyme (44). A rapid, pre-steady-state release of nitro-catechol ($P_1$ in Eq. 6)

$$ E + S \longleftrightarrow ES \longrightarrow ES' \longrightarrow E + P_2 \qquad (6) $$
$$ + $$
$$ P_1 $$

is followed by a slower, linear release of nitrocatechol and sulfate ($P_2$). The sulfonated "enzyme" (ES') probably has the sulfonyl group joined covalently to a nitrogen of an imidazole ring. As it happens, a histidine residue in the active center of the natural enzyme is also thought to play a part in the catalysis (45). Remarkably, the synthetic "enzyme" is a 100-fold better catalyst than the natural enzyme, and $10^{12}$-fold better than free imidazole in solution.

## *β*-Galactosidase [EC 3.2.1.23]

Of the many known glycosidases, *β*-galactosidase is one of the most widely studied. It catalyzes the hydrolysis of terminal, nonreducing *β*-D-galactose residues in *β*-galactosides to *β*-D-galactose, the steric configuration at the anomeric carbon being retained (46),

$$ \text{(galactoside)} + H_2O \longrightarrow \text{(galactose)} + ROH \qquad (7) $$

(42) S. J. Benkovic and J. M. Feder, *JACS* 94, 8928–8929 (1972).

(43) G. R. J. Burns, E. Galanopoulou, and C. H. Wynn, *Biochem. J.* 167, 223–227 (1977).

(44) H. C. Kiefer, W. I. Congdon, I. S. Scarpa, and I. M. Klotz, *PNAS* 69, 2155–2159 (1972).

(45) A. Jerfy and A. B. Roy, *BBA* 175, 355–364 (1969); G. D. Lee and R. L. Van Etten, *Arch. Biochem. Biophys.* 171, 424–434 (1975).

(46) K. Wallenfels and G. Kurz, *Biochem. Z.* 335, 559–572 (1962).

where R may be an oligosaccharide, or an aryl or alkyl group. Like so many other hydrolases, β-galactosidase can catalyze transferase reactions as well as hydrolysis. In the presence of a suitable hydroxyl-bearing acceptor the enzyme transfers the β-galactosyl group to the acceptor, the latter competing favorably with water. Again, the configuration of C-1 is retained in the product (47).

$$\text{(galactosyl–OR)} + R'OH \longrightarrow \text{(galactosyl–OR')} + ROH \quad (8)$$

The stereochemical consequences of β-galactosidase action exclude the possibility of a single-displacement reaction between the reactant molecules of reactions 7 and 8. This because a single displacement must lead to inversion of configuration at the anomeric carbon. Retention of configuration is most easily harmonized with a reaction pathway requiring a sequence of two $S_N 2$ (Walden) displacements, as in a double displacement in which a nucleophilic group of the enzyme acts upon substrate to form the intermediary α-galactosylenzyme.

$$\text{(galactosyl–OR, }\beta\text{)} + E \longleftarrow \text{(galactosyl–}E, \alpha\text{)} + ROH$$

$$\text{(galactosyl–}E, \alpha\text{)} \xrightarrow{H_2O} \text{(galactosyl–OH, }\beta\text{)} + E$$

$$\text{(galactosyl–}E, \alpha\text{)} \xrightarrow{R'OH} \text{(galactosyl–OR', }\beta\text{)} + E$$

(47) K. Wallenfels and O. P. Malhotra, *Advan. Carbohyd. Chem.* 16, 239–298 (1961); C. H. Kuo and W. W. Wells, *JBC* 253, 3550–3556 (1978).

Justification for this point of view lies in the finding that during the hydrolysis of eight aryl $\beta$-galactosides in the presence of 0.25 $M$ methanol, the products ($\beta$-galactose and methyl $\beta$-galactoside) were formed in a constant ratio (48). With 0.17 $M$ ethanol as acceptor, the products ($\beta$-galactose and ethyl $\beta$-galactoside) again appeared in a constant (but different) ratio. The fact that a series of galactosides with different leaving groups can react with a pair of competing acceptors to give a constant ratio of products points persuasively to the existence of a common intermediate, which in the present case is reasonably assigned the structure of $\alpha$-galactosylenzyme (49). It has been argued, however, that at the moment of reaction with acceptor the $\alpha$-galactosylenzyme may dissociate to an ion-pair in which the galactosyl cation is the immediate reacting entity (50).

More evidence for the existence of a galactosyl-enzyme intermediate comes from the action of the enzyme on $o$-nitrophenyl-$\beta$-D-galactoside in 50% aqueous dimethylsulfoxide at low temperature ($-22°C$) (51). Under these conditions, turnover of the enzyme is negligible; but a "burst" of $o$-nitrophenol, stoichiometric with enzyme concentration, is readily realized. The "burst" is regarded as direct evidence for the accumulation of a galactosyl-enzyme intermediate, in accord with the general equation,

$$\text{E} + \text{S} \underset{}{\overset{k_1}{\rightleftarrows}} \text{ES} \xrightarrow{k_2} \text{EG} \xrightarrow{k_3} \text{E} + \text{P}_2$$
$$+$$
$$\text{P}_1$$

where EG (the galactosylenzyme) and $o$-nitrophenol ($\text{P}_1$) evolve from the Michaelis complex (ES) at a rate ($k_2$) which (at $-22°C$) is vastly greater than the rate of hydrolysis of EG ($k_3$) to free enzyme and galactose ($\text{P}_2$). Since the galactosylenzyme breaks down so slowly, the "burst" of $o$-nitrophenol is a measure of the active site concentration. Similar "burst" kinetics have also been found in completely aqueous solutions and at more elevated temperatures (11°C) (52).

It is noteworthy that $\beta$-galactosidase also catalyzes the addition of water to D-galactal to yield 2-deoxy-$\beta$-D-galactose (53).

---

(48) T. M. Stokes and I. B. Wilson, *B* 11, 1061–1064 (1972).

(49) G. van der Groen, J. Wouters-Leysen, M. Yde, and C. K. De Bruyne, *EJB* 38, 122–129 (1973).

(50) M. L. Sinnott and I. J. L. Souchard, *Biochem. J.* 133, 89–98 (1973).

(51) A. L. Fink and K. J. Angelides, *Biochem. Biophys. Res. Commun.* 64, 701–708 (1975).

(52) P. J. Deschavanne, O. M. Viratelle, and J. M. Yon, *PNAS* 75, 1892–1896 (1978).

(53) J. Lehman and E. Schröter, *Carbohyd. Res.* 23, 359–368 (1972).

The somewhat unusual kinetic manifestations attending this process suggest that it involves a chemical reaction between substrate and enzyme, presumably the formation and hydrolysis of a 2-deoxygalactosyl-enzyme intermediate (54).

## NADase [EC 3.2.2.5]

This enzyme catalyzes the hydrolysis of the $N$-glycosyl bond of NAD.

ADPribose + $\begin{array}{c}\text{C}\!\!=\!\!\text{O}\\ \text{NH}_2\end{array}$ + H$^+$

The reaction is strongly inhibited by nicotinamide, which is of course a component of NAD and a product of its hydrolysis. At concentrations of nicotinamide which inhibit hydrolysis, it was found that [$^{14}$C]nicotinamide is easily incorporated into NAD in an exchange reaction (55, 56). It was also established that NAD synthesis from nicotinamide and ADPribose in the presence of enzyme does not occur (55). These observations lend force to the idea that the cleavage of NAD requires the formation of an intermediary ADPribosylenzyme, in which the enzyme is covalently linked to C-1 of the ribosyl moiety.

$$\underset{\text{(NAD)}}{\overset{+}{\text{ADPribosyl-nicotinamide}} + \text{E}} \longleftrightarrow \text{ADPribosyl-E} + \text{nicotinamide}$$

$$\downarrow \text{H}_2\text{O}$$

$$\text{ADPribose} + \text{E} + \text{H}^+$$

(54) D. F. Wentworth and R. Wolfenden, *B* 13, 4715–4720 (1974).
(55) L. J. Zatman, N.O. Kaplan, and S. P. Colowick *JBC* 200, 197–212 (1953); B. M. Anderson, C. J. Ciotti, and N. O. Kaplan, *JBC* 234, 1219–1225 (1959).
(56) F. Schuler, P. Travo, and M. Pascal, *EJB* 69, 593–602 (1976).

As with many hydrolases, NADase also catalyzes transferase reactions. In 5 $M$ methanol the methylglycoside of ADPribose is formed to the exclusion of the hydrolytic product (57).

The transfer reaction proceeds with over 99% retention of the $\beta$-configuration at C-1 of the ribosyl group. This finding, too, argues for the double-displacement mechanism in which ADPribosylenzyme participates in the catalysis. The catalytic group of the enzyme—perhaps a carboxylate ion (58), as in sucrose phosphorylase—probably links in the $\alpha$-configuration to the anomeric carbon of the ADPribosyl moiety. That the ADPribosylenzyme may exist in the mechanistically equivalent form of an ion-pair has been suggested (59).

## Carboxypeptidase A [EC 3.4.17.1]

Carboxypeptidase A is well established as a metalloprotease, with one atom of zinc per molecule of enzyme. It is an exopeptidase with a bias toward cleavage of amino acids with aromatic rings or large aliphatic side chains in the C-terminal position. The enzyme also possesses a considerable esterase activity, acting, for instance, on $O$-(trans-p-chlorocinnamoyl)-L-$\beta$-phenyl-lactate as follows (60):

(57) M. Pascal and F. Schuler, *FEBS lett.* 66, 107–109 (1976).

(58) F. Schuler and P. Travo, *EJB* 65, 247–255 (1976); F. Schuler, M. Pascal, and P. Travo, *EJB* 83, 205–214 (1978).

(59) H. G. Bull, J. P. Ferraz, E. H. Cordes, A. Ribbi, and R. Apitz-Castro, *JBC* 253, 5186–5192 (1978); F. Schuler, P. Travo, and M. Pascal, *Bioorg. Chem.* 8, 83–90 (1979).

(60) M. W. Makinen, K. Yamamura, and E. T. Kaiser, *PNAS* 73, 3882–3886 (1976).

$$Cl\text{—}C_6H_4\text{—}\underset{H}{\overset{H}{C}}\text{=}C\text{—}\overset{O}{\overset{\|}{C}}\text{—}O\text{—}CH(\text{—}CH_2C_6H_5)\text{—}COO + H_2O \xrightarrow{Zn^{2+}}$$

(9)

$$Cl\text{—}C_6H_4\text{—}\underset{H}{\overset{H}{C}}\text{=}C\text{—}COO + HO\text{—}CH(\text{—}CH_2C_6H_5)\text{—}COO$$

X-ray diffraction studies on crystalline carboxypeptidase A show that the $Zn^{2+}$ is liganded to three amino acid residues of the enzyme (two histidines and the carboxyl group of a glutamyl residue) and a molecule of water (61). Moreover, an X-ray investigation of the enzyme—glycyl-L-tyrosine complex reveals that the carbonyl oxygen of the peptide bond is coordinated to the $Zn^{2+}$, having displaced the water (62). This fact prompted the idea that $Zn^{2+}$ participates in catalysis by coordination of the carbonyl oxygen, thus polarizing the carbonyl bond and making its carbon more prone to reaction with a nucleophile (61). Spectral measurements on carboxypeptidase A in which the $Zn^{2+}$ is replaced by $Co^{2+}$ lead to the same conclusion about the polarizing function of the metal acting as a Lewis acid (63). Oxidation of the $Co^{2+}$ to $Co^{3+}$, however, causes the complete loss of peptidase as well as esterase activity. $Co^{3+}$, unlike $Co^{2+}$, is inert to substitution reactions. The loss of enzymic activity upon oxidation of the $Co^{2+}$-enzyme may thus be due to the failure of substrate to enter the inner coordination sphere of the metal (64).

The nucleophile most likely to react with the polarized carbonyl of substrate is the carboxylate group of Glu 270 in the active center of the enzyme. From the X-ray work it is clear that this carboxylate ion is ideally close to the hydrolysable bond of the substrate. The consequence of such nucleophilic interaction is the formation of an anhydride intermediate

(61) W. N. Lipscomb, *Chem. Soc. Rev.* 1, 319–336 (1972).

(62) W. N. Lipscomb, G. N. Reeke, Jr., J. A. Hartsuck, F. A. Quiocho, and P. H. Bethge, *Philos. Trans. R. Soc. London* B257, 177–214 (1970).

(63) S. A. Latt and B. L. Vallee, *B* 10, 4263–4270 (1971).

(64) M. M. Jones, J. B. Hunt, C. B. Storm, P. S. Evans, F. W. Carson, and W. J. Pauli, *Biochem. Biophys. Res. Commun.* 75, 253–258 (1977); H. E. Van Wart and B. L. Vallee, *Biochem. Biophys. Res. Commun.* 75, 732–738 (1977); idem., *B* 17, 3385–3394 (1978).

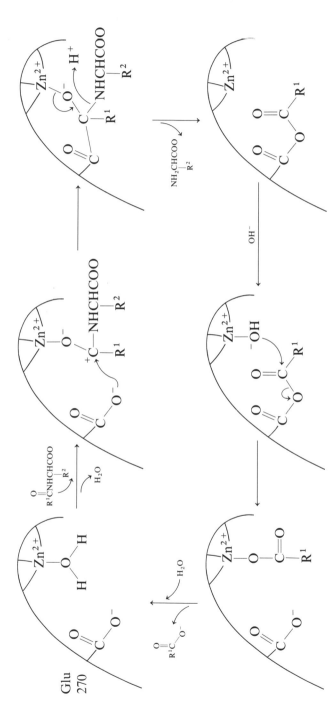

**Fig. 2.** A proposed mechanism of carboxypeptidase A action. Glu 270 participates in catalysis as a mixed anhydride with the acyl fragment of the substrate. Prosthetic zinc ion may act as a Lewis acid to activate substrate and to activate water. Tyr 248 (not shown) may supply the proton to the amino group of the C-terminal amino acid as it leaves the active center.

composed of the carboxyls of Glu 270 and the penultimate acid residue of the polypeptide (or ester) substrate (Fig. 2). Just such an intermediate can in fact be isolated, under special conditions, when carboxypeptidase A acts upon O-(*trans-p*-chlorocinnamoyl)-L-*β*-phenyllactate (60). In an organic-aqueous solvent at low temperature ($-60°C$), *trans-p*-chlorocinnamoylenzyme is revealed as an intermediate of reaction 9.

Hydrolysis of the anhydride function of the acyl enzyme requires participation of the zinc ion (65). Accordingly, Fig. 2 proposes that a zinc-coordinated hydroxide ion combines with the sensitive carbonyl of the acyl enzyme to disrupt the anhydride function, and yield a zinc-coordinated carboxylate ion (66). A water molecule ultimately displaces the carboxylate ion from the metal. In harmony with this picture is the observation that during inhibition of $Mn^{2+}$-carboxypeptidase A by a variety of carboxylate salts, the inhibitors are bound to the enzyme with their carboxylate groups in the first coordination shell of the $Mn^{2+}$, consistent with their character as competitive inhibitors (67). The mechanism depicted in Fig. 2 assigns a double function to the protein-bound $Zn^{2+}$. It is a Lewis acid in the first phase of the catalytic cycle, promoting formation of the acyl-enzyme intermediate. Then, to complete the cycle, the $Zn^{2+}$ activates a water molecule for hydrolysis of the acid anhydride. On this view, the substrate and its fragments form three different covalent bonds with holoenzyme in the course of one catalytic cycle of hydrolysis.

# Chymotrypsin [EC 3.4.21.1]

What may well be the world's most studied enzyme is the serine protease, chymotrypsin. A unique serine residue (Ser 195) is known from X-ray study of the crystalline enzyme to be in the active center as part of a catalytic "triad" which includes Asp 102 and His 57 (68). This serine—and none other—reacts stoichiometrically with diisopropyl phosphorofluoridate, resulting in

---

(65) L. C. Kuo, M. W. Makinen, and J. J. Dymowski, *Fed. Proc.* 37, 1696 (1978); M. W. Makinen, L. C. Kuo, J. J. Dymowski, and S. Jaffer, *JBC* 254, 356–366 (1979).

(66) Something analogous happens in the action of carbonic anhydrase when zinc-coordinated hydroxide reacts with carbon dioxide to yield zinc-coordinated carbonate (p. 174). Metal-coordinated hydroxide ions are coming to be recognized as potent nucleophiles at neutral pH [D. A. Buckingham and L. M. Engelhardt, *JACS* 97, 5915–5917 (1975); P. Woolley, *Nature (London)* 258, 677–682 (1975)]. As such they may be regarded as a species of "activated water." By coordinating to a metal ion a water molecule can have its protons more easily stripped away than otherwise, generating in the extreme case what may be a negatively charged oxygen atom linked to a metal. In the hydrophobic climate of an active center, this ought to be a potent nucleophile indeed.

(67) G. Navon, R. G. Shulman, B. J. Wyluda, and T. Yamana, *PNAS* 60, 86–91 (1968).

(68) D. M. Blow, J. J. Birktoft, and B. S. Hartley, *Nature (London)* 221, 337–340 (1969).

a completely inactive enzyme (69). When the enzyme at neutral pH acts upon the pseudosubstrate, p-nitrophenyl acetate, there is a "burst" release of p-nitrophenol in an amount equivalent to that of the enzyme, followed by a further slow (zero-order) release of p-nitrophenol (70). This biphasic release of p-nitrophenol is best described as a two-step chemical reaction, as expressed earlier in Eq. 6. Again, the ES' of Eq. 6 is considered to be the product (in this case the acetyl enzyme) of the stoichiometric reaction of the enzyme with substrate. In acid solution (pH 5), the "burst" release of p-nitrophenol is still observed, but the subsequent zero-order reaction is slowed essentially to zero. Low pH, then, permits the isolation of acetyl chymotrypsin (71). On degradation of this intermediate it becomes apparent that the same serine is acetylated by p-nitrophenyl acetate as is phosphorylated by diisopropyl phosphorofluoridate (72). Equation 6, as it applies to the action of chymotrypsin on p-nitrophenyl acetate, may therefore be written as follows:

$$\text{E-OH} + \text{CH}_3\overset{\displaystyle O}{\overset{\|}{\text{C}}}-\text{O}-\langle\bigcirc\rangle-\text{NO}_2 \longrightarrow$$

$$\begin{bmatrix}\text{Michaelis}\\\text{complex}\end{bmatrix} \xrightarrow{\text{HO}-\langle\bigcirc\rangle-\text{NO}_2} \text{E-O}-\overset{\displaystyle O}{\overset{\|}{\text{C}}}-\text{CH}_3 \xrightarrow{\text{H}_2\text{O}}$$

$$\text{E-OH} + \text{CH}_3\overset{\displaystyle O}{\overset{\|}{\text{C}}}-\text{OH}$$

A notable feature of chymotryptic catalysis is revealed in the chemical properties of the acetyl chymotrypsin prepared from p-nitrophenyl acetate at pH 5.5. The acetyl enzyme reacts readily with hydroxylamine at pH 5.5 to form one equivalent of acethydroxamic acid (73). By this criterion, the serine bond of the acetyl enzyme is energy-rich. But when acetyl chymotrypsin is denatured (reversibly) in 8 M urea at pH 3.0 and then reacted with hydroxylamine at pH 5.5, no hydroxamate is formed. In the denatured condition, therefore, the serine ester bond is energy-poor. (p-Nitrophenyl acetate itself reacts normally with hydroxylamine in 8 M urea.) Upon dilution of the urea, however, the acetyl enzyme recovers its normal properties and is again energy-rich and enzymically active. Clearly, the reactivity of Ser 195 and its esters depends on the integrity of the total protein structure of the enzyme. In this sense the native enzyme may be regarded as a transducer (or

(69) A. K. Balls and E. F. Jansen, *Adv. Enzymol.* 13, 321–343 (1952); N. K. Schaffer, S. C. May, Jr., and W. H. Summer, *JBC* 206, 201–207 (1954).
(70) B. S. Hartley and B. A. Kilby, *Biochem. J.* 56, 288–297 (1954).
(71) A. K. Balls and F. L. Aldrich, *PNAS* 41, 190–196 (1955); A. K. Balls and H. N. Wood, *JBC* 219, 245–256 (1956).
(72) R. A. Oosterbaan and M. E. van Andrichem, *BBA* 27, 423–425 (1958).
(73) G. H. Dixon, W. J. Dreyer, and H. Neurath, *JACS* 78, 4810 (1956).

reservoir) of chemical energy (p. 20). Parallel observations on [acyl-carrier-protein]malonyltransferase were remarked upon earlier (p. 76).

The acyl-enzyme mechanism is also operative in the hydrolysis of simple peptides and anilides, which are of course closer in structure than esters to the natural polypeptide substrates of the enzyme (74). Application of the partition criterion (4) reveals that a common intermediate mediates the hydrolysis of the ester, anilide, and peptide derivatives of $N$-acetyl-L-phenylalanine. A constant product ratio is found when these substrates are hydrolyzed by chymotrypsin in the presence of added acceptor nucleophiles which can compete effectively with water in the reaction. Thus, at a constant concentration of an acceptor (e.g., L-alaninamide), the same ratio of $N$-acetyl-L-phenylalanylalanylamide to $N$-acetyl-L-phenylalanine is found with the methyl ester, the $p$-trimethylammonium anilide, the dimethylamino-anilide, and the L-alaninamide peptide of $N$-acetyl-L-phenylalanine (74). The reaction with the last substrate is of course an exchange reaction.

Like some other hydrolases, chymotrypsin catalyzes an oxygen exchange between water and the carboxylic acids that are products of its hydrolytic activity (75). The exchange is believed to proceed over a pathway which includes the acyl-enzyme intermediate (76):

$$\text{E} + \text{R}-\overset{\overset{\displaystyle O}{\|}}{\text{C}}-\overset{*}{\text{O}}\text{H} \longrightarrow \text{R}-\overset{\overset{\displaystyle O}{\|}}{\text{C}}-\overset{*}{\text{O}}\text{H}\cdot\text{E} \xrightarrow{-H_2O^*} \text{R}-\overset{\overset{\displaystyle O}{\|}}{\text{C}}-\text{E} \xrightarrow{+H_2O}$$

$$\text{R}-\overset{\overset{\displaystyle O}{\|}}{\text{C}}-\text{OH}\cdot\text{E} \longrightarrow \text{R}-\overset{\overset{\displaystyle O}{\|}}{\text{C}}-\text{OH} + \text{E}$$

The acyl enzyme is easily prepared by incubating chymotrypsin with the chromophoric $N$-(2-furyl)acryloyl-L-tryptophan (77). The $k_{\text{obs}}$ for its decomposition is essentially the same as the $k_{\text{cat}}$ for the hydrolysis of the ester, $N$-(2-furyl)acryloyl-L-tryptophan methyl ester. The conclusion seems inescapable that the acyl enzyme formed from the acid (at low pH) is the kinetically significant intermediate in the hydrolysis of the ester.

All of the detailed mechanistic studies of chymotrypsin action have been made on pseudosubstrates of various kinds, each acting as a "model" of a protein substrate. It is now clear, however, that the basic reaction pathway established for such "models" applies as well to an authentic protein substrate. Thus, chymotrypsin, acting on the radioactively labeled (and slightly modified) B-chain of insulin, splits the polypeptide with the formation of an acyl enzyme, which can be isolated in denatured condition (78).

(74) J. Fastrez and A. R. Fersht, *B* 12, 2025–2034 (1973).

(75) D. B. Sprinson and D. Rittenberg, *Nature (London)* 167, 484 (1951); D. Doherty and F. Vaslow, *JACS* 74, 931–936 (1952); M. L. Bender and K. C. Kemp, *JACS* 79, 116–120 (1957).

(76) M. L. Bender, *JACS* 84, 2582–2590 (1962).

(77) C. G. Miller and M. L. Bender, *JACS* 90, 6850–6852 (1968).

(78) U. Dahlqvist and S. Wåhlby, *BBA* 391, 410–414 (1975).

While there is general agreement that Ser 195 is acylated during chymotrypsin catalysis, there are signs that the detailed mechanism of this serine protease (and probably all serine proteases) may be more complex than was hitherto thought. In particular, some hints are accumulating that the acyl group binds to His 57 as well as to Ser 195 during the catalytic cycle. The kinetics of acylation of chymotrypsin by a series of substituted phenyl esters of benzoic and acetic acid yield a Hammett $\rho$ value that points to an imidazole of the enzyme as initiating the nucleophilic attack upon substrate (79).

Kinetic solvent isotope effects support this idea (80). And further in accord with the same idea is the $^{31}P$ NMR spectrum of diisopropyl phosphoryl chymotrypsin, which displays two peaks. These have a ratio of intensities which is dependent upon the pH of the solution, and are assigned to two interconverting isomers of the enzyme complex. It is suggested that the isomerization entails phosphoryl transfer between the hydroxyl of Ser 195 and the imidazole nitrogen of His 57 (81).

Like the acylation of chymotrypsin, the deacylation of the acyl enzyme may also be more complex than was formerly thought. Kinetic studies were made on the acyl enzymes prepared from the homologous series of *p*-nitrophenyl esters of 2-(5-*n*-alkyl)furoic acid.

(79) C. D. Hubbard and J. F. Kirsch, *B* 11, 2483–2493 (1972); N. Shimamoto, *J. Biochem.* (*Tokyo*) 80, 961–968 (1976); C. D. Hubbard and T. S. Sharpe, *JBC* 252, 1633–1638 (1977); N. Shimamoto, *J. Biochem.* (*Tokyo*) 82, 185–193 (1977).
(80) J. F. Kirsch and V. I. Zannis, *Fed. Proc.* 38, 474 (1979).
(81) D. G. Gorenstein and J. B. Findlay, *Biochem. Biophys. Res. Commun.* 72, 640–645 (1976).

The Arrhenius plots for the hydrolyses of the furoyl chymotrypsins exhibit a sharp discontinuity. It is conjectured that the deacylation involves a minimum of two elementary steps, which are differently affected by temperature. Instead of invoking protein conformational changes, it is proposed that "the furoyl residue may lie in more than one position within the active site region and still undergo hydrolysis" (82). Not stated is which amino acid(s) in the active site the furoyl residue might link to enroute from Ser 195 to reaction with water.

In sum, there is a consensus that chymotrypsin and the other serine proteases effect hydrolysis through an acyl-enzyme intermediate—that is, by double-displacement catalysis. But newer developments are suggesting a pathway more complex than this; possibly a *multidisplacement* pathway, in which the acyl group joins covalently not only to Ser 195 but to one or more other catalytic groups in the active center, in a fully organized "surface walk."

## Papain [EC 3.4.22.2]

The cysteine proteases form a group of enzymes which depend for their activity on the nucleophilic action of a thiol function in the active center. Of the several familiar thiol proteases, papain has probably had the most searching scrutiny, and is presented here as representative of this group of enzymes. Papain cleaves a variety of peptide bonds, but it also cleaves esters (83) and thiolesters (84). Transamidations of the type

$$\text{Carbobenzoxyglycinamide} + H_2\text{N-R} \longleftarrow$$
$$\text{carbobenzoxyglycyl-NH-R} + NH_3$$

are also catalyzed by papain (85), as are a variety of transesterification reactions (83, 86). The multiple activities of papain are readily accounted for in chemical terms by a mechanism which includes an acyl-enzyme intermediate (87). That an acyl enzyme does indeed mediate papain-catalyzed hydrolysis of esters and amides first became apparent in the observation that $\alpha$-$N$-benzoyl-L-arginine ethyl ester and $\alpha$-$N$-benzoyl-L-argininamide are hydrolyzed with very nearly the same maximum velocity (88). A rate-determining deacylation of a common intermediate—the $\alpha$-$N$-benzoyl-L-argininyl papain—in all probability accounts for this phenomenon. Similar kinetic evidence for a common intermediate stems from the papain-catalyzed

(82) J. E. Baggott and M. H. Klapper, *B* 15, 1473–1481 (1976).

(83) A. N. Glazer, *JBC* 241, 3811–3817 (1966).

(84) R. M. Metrione and R. B. Johnston, *B* 3, 482–485 (1964).

(85) M. J. Mycek and J. S. Fruton, *JBC* 226, 165–171 (1957).

(86) A. W. Lake and G. Lowe, *Biochem. J.* 101, 402–410 (1966).

(87) E. L. Smith, B. J. Finkle, and A. Stockell, *Discuss Faraday Soc.* 20, 96–104 (1955).

(88) E. L. Smith and M. J. Parker, *JBC* 233, 1387–1391 (1958); J. R. Whitaker and M. L. Bender, *JACS* 87, 2728–2737 (1965).

hydrolysis of various esters of carbobenzoxyglycine (89), of alkyl and aryl esters of hippuric acid (90), of aryl and alkyl esters of carbobenzoxy-L-lysine (91), and of a series of N-methanesulfonylglycine esters (92). In a similar vein, the partitioning ratio ($k_{\text{ethanol}}/k_{\text{water}}$) for the papain-catalyzed hydrolysis and ethanolysis of ethylhippurate and p-nitrophenyl hippurate proved to be independent of the substrate (93). Also consonant with the acyl-enzyme pathway for papain catalysis is the "burst" release of p-nitrophenol which precedes the steady state in the hydrolysis of N-carbobenzoxy-L-tyrosine p-nitrophenyl ester (94).

But even more convincing evidence for the acyl-enzyme intermediate comes from the actual isolation of it. By incubating methyl thionohippurate with papain at pH 6.0 for 30 sec and then adjusting the pH to 2.5, the acyl enzyme is generated and stabilized against hydrolysis (95).

The wavelength and extinction of the ultraviolet absorption of the thionohippuryl papain signaled the presence of a dithioester function, affirming the acylation of an enzymic sulfhydryl group. By reacting papain with an excess of *trans*-cinnamoyl imidazole at pH 3.4 followed by gel filtration, *trans*-cinnamoyl papain is readily isolated, and its deacylation can be studied independently (96).

(89) J. F. Kirsch and M. Igleström, *B* 5, 783–791 (1966).
(90) G. Lowe and A. Williams, *Biochem. J.* 96, 199–204 (1965).
(91) M. L. Bender and L. J. Brubacher, *JACS* 88, 5880–5889 (1966).
(92) E. C. Lucas and A. Williams, *B* 8, 5125–5135 (1969).
(93) A. C. Henry and J. F. Kirsch, *B* 6, 3536–3544 (1967).
(94) M. L. Bender, M. L. Begué-Canton, R. L. Blakeley, L. J. Brubacher, J. Feder, C. R. Gunter, F. J. Kézdy, J. V. Killheffer, Jr., T. H. Marshall, C. G. Miller, R. W. Roeske, and J. K. Stoops, *JACS* 88, 5890–5913 (1966).
(95) G. Lowe and A. Williams, *Biochem. J.* 96, 189–193 (1965).
(96) L. J. Brubacher and M. L. Bender, *JACS* 88, 5871–5880 (1966).

Spectroscopic measurements made on *trans*-cinnamoyl papain revealed the existence of a thioester bond, confirming the acylation of an enzymic thiol. And at subzero temperatures in organic-aqueous media, when turnover of the enzyme is extremely slow, the $N^x$-carbobenzoxy-L-lysyl papain can be prepared quantitatively from the *p*-nitrophenylester (97) or *p*-nitrophenyl-anilide (98) of the substrate.

The enzymic thiol in question is now known to belong to Cys 25, with His 159 as its very close spatial neighbor (99). Together, these two residues are responsible for splitting the substrate. The chemical reaction is facilitated by the preexistence of the Cys 25–His 159 pair as a thiolate–imidazolium ion-pair, in which the thiol proton is already on His 159, the $pK_a$ of the thiol proton being about 4 (100). The preexisting thiolate ion is therefore a potent nucleophile, and accounts for the high activity of Cys 25 even in the acid pH range. The imidazolium ion is thought to promote acyl-enzyme formation by providing general acid catalysis.

## Pepsin [EC 3.4.23.1]

The acid proteases form a group of enzymes whose hydrolytic activity is maximal in the pH range 1 to 5. Of such enzymes, pepsin has doubtless received the most extensive study. Besides the hydrolysis of the peptide bond, pepsin also catalyzes transpeptidation reactions some of which are distinguished by the transfer of the amino portion of the peptide function to a free carboxyl acceptor.

$$
\begin{array}{c} \quad\quad O \\ \quad\quad \| \\ R^1\!-\!C\!-\!N\!-\!R + R^2\!-\!COOH \rightleftharpoons R^2\!-\!C\!-\!N\!-\!R + R^1\!-\!COOH \quad (10)\\ \quad\quad H \quad\quad\quad\quad\quad\quad\quad\quad\quad\quad\quad II \end{array}
$$

Such transfer reactions, it was early proposed, proceed over a pathway which includes an intermediate formed by covalent union of the enzyme with the amino nitrogen of the transferring amino acid (101). Support for an amino-enzyme intermediate in pepsin-catalyzed reactions comes from experiments on isotope exchange. When *N*-benzyloxycarbonyl-Tyr-Tyr is incubated with labeled *N*-benzyloxycarbonyl-Tyr, extensive labeling of the residual *N*-acyldipeptide is observed, while incubation with labeled tyrosine gives no exchange (102). More recently it was found that when pepsin is incubated

(97) A. L. Fink and K. J. Angelides, *B* 15, 5287–5293 (1976).
(98) K. J. Angelides and A. L. Fink, *B* 18, 2355–2368 (1979).
(99) J. Drenth, J. N. Jansonius, R. Koekoek, H. M. Swen, and B. G. Wolthers, *Nature (London)* 218, 929–932 (1968).
(100) L. Polgár, *FEBS lett.* 47, 15–18 (1974); M. Shipton, M. P. J. Kierstan, J. P. G. Malthouse, T. Stuchbury, and K. Brocklehurst, *FEBS lett.* 50, 365–368 (1975); S. D. Lewis, F. A. Johnson, and J. A. Shafer, *B* 15, 5009–5017 (1976); *idem.*, *B* 20, 48–51 (1981).
(101) H. Neumann, Y. Levin, A. Berger, and E. Katchalski, *Biochem. J.* 73, 33–41 (1959).
(102) J. S. Fruton, S. Fujii, and M. H. Knappenberger, *PNAS* 47, 759–761 (1961).

with $N$-benzyloxycarbonyl-Glu-$[^{14}C]$Tyr ethyl ester, $[^{14}C]$Tyr ethyl ester is bound to the enzyme; but no radioactivity binds to protein when $N$-benzyloxycarbonyl-$[^{14}C]$Glu-Tyr ethyl ester is the substrate (103). The same enzyme-$[^{14}C]$Tyr ethyl ester can be prepared by simply incubating the enzyme at pH 5 with $[^{14}C]$Tyr ethyl ester (104). Denaturation of the protein did not release the label. Treatment of the inactivated, labeled protein with dinitrofluorobenzene followed by total acid hydrolysis led to the eventual isolation of $O$-dinitrophenyl-$[^{14}C]$tyrosine. Failure to find $O,N$-di-dinitrophenyl-$[^{14}C]$Tyr is accounted for by covalent binding of the tyrosine to the protein through the tyrosyl amino group. An enzymic carboxyl is the point of attachment of the amino group (105).

The kinetics of pepsin-catalyzed transpeptidation were investigated with $N$-acetyl-Phe-Tyr and $N$-acetyl-Tyr-Tyr as tyrosyl donors to the chromophoric $N$-benzyloxycarbonyl-$p$-nitro-L-phenylalanine as acceptor. Under conditions in which hydrolysis is negligible, it is found that the rates of transfer of the tyrosyl group from the two donors to the acceptor are essentially identical. This result, too, accords with the proposition that E—Tyr is an intermediate of the reaction (106).

Pepsin reacts stoichiometrically with active-site-directed irreversible inhibitors of the diazoketone and diazoacetamide type, or with 1,2-epoxy-3-($p$-nitrophenoxy) propane, with concurrent loss of all catalytic activity. These reagents esterify one or the other $\beta$-carboxyl group of a unique pair of aspartate residues—Asp 32 and Asp 215 (107)—which are in close proximity to each other within the active center of the enzyme (108). One of these carboxyls could, in principle, unite in amide linkage with the amino group of the transferring amino acid in the transpeptidation reactions cited above (109). Thus, a minimal mechanism for reaction 10 could be written as

$$E-\overset{\overset{\displaystyle O}{\|}}{C}-OH + R^1-\overset{\overset{\displaystyle O}{\|}}{C}-\underset{\underset{\displaystyle H}{|}}{N}-R \; \longleftrightarrow \; E-\overset{\overset{\displaystyle O}{\|}}{C}-\underset{\underset{\displaystyle H}{|}}{N}-R + R^1-\overset{\overset{\displaystyle O}{\|}}{C}-OH \quad (11)$$

$$E-\overset{\overset{\displaystyle O}{\|}}{C}-\underset{\underset{\displaystyle H}{|}}{N}-R + R^2-\overset{\overset{\displaystyle O}{\|}}{C}-OH \; \longleftrightarrow \; E-\overset{\overset{\displaystyle O}{\|}}{C}-OH + R^2-\overset{\overset{\displaystyle O}{\|}}{C}-\underset{\underset{\displaystyle H}{|}}{N}-R \quad (12)$$

(103) C. Godin and C. Y. Yuan, *J. Chem. Soc. (Chem. Commun.)* 84 (1970).

(104) L. M. Ginodman, N. G. Lutsenko, T. N. Barshevskaya, and V. V. Somova, *B* (Engl. transl.) 36, 510–518 (1971).

(105) T. Valueva, L. M. Ginodman, T. N. Barshevskaya, and F. T. Guseinov, *B* (Engl. transl.) 38, 355–360 (1973).

(106) V. K. Antonov, L. D. Rumsh, and A. G. Tikhodeeva, *FEBS lett.* 46, 29–33 (1974).

(107) J. Sodek and T. Hofmann, *Can. J. Biochem.* 48, 1014–1016 (1970); R. S. Bayliss, J. R. Knowles, and G. B. Wybrandt, *Biochem. J.* 113, 377–386 (1969); K. C. S. Chen and J. Tang, *JBC* 247, 2566–2574 (1972).

(108) I. N. Hsu, L. T. J. Debaere, M. N. G. James, and T. Hofmann, *Nature (London)* 266, 140–145 (1977).

(109) J. R. Knowles, *Philos. Trans. R. Soc. London* B257, 135–146 (1970).

It is certain, however, that the chemical pathway of pepsin-catalyzed transpeptidation is more complex than is indicated in Eqs. 11 and 12 (110). In this connection we note here the striking chemical parallel between transpeptidation by pepsin (Eq. 10) and the reaction catalyzed by coenzyme A transferase (p. 104), which can be generalized as follows:

$$R^1\text{—}\overset{\overset{\displaystyle O}{\|}}{C}\text{—SCoA} + R^2\text{—}\overset{\overset{\displaystyle O}{\|}}{C}\text{—OH} \longleftrightarrow R^2\text{—}\overset{\overset{\displaystyle O}{\|}}{C}\text{—SCoA} + R^1\text{—}\overset{\overset{\displaystyle O}{\|}}{C}\text{—OH} \quad (13)$$

Reactions 10 and 13 are alike in that a transfer of something (an amino acid or a coenzyme A molecule, as the case may be) takes places between a pair of carboxylic acid substrates. There is therefore justification for proposing a chemical mechanism for amino transpeptidation analogous to the one that holds for coenzyme A transferase. The counterpart to the amino enzyme of transpeptidation is the coenzyme A enzyme which mediates reaction 13. It is known that the transferring coenzyme A is joined transiently, as a thiolester, to a carboxyl in the active center of coenzyme A transferase. Also established for the latter enzyme is the concurrent transfer—but in the "opposite" direction—of an oxygen atom between the carboxyl groups of the two carboxylic acid substrates. Such transfer is of course predicted by Eq. 13. It actually takes place indirectly via interchange of the transferring oxygen atom with an oxygen of the enzyme (see Fig. 12, p. 105). Expectation is that amino transpeptidation (Eq. 10) must similarly require an oxygen transfer concurrent with the transfer of the amino acid, and in the "opposite" direction. Complete oxygen transfer between substrate carboxylic acids during transpeptidation has still to be demonstrated. Nonetheless, pepsin does catalyze the exchange of $^{18}O$ between the carboxyl group of an acyl amino acid and water (111). Moreover, incubating pepsin with $H_2{}^{18}O$ at pH 4 leads to the incorporation of $^{18}O$ into a carboxyl group in the active center (112); and, more to the point, the enzymic carboxyl, so labeled, can transfer its label to the carboxyl of N-acetylphenylalanine, when the latter is present as a virtual substrate (113). This oxygen-transferring capacity of pepsin—so far as it has been studied—parallels that of coenzyme A transferase, and constitutes another point of chemical kinship between the two enzymes.

The lower part of Fig. 3 proposes a chemical mechanism for pepsin-catalyzed amino transpeptidation which repeats in its essentials the one proposed (on stronger grounds) for coenzyme A transferase. In it the amino enzyme shares place with an acyl enzyme as intermediates of the reaction. The amino enzyme has the composition of an amide, while the acyl enzyme

(110) M. S. Silver, M. Stoddard, and M. H. Kelleher, JACS 98, 6684–6690 (1976).

(111) N. Sharon, V. Grisaro, and H. Neumann, Arch. Biochem. Biophys. 97, 219–221 (1962).

(112) L. S. Shkarenkova, L. M. Ginodman, L. V. Kozlov, and V. N. Orekhovich, B (Engl. transl.) 33, 131–137 (1968); L. M. Ginodman, T. A. Valueva, L. V. Kozlov, and L. S. Shkarenkova, B (Engl. transl.) 34, 171–172 (1969).

(113) L. M. Ginodman and L. S. Shkarenkova, FEBS Meet. 5th, 1968, p. 274.

Hydrolysis

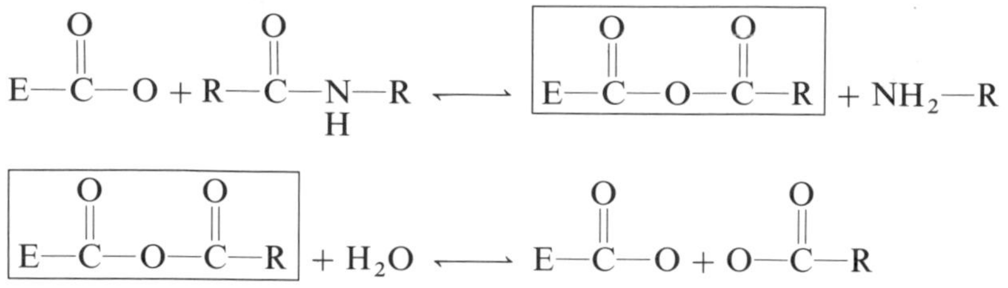

Transpeptidation (acyl transfer)

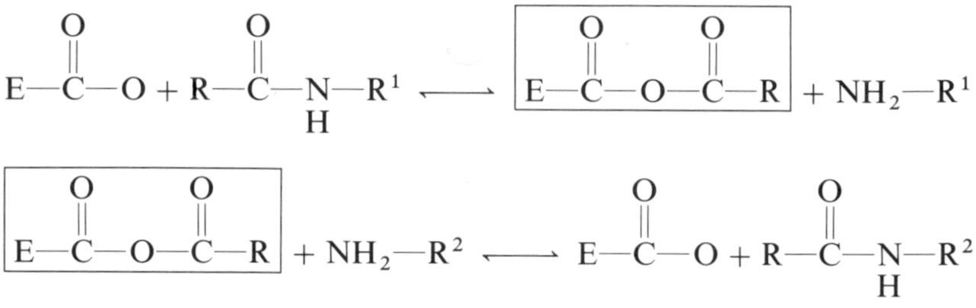

Transpeptidation (amino transfer)

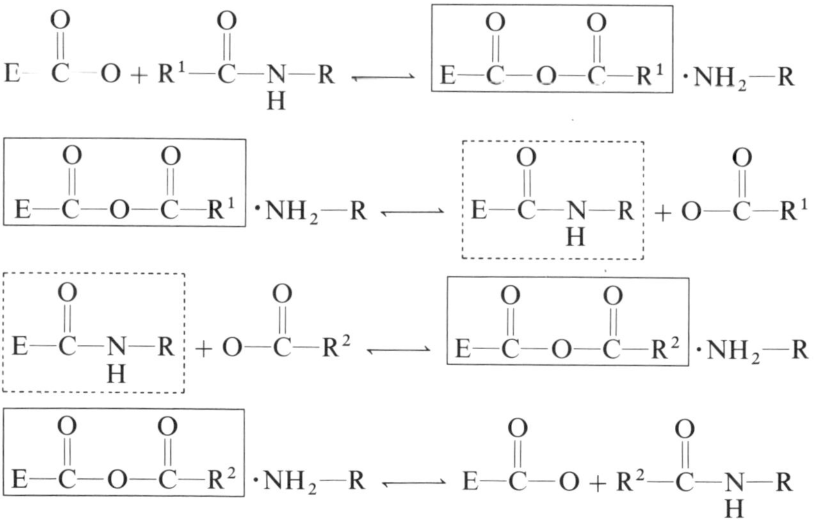

**Fig. 3.** Proposed mechanism of pepsin-catalyzed hydrolysis and transpeptidation. The key intermediate in each case is the acyl enzyme—a mixed anhydride. Whether transpeptidation takes the route of acyl or amino transfer may be a matter of the relative ease of desorption of the products of peptide cleavage. Thus, if the amino product of cleavage is less easily desorbed, amino transfer may be favored. The acyl enzyme is framed in solid lines, the amino enzyme in dashed lines.

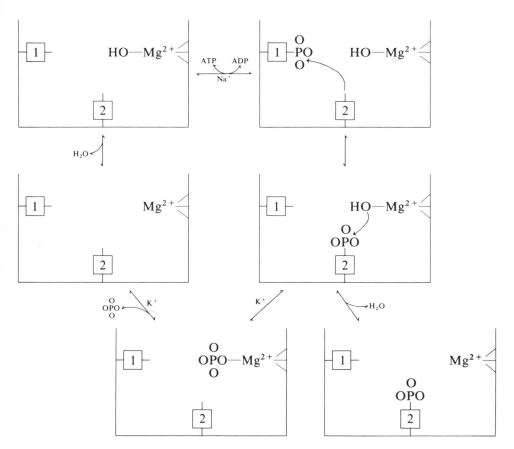

**Fig. 6.** Proposed mechanism for the ion-translocating ATPase reaction. The sites numbered 1 and 2 are nucleophiles (at least one of which is a carboxyl group) which join sequentially to the transferring phosphoryl group on its route to reaction with $Mg^{2+}$-activated water. When the phosphoryl group is on site 1 it is reactive with ADP; when on site 2, it is not. Site 1 catalyzes an ADP–ATP exchange, and site 2 catalyzes hydrolysis with $Mg^{2+}$-activated water, and oxygen exchange between ortho-phosphate and water.

Nuclear relaxation and kinetic studies have been made on (Na,K)ATPase in which the $Mg^{2+}$ in the active center is replaced by $Mn^{2+}$, and $K^+$ (or $Na^+$) is replaced by thallium-205. The divalent and monovalent ions are found to be at such a distance from each other as to easily coordinate a phosphate ion between them (135, 136). Phosphate, so bound, is thought to be activated sufficiently to enable its reversible transfer as a phosphoryl group to the $\beta$-carboxyl of the nearby aspartyl residue (Fig. 6) (135). These considerations account for the frequently observed phosphorylation of the

(135) C. M. Grisham, R. K. Gupta, R. E. Barnett, and A. S. Mildvan, *JBC* 249, 6738–6744 (1974).
(136) C. M. Grisham and A. S. Mildvan, *J. Supramol. Struct.* 3, 304–313 (1975).

enzyme with $P_i$ and $Mg^{2+}$ (137), and indeed for the overall reversal of ATP hydrolysis (138). A rapid $Mg^{2+}$- plus $K^+$-dependent exchange of water oxygens with those of $P_i$ (139) as well as a $Na^+$- plus $Mg^{2+}$-dependent ADP–ATP exchange (140) are also accommodated by the reaction mechanism of Fig. 6, which depicts the hydrolytic water as being activated by coordination to the prosthetic $Mg^{2+}$ ion.

Like the (Na,K)ATPase, the $Ca^{2+}$- (and $Mg^{2+}$-)dependent ATPase of sarcoplasmic reticulum catalyzes the hydrolysis of ATP with a phosphoenzyme figuring prominently as a component of the catalytic pathway (141). The ATP cleavage in this case powers the transport of calcium ions into sarcoplasmic reticulum vesicles against a concentration gradient (142). The process being reversible, ATP can be synthesized from $P_i$ and ADP during the outflow of $Ca^{2+}$ from the vesicles (143). Indeed, ATP can be synthesized from $P_i$ and ADP, under some conditions, even in the absence of an ion gradient (144–146). Such synthesis proceeds through a $Mg^{2+}$-dependent phosphorylation of the enzyme by $P_i$, reversing what appears to be the late, $Ca^{2+}$-independent steps in the forward hydrolytic process (144–146). Phosphorus in the phosphoenzyme is in dynamic equilibrium with $P_i$ of the medium (145). Concurrent with the formation of phosphoenzyme there is a rapid exchange of the $P_i$ oxygens with the oxygens of water (147). These chemical analogies with the (Na,K)ATPase point to close mechanistic parallels between the two ATPases (Fig. 6), including the participation of two phosphoenzyme forms in the catalytic cycle, one of which is reactive with ADP to yield ATP, while the other is not (148). At the time of writing it is

(137) G. E. Lindenmayer, A. H. Laughter, and A. Schwartz, *Arch. Biochem. Biophys.* 127, 187–192 (1968); R. W. Albers, G. J. Koval, and G. J. Siegel, *Mol. Pharmacol.* 4, 324–336 (1968); A. K. Sen, T. Tobin, and R. L. Post, *JBC* 244, 6596–6604 (1969); W. F. Dudding and C. G. Winter, *BBA* 241, 650–660 (1971); R. L. Post, G. Toda, and F. N. Rogers, *JBC* 250, 691–701 (1975).

(138) K. Taniguchi and R. L. Post, *JBC* 250, 3010–3018 (1975).

(139) E. G. Skvortsevich, N. S. Panteleeva, and L. N. Pisareva, *Doklady* (Engl. transl.) 206, 363–365 (1972); A. S. Dahms and P. D. Boyer, *JBC* 248, 3155–3162 (1973).

(140) S. Fahn, G. J. Koval, and R. W. Albers, *JBC* 241, 1882–1889 (1966); W. L. Stahl, *Arch. Biochem. Biophys.* 120, 230–231 (1967); R. Blostein, *JBC* 243, 1957–1965 (1968).

(141) T. Yamamoto and Y. Tonomura, *J. Biochem.* (*Tokyo*) 62, 558–575 (1967); A. Martonosi, *Biochem. Biophys. Res. Commun.* 29, 753–757 (1967); T. Yamamoto and Y. Tonomura, *J. Biochem.* (*Tokyo*) 64, 137–145 (1968); M. Makinose, *EJB* 10, 74–82 (1969); R. L. Coffey, E. Lagwinska, M. Oliver, and A. Martonosi, *Arch. Biochem. Biophys.* 170, 37–48 (1975).

(142) S. Ebashi and F. Lipmann, *J. Cell. Biol.* 14, 389–400 (1962).

(143) M. Makinose and W. Hasselbach, *FEBS lett.* 12, 271–272 (1971); R. Panet and Z. Selinger, *BBA* 255, 34–42 (1972).

(144) H. Masuda and L. de Meis, *B* 12, 4581–4585 (1973).

(145) T. Kanazawa, *JBC* 250, 113–119 (1975).

(146) A. F. Knowles and E. Racker, *JBC* 250, 1949–1951 (1975).

(147) T. Kanazawa and P. D. Boyer, *JBC* 248, 3163–3172 (1973).

(148) M. Shigekawa, J. P. Dougherty, and A. M. Katz, *JBC* 253, 1442–1450 (1978); M. Shigekawa and J. P. Dougherty, *JBC* 253, 1451–1464 (1978); Y. Nakamura, *J. Biochem.* (*Tokyo*) 88, 177–181 (1980).

not clear whether the two phosphoenzymes are conformational or positional isomers. In Fig. 6 we hazard the guess that they are positional isomers, and that the phosphoryl group is transferred from ATP to water with net steric inversion in a triple-displacement reaction. Of interest in this connection is the finding that the ATPase of myosin and of beef heart mitochondria catalyzes just such a steric inversion on phosphorus (149). A phosphoenzyme involvement in myosin ATPase action has been proposed (150) and disputed (151).

The electrogenic, proton-translocating ATPase of *Neurospora* plasma membrane also hydrolyzes ATP via a phosphoenzyme intermediate (152). The phosphoryl group is released from the protein by hydroxylamine, and, as in the above-cited ATPases, is evidently present as an acyl phosphate.

**Table 4.1.** Hydrolases Known to Act by Covalent Catalysis

| EC no. | Familiar name of enzyme[a] | Criteria[b] | References |
|--------|----------------------------|-------------|------------|
| 3.1.1.1 | Carboxylesterase | K, D, B, M | (153) |
| 3.1.1.3 | Lipase | I, D, B | (154) |
| 3.1.1.5 | Lysophospholipase | M, D | (155) |
| 3.1.1.6 | Acetylesterase | D | (156) |
| 3.1.1.7 | Acetylcholinesterase [Ca] | K, D, B, M | (157) |
| 3.1.1.8 | Cholinesterase | K, D, B | (158) |
| 3.1.1. | Cutinase | D | (159) |
| 3.1.2. | Fibrinoligase [Ca] | $V_{max}$ | (160) |
| 3.1.2 | Acyl-CoA hydrolase | D | (161) |
| 3.1.3.1 | Alkaline phosphatase [Zn] | $I^c$, K, B, $V_{max}$, M, C | (162) |
| 3.1.3.2 | Acid phosphatase | $I^c$, K, B, $V_{max}$ | (163) |
| 3.1.3.9 | Glucose-6-phosphatase | $I^c$, K, E, $V_{max}$ | (164) |
| 3.1.3.11 | Fructose-1,6-diphosphatase [Zn] | M | (165) |
| 3.1.3.13 | Bisphosphoglycerate phosphatase | M | (166) |
| 3.1.4.1 | Phosphodiesterase I | B, $V_{max}$, $I^c$, C | (167) |
| 3.1.6.1 | Arylsulfatase | B, $V_{max}$, M, K | (168) |
| 3.2.1.1 | α-Amylase [Ca] | C | (169) |
| 3.2.1.4 | Cellulase | C | (170) |
| 3.2.1.6 | Laminarinase | C | (171) |
| 3.2.1.10 | Isomaltase | C | (172) |
| 3.2.1.17 | Lysozyme | E, M, C | (173) |
| 3.2.1.20 | α-Glucosidase | $I^c$, C | (174) |
| 3.2.1.21 | β-Glucosidase | E, B, M, C | (175) |
| 3.2.1.22 | α-Galactosidase | E, C | (176) |
| 3.2.1.23 | β-Galactosidase | K, B, $V_{max}$, M, C | (177) |
| 3.2.1.24 | α-Mannosidase [Zn] | K, C | (178) |
| 3.2.1.26 | Invertase | E, C | (179) |
| 3.2.1.28 | α,α-Trehalase | C | (180) |
| 3.2.1.30 | β-N-Acetylglucosaminidase | M, C | (181) |
| 3.2.1.31 | β-Glucuronidase | K, C | (182) |

**Table 4.1.** Hydrolases Known to Act by Covalent Catalysis   (*Continued*)

| EC no. | Familiar name of enzyme[a] | Criteria[b] | References |
|--------|----------------------------|-------------|------------|
| 3.2.1.33 | Amylo-1,6-glucosidase | M, C | (183) |
| 3.2.1.37 | Exo-1,4-$\beta$-D-xylosidase | $V_{max}$, C | (184) |
| 3.2.1.48 | Sucrose $\alpha$-glucohydrolase | E, K, C | (185) |
| 3.2.1.59 | Endo-1,3-$\alpha$-D-glucanase | C | (186) |
| 3.2.2.5 | NADase | E, K, C | (187) |
| 3.2.3.1 | Thioglucosidase | C | (188) |
| 3.3.1.1 | Adenosylhomocysteinase [NAD] | I[e] | (189) |
| 3.4.14.1 | Dipeptidylpeptidase | D, M | (190) |
| 3.4.15.2 | Carboxyamidase | D | (191) |
| 3.4.16.1 | Carboxypeptidase C | D, B | (192) |
| 3.4.17.1 | Carboxypeptidase A [Zn] | I[e], K, M | (193) |
| 3.4.17.6 | D-Alanine carboxypeptidase | I[d], E, K, M | (194) |
| 3.4.21.1 | Chymotrypsin [Ca] | I, I[d], K, D, B, M | (195) |
| 3.4.21.3 | Metridium proteinase A | D | (196) |
| 3.4.21.4 | Trypsin [Ca] | I[c], I[d], D, B | (197) |
| 3.4.21.5 | Thrombin | I[d], D, B, $V_{max}$ | (198) |
| 3.4.21.6 | Prothrombinase | B | (199) |
| 3.4.21.7 | Plasmin | I[d], D, B, $V_{max}$ | (200) |
| 3.4.21.8 | Kallikrein | I[d], D, B | (201) |
| 3.4.21.9 | Enterokinase | D, B | (202) |
| 3.4.21.10 | Acrosin | B | (203) |
| 3.4.21.11 | Elastase | I[e], D, B, $V_{max}$, M | (204) |
| 3.4.21.12 | $\alpha$-Lytic proteinase | B, M | (205) |
| 3.4.21.14 | Subtilisin | I[d], D, B, M | (206) |
| 3.4.21.14 | *Aspergillus* alkaline proteinase | D, B | (207) |
| 3.4.21.14 | *Arthrobacter* serine proteinase | D | (208) |
| 3.4.21.14 | Thermophilic *Streptomyces* alkaline proteinase | D | (209) |
| 3.4.21.14 | Protease I | D | (210) |
| 3.4.21.14 | Thermomycolase | D | (211) |
| 3.4.21.14 | Proteinase K [Ca] | $V_{max}$ | (212) |
| 3.4.21.19 | Staphylococcal serine proteinase | K | (213) |
| 3.4.21.26 | Post-proline cleaving enzyme | D | (214) |
| 3.4.21.31 | Urokinase | D, B | (215) |
| 3.4.21. | Cocoonase | K, B, M | (216) |
| 3.4.21. | Cotton seed proteinase | D | (217) |
| 3.4.21. | Thrombocytin | D, B | (218) |
| 3.4.22.2 | Papain | I[d], K, B, $V_{max}$ | (219) |
| 3.4.22.3 | Ficin | I[d], K, B | (220) |
| 3.4.22.4 | Bromelain | I[d], K, M | (221) |
| 3.4.22.8 | Clostripain [Ca] | $V_{max}$ | (222) |
| 3.4.22.9 | Yeast proteinase B | D | (223) |
| 3.4.22.10 | Streptococcal proteinase | K, $V_{max}$ | (224) |
| 3.4.22.14 | Actinidin | K | (225) |
| 3.4.23.1 | Pepsin | I[c], E, K, M, Z | (226) |

**Table 4.1.** Hydrolases Known to Act by Covalent Catalysis   (*Continued*)

| EC no. | Familiar name of enzyme[a] | Criteria[b] | References |
|---|---|---|---|
| 3.4.23.3 | Gastricsin | Z | (227) |
| 3.4.23.4 | Rennin | Z | (228) |
| 3.4.23.5 | Cathepsin D | M, Z | (229) |
| 3.4.23.6 | Proteinase A | M, Z | (230) |
| 3.4.23.6 | Penicillopepsin | M, Z | (231) |
| 3.4.23.6 | Acid protease from *Rhizopus chinensis* | M, Z | (232) |
| 3.4.23.6 | *Endothia* acid proteinase | M, Z | (233) |
| 3.4.23. | Protease VI of *Mucor miehi* | Z | (234) |
| 3.4.24.4 | *Streptomyces griseus* proteinase | $I^d$, D, B, M | (235) |
| 3.4.24.4 | Thermolysin [Zn] | M | (236) |
| 3.4.99.19 | Renin | Z | (237) |
| 3.5.1.1 | Asparaginase | K, M | (238) |
| 3.5.1.2 | Glutaminase | E, K, M, G | (239) |
| 3.5.1.3 | $\omega$-Amidase | K, $V_{max}$ | (240) |
| 3.5.1.5 | Urease [Ni] | $V_{max}$ | (241) |
| 3.5.1.13 | Arylacylamidase | D | (242) |
| 3.5.1.14 | Aminoacylase [Zn] | B | (243) |
| 3.5.1.19 | Nicotinamidase | K, D, B | (244) |
| 3.5.1.22 | Pantothenase | K, E | (245) |
| 3.5.1.36 | N-Methyl-2-oxoglutaramate hydrolase | E, K, $V_{max}$ | (246) |
| 3.5.2.6 | Penicillinase | I, $I^{d,c,e}$, B, K | (247) |
| 3.5.4.4 | Adenosine deaminase | K, $V_{max}$ | (248) |
| 3.5.4.6 | AMP deaminase [Zn] | B | (249) |
| 3.6.1.1 | Inorganic pyrophosphatase [Mg] | $I^c$, M | (250) |
| 3.6.1.3 | ATPase [Mg or Ca] | I, E, M | (251) |
| 3.7.1.3 | Kynureninase [PLP] | M | (252) |

[a] Prosthetic groups are indicated in brackets. Abbreviation: PLP, pyridoxal-5'-P.

[b] The symbols mean the following:

I, the holoenzyme links covalently with the substrate or a fragment of it to form a chemically competent intermediate;

E, the enzyme catalyzes one or more exchange reactions consistent with the participation of a covalent enzyme–substrate intermediate;

K, the enzyme exhibits kinetic properties (exclusive of "burst" and $V_{max}$ kinetics) consistent with the participation of a covalent enzyme–substrate intermediate;

D, the enzyme is inactivated upon stoichiometric reaction with diisopropyl fluorophosphate, phenylmethylsulfonyl fluoride, or other seryl group titrant, with blockage of the seryl hydroxyl in the active center. Such titration and inactivation is regarded as diagnostic of the serine proteases, in which the active-site serine is acylated by substrate;

B, the enzyme displays "burst" kinetics; $V_{max}$, the enzyme acts on diverse substrates in conformity with the maximum velocity principle (W. P. Jencks, *Catalysis in Chemistry and Enzymology*, McGraw-Hill, New York, 1969, pp. 50–53);

M, miscellaneous data and derivative arguments that are peculiar to the enzyme in question;

C, the enzyme catalyzes a reaction in which steric configuration is retained at the carbon at which reaction occurs, thereby excluding the possibility of a single-displacement mechanism;

G, the enzyme is irreversibly inactivated by stoichiometric alkylation with a glutamine analogue; inferred from this is the participation of a glutamyl-enzyme intermediate;

Z, the enzyme is irreversibly inhibited by diazoacetyl norleucine methyl ester in the presence of $Cu^{2+}$ and by 1,2-epoxy-3-($p$-nitrophenoxy)propane. Stoichiometric reaction with these reagents is typical of the carboxyl groups in the active site of pepsin, and is characteristic of acid proteases in general. Enzymes that undergo such inhibition are believed to have the same chemical mechanism as pepsin [J. Tang, *Trends Biochem. Sci.* 1, 205–208 (1976)]. The Z criterion is thus analogous to the D (for serine proteases) and the G (for glutamine-utilizing enzymes) criteria.

$^c$ The covalent enzyme–substrate intermediate is isolated in denatured condition.

$^d$ The covalent enzyme–substrate intermediate is formed from an unnatural substrate.

$^e$ The covalent enzyme–substrate intermediate is identified spectroscopically.

(149) M. R. Webb and D. R. Trentham, *JBC* 255, 8629–8632 (1980); M. R. Webb, C. Grubmeyer, H. S. Penefsky, and D. R. Trentham, *JBC* 255, 11637–11639 (1980).

(150) N. Kinoshita, S. Kubo, H. Onishi, and Y. Tonomura, *J. Biochem.* (*Tokyo*) 65, 285–301 (1969).

(151) R. G. Wolcott and P. D. Boyer, *BBA* 303, 292–297 (1973).

(152) J. B. Dame and G. A. Scarborough, *B* 19, 2931–2937 (1980).

(153) Refs. (1–9).

(154) M. F. Maylié, M. Charles, and P. Desnuelle, *BBA* 276, 162–175 (1972); J. E. Fulton, Jr., N. L. Noble, S. Bradley, and W. M. Awad, *B* 13, 2320–2327 (1974); M. Sémériva, C. Chapus, C. Bovier-Lapierre, and P. Densnuelle, *Biochem. Biophys. Res. Commun.* 58, 808–813 (1974); C. Chapus, M. Sémériva, and P. Desnuelle, *Fed. Proc.* 36, 720 (1977).

(155) G. P. H. van den Heusden and H. van den Bosch, *Biochem. Biophys. Res. Commun.* 90, 1000–1006 (1979).

(156) A. J. J. Ooms, J. C. A. E. Breebaart-Hansen, and B. I. Ceulen, *Biochem. Pharmacol.* 15, 17–30 (1966).

(157) I. B. Wilson, F. Bergmann, and D. Nachmansohn, *JBC* 186, 781–790 (1950); I. B. Wilson, in *The Enzymes*, 2nd ed., P. D. Boyer, H. Lardy, and K. Myrback, eds., Academic Press, New York, 1960, Vol. 4, pp. 501–520; H. C. Froede and I. B. Wilson, *The Enzymes*, 3rd ed., P. D. Boyer, ed., Academic Press, New York, 1971, Vol. 5, pp. 87–114; Ref. (94).

(158) H. S. Jansz, D. Brons, and M. G. P. J. Warringa, *BBA* 34, 573–575 (1959); A. R. Main, E. Tarkan, J. L. Aull, and W. G. Soucie, *JBC* 247, 566–571 (1972); K. B. Augustinsson, T. Bartfai, and B. Mannervik, *Biochem. J.* 141, 825–834 (1974); O. Lockridge and B. N. La Du, *JBC* 253, 361–366 (1978).

(159) R. E. Purdy and P. E. Kolattukudy, *B* 14, 2832–2840 (1975).

(160) C. G. Curtis, P. Stenberg, K. L. Brown, A. Baron, K. Chen, A. Gray, I. Simpson, and L. Lorand *B* 13, 3257–3262 (1974).

(161) L. J. Libertini and S. Smith, *JBC* 253, 1393–1401 (1978).

(162) Refs. (10–26); M. Caswell and M. Caplow, *B* 19, 2907–2911 (1980).

(163) Refs. (28–32); R. Y. Hsu, W. W. Cleland, and L. Anderson, *B* 5, 799–807 (1966).

(164) Refs. (33–36).

(165) S. J. Benkovic, J. J. Villafranca, and J. J. Kleinschuster, *Arch. Biochem. Biophys.* 155, 458–463 (1973).

(166) K. Ikura, R. Sasaki, H Narita, E. Sugimoto, and H. Chiba, *EJB* 66, 515–522 (1976).

(167) Refs. (38) and (39).

(168) Refs. (40–44).

(169) R. Kuhn, *Ann. Chem.* 443, 1–71 (1925); G. Okada, D. S. Genghof, and E. J. Hehre, *Carbohyd. Res.* 71, 287–298 (1979).

(170) D. R. Whitaker, *Arch. Biochem. Biophys.* 53, 436–438 (1954).

(171) D. R. Clark, J. Johnson, Jr., K. H. Chung, and S. Kirkwood, *Carbohyd. Res.* 61, 457–477 (1978).

(172) G. Semenza, C. H. Curtius, J. Kolinska, and M. Muller, *BBA* 146, 196–204 (1967).

(173) D. C. Phillips, *Harvey Lect.* 66, 135–160 (1970–1971); M. A. Raftery and T. Rand-Meir,

*B* 7, 3281–3289 (1968); J. J. Pollock and N. Sharon, *B* 9, 3913–3925 (1970); R. F. Atkinson and T. C. Bruice, *JACS* 96, 819 825 (1974); P. R. Young and W. P. Jencks, *JACS* 99, 8238 8248 (1977).

(174) T. N. Palmer, *Biochem. J.* 124, 701–724 (1971); H. Y. L. Lai, L. G. Butler, and B. Axelrod, *Biochem. Biophys. Res. Commun.* 60, 635–640 (1974); S. Chiba, K. Hiromi, N. Minamiura, M. Ohnishi, T. Shinomura, K. Suga, T. Suganuma, A. Tanaka, S. Tomioka, and T. Yamamoto, *J. Biochem. (Tokyo)* 85, 1135–1141 (1979).

(175) M. J. Rabaté, *Bull. Soc. Chim. Biol.* 17, 572–601 (1935); E. M. Crook and B. A. Stone, *Biochem. J.* 65, 1–12 (1957); G. Legler, *Z. Physiol. Chem.* 349, 767–774 (1968); F. Dahlquist, T. Rand-Meir, and M. A. Raftery, *B* 8, 4214–4221 (1969); A. L. Fink and N. E. Good, *Biochem. Biophys. Res. Commun.* 58, 126–131 (1974); S. S. Raghavan, R. A. Mumford, and J. N. Kanfer, *Biochem. Biophys. Res. Commun.* 58, 99–106 (1974); G. Legler, K. R. Roeser, and H. K. Illig, *EJB* 101, 85–92 (1979); C. K. De Bruyne, G. M. Aerts, and R. L. De Gussem, *EJB* 102, 257–267 (1979); J. P. Weber and A. L. Fink, *JBC* 255, 9030–9032 (1980).

(176) Y. T. Li and M. R. Shetlar, *Arch. Biochem. Biophys.* 108, 301–313 (1964); P. M. Dey, *BBA* 191, 644–652 (1969).

(177) Refs. (46–54); S. Rosenberg and J. F. Kirsch, *Fed. Proc.* 37, 1296 (1978); P. J. Deschavanne, O. M. Viratelle, and J. M. Yon, *PNAS* 75, 1892–1896 (1978).

(178) J. De Prijcker, A. Vervoort, and C. K. De Bruyne, *EJB* 47, 561–566 (1974); J. De Prijcker, C. K. De Bruyne, M. Claeyssens, and A. De Bruyne, *Carbohyd. Res.* 43, 380–382 (1975); J. De Prijcker, A. De Bock, and C. K. De Bruyne, *Carbohyd. Res.* 60, 141–153 (1978).

(179) J. Edelmen, *Biochem. J.* 57, 22–33 (1954).

(180) J. Labat, F. Baumann, and J. E. Courtois, *C.R. Acad. Sci. (Paris), Sect. D*, 274, 1967–1969 (1972).

(181) D. H. Leaback and P. G. Walker, *Biochem. J.* 104, 70P (1967); D. H. Leaback, *Biochem. Biophys. Res. Commun.* 32, 1025–1030 (1968).

(182) W. H. Fishman and S. Green, *JBC* 225, 435–452 (1957); C. C. Wang and O. Touster, *JBC* 247, 2650–2656 (1972); R. Niemann and E. Buddecke, *BBA* 567, 196–206 (1979).

(183) T. E. Nelson and J. Larner, *BBA* 198, 538–545 (1970); T. E. Nelson, R. C. White, and B. K. Gillard, *Fed. Proc.* 32, 627 (1973); B. K. Gillard and T. E. Nelson *B* 16, 3978–3987 (1977).

(184) F. Deleyn, M. Claeyssens, J. Van Beeumen, and C. K. De Bruyne, *Can. J. Biochem.* 56, 43–50 (1978); F. Deleyn, M. Claeyssens, and C. K. De Bruyne, *Can. J. Biochem.* 58, 5–8 (1980).

(185) A. Dahlqvist and B. Bergström, *Acta Chem. Scand.* 13, 1659–1667 (1959); J. A. Carnie and J. W. Porteous, *Biochem. J.* 85, 450–456 (1962); G. Semenza and A. K. von Balthazar, *EJB* 41, 149–162 (1974); A. Cogoli and G. Semenza, *JBC* 250, 7802–7809 (1975); B. Zagalak and H. C. Curtius, *Biochem. Biophys. Res. Commun.* 62, 503–509 (1975); R. E. Huber and R. D. Mathison, *Can. J. Biochem.* 54, 153–164 (1976).

(186) S. Hasegawa, J. H. Nordin, and S. Kirkwood, *JBC* 244, 5460–5470 (1969).

(187) Refs. (55–57); K. Ueda, M. Fukushima, H. Okayama, and O. Hayaishi, *JBC* 250, 7541–7546 (1975).

(188) M. G. Ettlinger, G. P. Dateo, B. W. Harrison, T. J. Mabry, and C. P. Thompson, *PNAS* 47, 1875–1880 (1961).

(189) J. L. Palmer and R. H. Abeles, *JBC* 251, 5817–5819 (1976); idem., *JBC* 254, 1217–1226 (1979); R. H. Abeles, A. H. Tashjian, Jr., and S. Fish, *Biochem. Biophys. Res. Commun.* 95, 612–617 (1980).

(190) C. P. Heinrich and J. S. Fruton, *B* 7, 3556–3565 (1968); A. J. Kenny, A. G. Booth, S. G. George, J. Ingram, D. Kershaw, E. J. Wood, and A. R. Young, *Biochem. J.* 157, 169–182 (1976).

(191) W. H. Simmons and R. Walter, *B* 19, 39–48 (1980).

(192) R. Hayashi, S. Moore, and W. H. Stein, *JBC* 248, 8366–8369 (1973); Y. Nakagawa and E. T. Kaiser, *Biochem. Biophys. Res. Commun.* 61, 730–734 (1974); Y. Bai, R. Hayashi, and T. Hata, *J. Biochem. (Tokyo)* 78, 617–626 (1975).

(193) Refs. (60–65, 67).

(194) J. J. Pollock, J. M. Ghysen, R. Linder, M. R. J. Salton, H. R. Perkins, M. Nieto, M. Leyh-Bouille, J. M. Frére, and K. Johnson, *PNAS* 69, 662–666 (1972); J. N. Umbreit and

J. L. Strominger, *JBC* 248, 6767–6771 (1973); R, R. Yocum, P. M. Blumberg, and J. L. Strominger, *JBC* 249, 4863–4871 (1974); P. M. Blumberg, R. R. Yocum, E. Willoughby, and J. L. Strominger *JBC* 249, 6828–6835 (1974); S. Hammarström and J. L. Strominger, *PNAS* 72, 3463–3467 (1975); J. M. Frére, C. Duez, J. M. Ghysen, and J. Vandekerkhove, *FEBS lett.* 70, 257–260 (1977); N. Georgopapadakou, S. Hammarström, and J. L. Strominger, *PNAS* 74, 1009–1012 (1977); T. Nishino, J. W. Kozarich, and J. L. Strominger, *JBC* 252, 2934–2939 (1977); R. R. Yocum, J. R. Rasmussen, and J. L. Strominger, *JBC* 255, 3977–3986 (1980).
(195) Refs. (69–82); A. L. Fink, *Arch. Biochem. Biophys.* 155, 473–474 (1973); *idem.*, *B* 15, 1580–1586 (1976).
(196) D. Gibson and G. H. Dixon, *Nature (London)* 222, 753–756 (1969).
(197) G. H. Dixon, D. L. Kaufman, and H. Neurath, *JACS* 80, 1260–1261 (1958); M. L. Bender and E. T. Kaiser, *JACS* 84, 2556–2561 (1962); Ref. (94); J. S. Huang and I. E. Liener, *B* 16, 2474–2478 (1977); M. W. Hunkapiller, M. D. Forgac, E. H. Yu, and J. H. Richards, *Biochem. Biophys. Res. Commun.* 87, 25–31 (1979).
(198) J. A. Gladner and K. Laki, *JACS* 80, 1263–1264 (1958); E. F. Curragh and D. T. Elmore, *Biochem. J.* 93, 163–171 (1964); F. J. Kézdy, L. Lorand, and K. D. Miller, *B* 4, 2302–2308 (1965); J. B. Baird and D. T. Elmore, *FEBS lett.* 1, 343–345 (1968); T. Chase and E. Shaw, *B* 8, 2212–2224 (1969).
(199) D. V. Roberts, R. W. Adams, D. T. Elmore, G. W. Jameson, and W. S. A. Kyle, *Biochem. J.* 123, 41P (1971); R. L. Smith, *JBC* 248, 2418–2423 (1973).
(200) W. R. Grosskopf, L. Summaria, and K. C. Robbins, *JBC* 244, 3590–3597 (1969); T. Chase and E. Shaw, *B* 8, 2212–2224 (1969); U. Christensen, *BBA* 397, 459–467 (1975).
(201) F. Fiedler, B. Müller, and E. Werle, *FEBS lett.* 22, 1–4 (1972); *idem.*, *FEBS lett.* 24, 41–44 (1972); M. Zuber and E. Sachse, *B* 13, 3098–3110 (1974); S. Sampaio, S. C. Wong, and E. Shaw, *Arch. Biochem. Biophys.* 165, 133–139 (1974).
(202) S. Maroux, J. Baratti, and P. Desnuelle, *JBC* 246, 5031–5039 (1971); J. J. Liepnieks and A. Light, *Fed. Proc.* 35, 1460 (1976); S. Baratti and S. Maroux, *BBA* 452, 488–496 (1976); J. J. Liepnieks and A. Light, *JBC* 254, 1677–1683 (1979).
(203) C. R. Brown, Z. Andani, and E. F. Hartree, *Biochem. J.* 149, 147–154 (1975); W. D. Schleuning, R. Hell, H. Schiessler, and H. Fritz, *Z. Physiol. Chem.* 356, 1915–1921 (1975).
(204) M. A. Naughton, F. Sanger, B. S. Hartley, and D. C. Shaw, *Biochem. J.* 77, 149–163 (1960); M. L. Bender and T. H. Marshall, *JACS* 90, 201–207 (1968); D. M. Shotton and H. C. Watson, *Nature (London)* 225, 811–816 (1970); E. J. Breaux and M. L. Bender, *Biochem. Biophys. Res. Commun.* 70, 235–240 (1976); A Gertler, Y. Weiss, and Y. Burstein, *B* 16, 2709–2716 (1977); A. L. Fink and P. Meehan, *PNAS* 76, 1566–1569 (1979).
(205) H. Kaplan and D. R. Whitaker, *Can. J. Biochem.* 47, 305–316 (1960); M. W. Hunkapiller, S. H. Smallcombe, D. R. Whitaker, and J. H. Richards, *B* 12, 4732–4742 (1973); M. W. Hunkapiller, M. D. Forgac, and J. H. Richards, *B* 15, 5581–5588 (1976).
(206) F Sanger and D. C. Shaw, *Nature (London)* 187, 872–873 (1960); S. A. Bernhard, S. J. Lau, and H. Noller, *B* 4, 1108–1118 (1965); J. D. Robertus, J. Kraut, R. A. Alden, and J. J. Birktoft, *B* 11, 4293–4303 (1972); Ref. (94); D. A. Matthews, R. A. Alden, J. J. Birktoft, S. T. Freer, and J. Kraut, *JBC* 252, 8875–8883 (1977); M. Philipp, I.-H. Tsai, and M. L. Bender, *B* 18, 3769–3773 (1979).
(207) O. Mikeš, J. Turková, N. B. Toan, and F. Šorm, *BBA* 178, 112–117 (1969); S. van Heyningen, *EJB* 28, 432–437 (1972).
(208) S. Wählby, *BBA* 151, 409–413 (1968).
(209) K. Mizusawa and F. Yoshida, *JBC* 248, 4417–4423 (1973).
(210) M. Pacaud, *EJB* 82, 439–451 (1978).
(211) G. Voordouw, G. M. Gaucher, and R. S. Roche, *Can. J. Biochem.* 52, 981–990 (1974).
(212) E. Kraus and U. Femfert, *Z. Physiol Chem.* 357, 937–947 (1976).
(213) J. Houmard, *EJB* 68, 621–627 (1976).
(214) T. Yoshimoto, R. C. Orlowski, and R. Walter, *B* 16, 2942–2948 (1977); T. Yoshimoto, R. Walter, and D. Tsuru, *JBC* 255, 4786–4792 (1980).
(215) D. Petkov and I. Stoinova, *BBA* 370, 546–555 (1974); D. Petkov, E. Christova, I. Pojar-

lieff, and N. Stambolieva, *EJB* 51, 25–32 (1975); M. E. Soberano, E. B. Ong, A. J. Johnson, M. Levy, and G. Schoellmann, *BBA* 445, 763–773 (1976).

(216) F. C. Kafatos, J. H. Law, and A. M. Tartakoff, *JBC* 242, 1488–1494 (1967).

(217) J. N. Ihle, *Fed. Proc.* 30, 1184 (1971).

(218) E. P. Kirby, S. Niewiarowski, K. Stocker, C. Kettner, E. Shaw, and T. M. Brudzynski, *B* 18, 3564–3570 (1979).

(219) Refs. (87–98).

(220) G. Lowe and A. Williams, *Biochem. J.* 96, 189–193 (1965); M. R. Holloway, E. Antonini, and M. Brunori, *EJB* 24, 332–341 (1971); M. R. Holloway and M. J. Hardman, *EJB* 32, 537–546 (1973).

(221) T. Inagami and T. Murachi, *B* 2, 1439–1444 (1963); K. Brocklehurst, E. M. Crook, and C. W. Wharton, *Chem. Commun.* 1185–1187 (1967); C. W. Wharton, *Biochem. J.* 143, 575–586 (1974).

(222) P. W. Cole, K. Murakami, and T. Inagami, *B* 10, 4246–4252 (1971).

(223) R. E. Ulane and E. Cabib, *JBC* 251, 3367–3376 (1976).

(224) A. A. Kortt and T. Y. Liu, *B* 12, 328–337 (1973).

(225) M. J. Boland and M. J. Hardman, *EJB* 36, 575–582 (1973).

(226) Refs. (101–116).

(227) J. Tang, *JBC* 246, 4510–4517 (1971).

(228) W. J. Chang and K. Takahashi, *J. Biochem.* (*Tokyo*) 76, 467–474 (1974).

(229) H. Keilova, *FEBS lett.* 6, 312–314 (1970); M. Cunningham and J. Tang, *JBC* 251, 4528–4536 (1976); A. Moriyama and K. Takahashi, *J. Biochem.* (*Tokyo*) 83, 441–451 (1978).

(230) F. Meussdoerffer, P. Tortora, and H. Holzer, *JBC* 255, 12087–12093 (1980).

(231) J. Sodek and T. Hofmann, *Can. J. Biochem.* 48, 1014–1016 (1970); T. T. Wang and T. Hofmann, *Can. J. Biochem.* 55, 286–294 (1977).

(232) P. Sepulveda, K. W. Jackson, and J. Tang, *Biochem. Biophys. Res. Commun.* 63, 1106–1112 (1975); E. Subramanian, I. D. A. Swan, and D. R. Davies, *Biochem. Biophys. Res. Commun.* 68, 875–880 (1976).

(233) C. H. Wong, T. J. Lee, T. Y. Lee, T. H. Lu, and I. H. Yang, *B* 18, 1638–1640 (1979).

(234) W. S. Rickert and P. A. McBride-Warren, *BBA* 480, 262–274 (1977).

(235) S. Wählby and L. Engström, *BBA* 151, 402–408 (1968); S. Wählby, *Acta Chem. Scand.* 24, 703–704 (1970); C.-A. Bauer, B. Löfqvist, and G. Pettersson, *EJB* 41, 45–49 (1974); G. D. Brayer, L. T. J. Delbaere, M. N. G. James, C.-A. Bauer, and R. C. Thompson, *PNAS* 76, 96–100 (1979).

(236) W. L. Bigbee and F. W. Dahlquist, *B* 13, 3542–3549 (1974); M. K. Pangburn and K. A. Walsh, *B* 14, 4050–4054 (1975); W. R. Kester and B. W. Matthews, *B* 16, 2506–2516 (1977); N. Nishino and J. C. Powers, *B* 17, 2846–2850 (1978); M. C. Bolognesi and B. W. Matthews, *JBC* 254, 634–639 (1979).

(237) T. Inagami, K. Misono, and A. M. Michelakis, *Biochem. Biophys. Res. Commun.* 56, 503–509 (1974); K. S. Misono and T. Inagami, *B* 19, 2616–2622 (1980).

(238) R. C. Jackson and R. E. Handschumacher, *B* 9, 3585–3590 (1970); M. Ehrman, H. Cedar, and J. H. Schwartz, *JBC* 246, 88–94 (1971); J. B. Howard and F. H. Carpenter, *JBC* 247, 1020–1030 (1972); J. O. Westerik and R. Wolfenden, *JBC* 249, 6351–6353 (1974); P. C. Dunlop, G. M. Meyer, and R. J. Roon, *JBC* 255, 1542–1546 (1980).

(239) Refs. (117–121).

(240) L. B. Hersh, *B* 11, 2251–2256 (1972).

(241) J. Bennett and E. A. Wren, *BBA* 482, 421–426 (1977).

(242) E. Heymann and H. Rix, *Int. J. Pept. Protein Res.* 11, 59–64 (1978).

(243) J. C. Lugay and J. N. Aronson, *BBA* 191, 397–414 (1969).

(244) S. Su, L. Albizati and S. Chaykin, *JBC* 244, 2956–2965 (1969); L. Albizati and J. L. Hedrick, *B* 11, 1508–1517 (1972); S. S. Gillam, J. G. Watson, and S. Chaykin, *Arch. Biochem. Biophys.* 157, 268–284 (1973); R. E. A. Gadd and W. J. Johnson, *Int. J. Biochem.* 5, 397–408 (1975).

(245) R. K. Airas, *B* 17, 4932–4938 (1978).

(246) L. B. Hersh, *JBC* 245, 3526–3535 (1970); L. B. Hersh, *JBC* 246, 6803–6806 (1971).

(247) R. Virden, A. F. Bristow, and R. H. Pain, *Biochem. J.* 149, 397–401 (1975); J. Fisher, J. G. Belasco. S. Khosla, and J. R. Knowles, *B* 19, 2895–2901 (1980).

(248) R. Wolfenden, *JACS* 88, 3157–3158 (1966); R. Wolfenden and J. F. Kirsch, *JACS* 90, 6849–6850 (1968); R. Wolfenden, *B* 8, 2409–2412 (1969); B. A. Orsi, N. McFerran, A. Hill, and A. Bingham, *B* 11, 3386–3392 (1972).

(249) A. C. Quenelle and W. R. Melander, *Arch. Biochem. Biophys.* 170, 601–607 (1975).

(250) See discussion of pyrophosphatase on p. 143.

(251) Refs. (125–152).

(252) M. Moriguchi and K. Soda, *B* 12, 2974–2980 (1973); K. Tanizawa and K. Soda, *J. Biochem.* (*Tokyo*) 86, 1199–1209 (1979).

# Chapter 5
# Lyases

Lyases are enzymes that cleave C—C, C—O, C—N, and other bonds in such manner that the cleavage event does not involve hydrolysis or oxidation. In the cleavage direction the reaction is often upon one substrate only. During cleavage a molecule is eliminated from the substrate, leaving an unsaturated residue. The eliminated fragment is usually linked covalently to the (holo)enzyme during the cleavage process, and is ultimately transferred to an acceptor, often a proton. The decarboxylation of acetoacetate to acetone and carbon dioxide is a familiar example of a lyase reaction in which a C—C bond is cleaved. The present chapter gives a sampling of lyase reactions, showing the diverse chemical mechanisms by which they are (thought to be) catalyzed. Table 5.1 at the end of the chapter lists 56 lyases for which there is evidence of a covalent enzyme–substrate intermediate in their catalytic cycles. They comprise 23% of all the lyases recognized officially by the Enzyme Commission; and, on grounds of chemical analogy, they exemplify, in aggregate, over 90% of the EC lyases (cf. Chapter 8).

## Pyruvate Decarboxylase [EC 4.1.1.1]

One of the decarboxylases to come under early study was the thiamine-pyrophosphate-requiring enzyme which catalyzes the decarboxylation of pyruvate.

$$CH_3COCOO^- + H^+ \xrightarrow{Mg^{2+}} CH_3CHO + CO_2$$

The chemical action takes place on C-2 of the thiazolium ring of thiamine (Fig. 1). α-Hydroxyethyl-TPP has been isolated from E. coli cells (1) and from

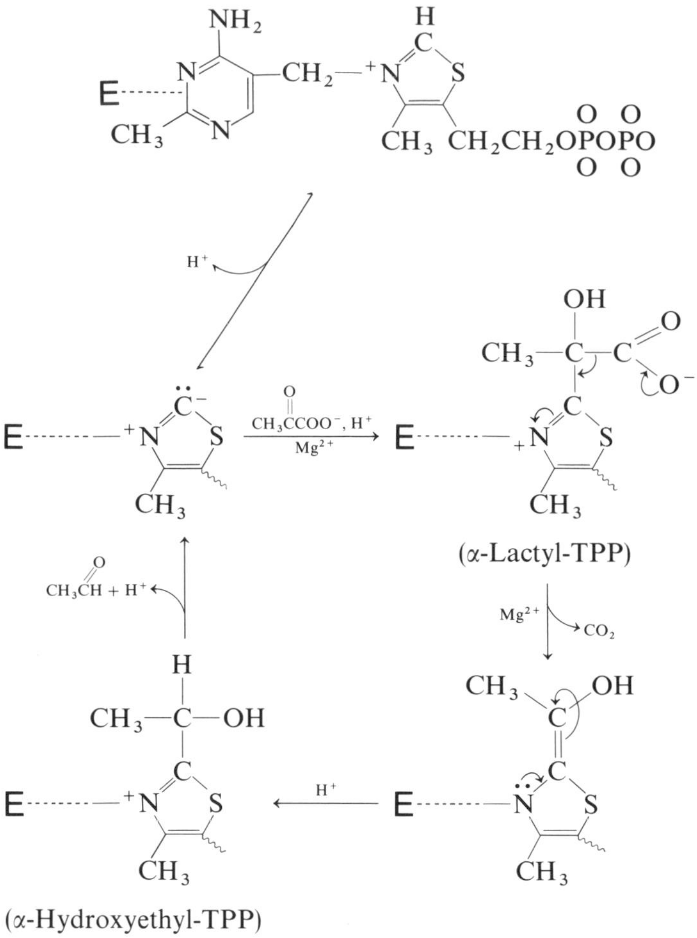

**Fig. 1.** The decarboxylation of pyruvate by pyruvate decarboxylase, showing the co-enzymic function of thiamine pyrophosphate (TPP). Enzyme-bound α-lactyl-TPP and α-hydroxyethyl-TPP are sequential intermediates of the reaction.

deproteinized incubation mixtures of pyruvate decarboxylase from wheat germ and from brewer's yeast (1, 2). α-Lactyl-TPP, the precursor of α-hydroxyethyl-TPP, could also be isolated after a very brief incubation (2). A chemical synthesis of α-hydroxyethyl-TPP was achieved by reacting acetaldehyde with TPP at pH 8.8 (3). The synthetic product, upon incubation with a substrate amount of apopyruvate decarboxylase from wheat germ, was transformed into acetaldehyde.

(1) G. L. Carlson and G. M. Brown, *JBC* 235, PC3 (1960).
(2) H. Holzer and K. Beaucamp, *BBA* 46, 225–243 (1961).
(3) L. O. Krampitz, I. Suzuki, and G. Greull, *Fed. Proc.* 20, 971–977 (1961).

# Acetoacetate Decarboxylase [EC 4.1.1.4]

While most enzymic decarboxylations require the participation of a prosthetic group of one kind or another, the decarboxylation of acetoacetate

$$CH_3COCH_2COO^- + H^+ \longrightarrow CH_3COCH_3 + CO_2$$

has no need of such assistance. In place of a prosthetic group, an $\varepsilon$-amino group of a lysyl residue in the active center of acetoacetate decarboxylase reacts directly with the substrate, forming a Schiff base with it (Fig. 2) (4). Upon loss of carbon dioxide, this gives way to a new Schiff base between enzyme and the product, acetone. The latter intermediate can be trapped when the enzyme is treated with sodium borohydride in the presence of acetoacetate, a process which inactivates the enzyme. Hydrolysis of the inactivated enzyme yields $\varepsilon$-$N$-isopropyllysine. With a unique lysyl residue acting thus as a nucleophile in the decarboxylation of acetoacetate, it is fitting that the amino group of the said lysyl residue has the remarkably low pK of 5.9 (5), four pK units lower than that of the normal $\varepsilon$-amino group of free lysine.

In keeping with the mechanism of Fig. 2, acetoacetate decarboxylase catalyzes the exchange of the carbonyl oxygen of its substrates with the

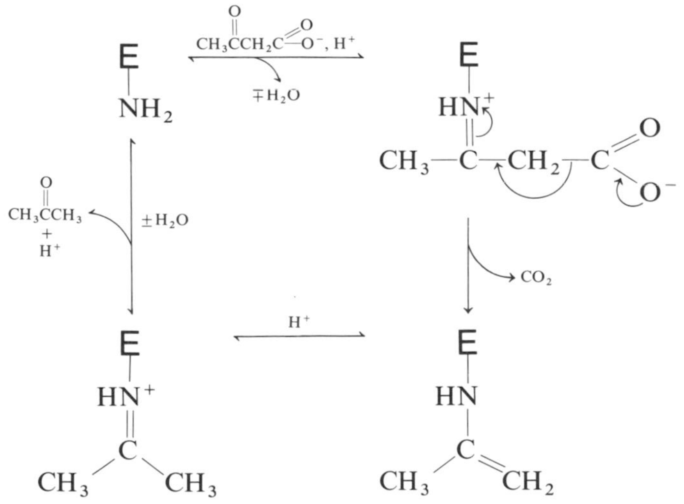

**Fig. 2.** The decarboxylation of acetoacetate by acetoacetate decarboxylase via Schiff base intermediates (4).

(4) S. Warren, B. Zerner, and F. H. Westheimer, B 5, 817–823 (1966); R. A. Laursen and F. H. Westheimer, JACS 88, 3426–3430 (1966).
(5) D. E. Schmidt, Jr., and F. H. Westheimer, B 10, 1249–1253 (1971).

oxygen of water (6). Moreover, the enzyme catalyzes proton-exchange reactions in acetone (7) and butanone (8). Of special interest is the stereospecific proton exchange which takes place at the prochiral 3-position of butanone, pointing to the presence in the active center of a specific catalytic group responsible for the abstraction and placement of a proton (8).

## Glutamate Decarboxylase [EC 4.1.1.15]

The enzymic decarboxylation of most amino acids is a pyridoxal-P-requiring process. Like the transaminases (Chapter 3), the amino acid decarboxylases grip the coenzyme by Schiff base bonding to the ε-amino group of an enzymic lysyl residue. Decarboxylation, like transamination, is thought to begin with a "transaldimination," in which the amino group of the substrate replaces the enzymic lysyl residue in linkage to the coenzyme (Fig. 3) (9). Loss of

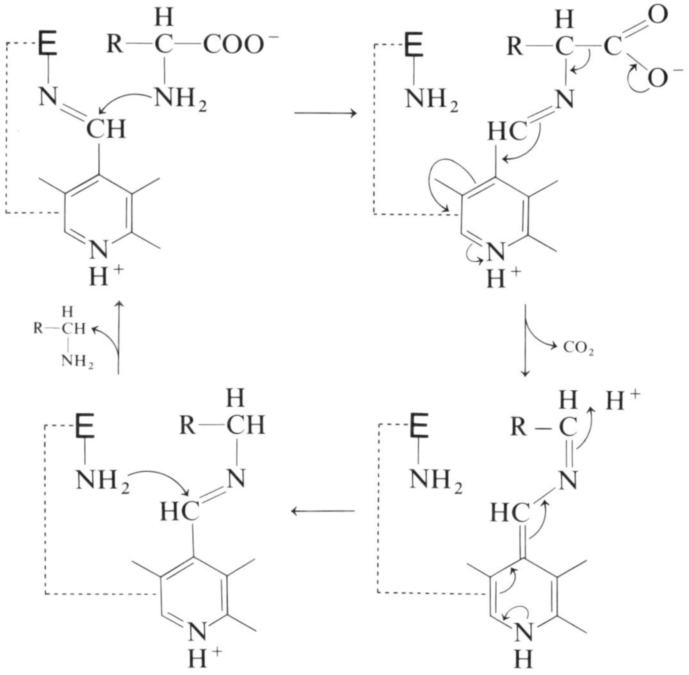

**Fig. 3.** Proposed mechanism for the α-decarboxylation of amino acids with participation of pyridoxal-P (9).

(6) G. A. Hamilton and F. H. Westheimer, *JACS* 81, 6332–6333 (1959).

(7) W. Tagaki and F. H. Westheimer, *B* 7, 901–905 (1968).

(8) G. Hammons, F. H. Westheimer, K. Nakaoka, and R. Kluger, *JACS* 97, 1568–1572 (1975).

(9) E. A. Boeker and E. E. Snell, in *The Enzymes*, 3rd ed., P. D. Boyer, ed., Academic Press, New York, 1972, Vol. 6, p. 217.

carbon dioxide, its replacement by a proton, and a second "transaldimination" complete the catalytic cycle. Though this decarboxylation pathway is widely accepted, and seems on general grounds almost certain to be valid, there is surprisingly little in the way of solid evidence to sustain it. Some of the best evidence for the catalytic pathway of Fig. 3 stems from work on glutamate decarboxylase of *E. coli* (10, 11), which catalyzes the decarboxylation of L-glutamate to give 4-aminobutyrate.

$$OOCCH_2CH_2\underset{\underset{NH_2}{|}}{C}HCOO^- + H^+ \longrightarrow CO_2 + OOCCH_2CH_2CH_2NH_2$$

Besides L-glutamate, however, the enzyme is also slowly active on α-methyl-DL-glutamate. Normal decarboxylation of this substrate yields 4-amino-valerate.

$$OOCCH_2CH_2\underset{\underset{NH_2}{|}}{\overset{\overset{CH_3}{|}}{C}}-COO^- + H^+ + E \cdot PLP \longrightarrow$$

$$OOCCH_2CH_2\underset{\underset{NH_2}{|}}{\overset{\overset{CH_3}{|}}{C}}H + CO_2 + E \cdot PLP$$

But small amounts of levulinic acid and pyridoxamine-P (PMP) also accumulate, in consequence of a small concurrent transamination with the coenzyme (Fig. 4).

$$OOCCH_2CH_2\underset{\underset{NH_2}{|}}{\overset{\overset{CH_3}{|}}{C}}-COO^- + H^+ + E \cdot PLP \longrightarrow$$

$$OOCCH_2CH_2COCH_3 + CO_2 + apoenzyme + PMP$$

Gradual inactivation of the enzyme ensues as its PMP dissociates. Addition of fresh PLP reactivates the enzyme. The appearance of PMP in the abnormal decarboxylation testifies to the formation of a Schiff base between substrate and holoenzyme during catalysis. Figure 4 accounts reasonably for these events. The carbanion shown within the brackets in Fig. 4 can be detected through the use of tetranitromethane (12). Protonation of the carbanion (in the case of L-glutamate) occurs with retention of configuration relative to the

(10) T. E. Huntley and D. E. Metzler, *Symposium on Pyridoxal Enzymes*, Maruzen, Tokyo, 1968, p. 81.
(11) V. S. Sukhareva and A. E. Braunshtein, *Mol. Biol.* (Engl. transl.) 5, 241–252 (1971).
(12) M. L. Mekhanik and Yu. M. Torchinskii, *Biochemistry* (Engl. transl.) 37, 1095–1097 (1972).

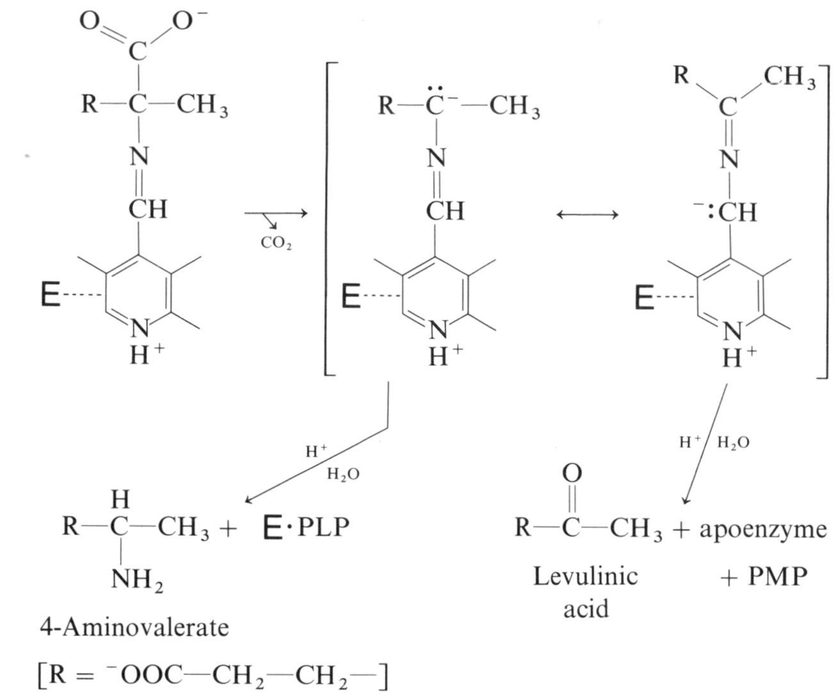

**Fig. 4.** Proposed mechanism for the normal and abnormal decarboxylation of α-methyl-DL-glutamate in the form of the enzyme-bound Schiff base with pyridoxal-P (11).

lost carbon dioxide, implying the presence in the active center of a catalytic group specifically sited for the placement of the proton (13).

## Histidine Decarboxylase [EC 4.1.1.22]

The decarboxylation of histidine to form histamine,

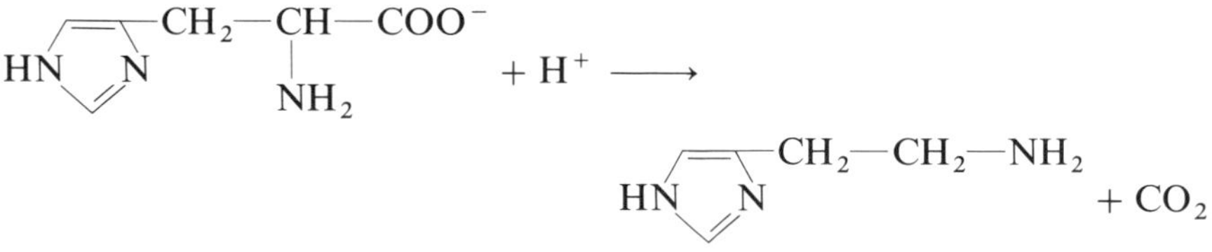

like the decarboxylation of other amino acids, proceeds by way of a Schiff base intermediate. But the prosthetic group of histidine decarboxylase is not

(13) H. Yamada and M. H. O'Leary, *B* 17, 669–672 (1978); E. Santaniello, M. G. Kienle, and A. Manzocchi, *J. Chem. Soc. Perkin Trans. 1* 1677–1679 (1979); M. Bouclier, M. J. Fung, and B. Lippert, *EJB* 98, 363–368 (1979).

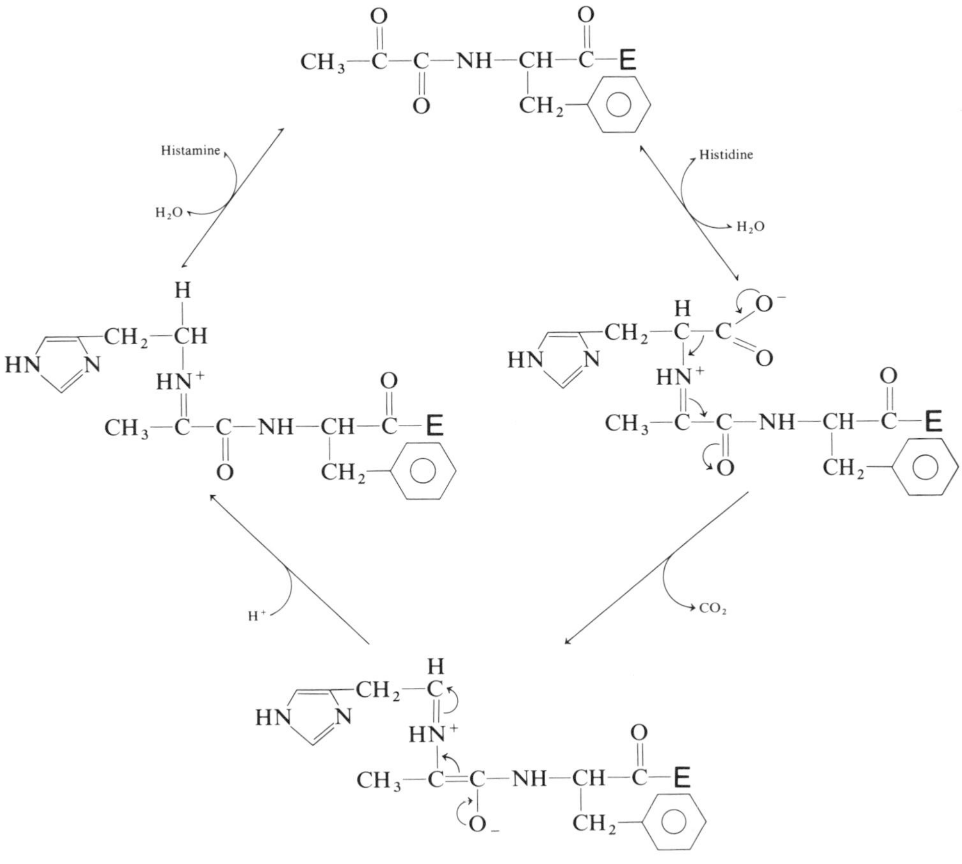

**Fig. 5.** Mechanism proposed for the action of histidine decarboxylase, showing the participation of the peptidically linked pyruvyl group (15).

pyridoxal-P; it is, rather, a covalently bound pyruvyl residue (14). The pyruvyl group is linked peptidically to a phenylalanyl residue of the enzyme. Reducing the enzyme at low temperature with sodium borohydride in the presence of labeled histidine fixes label to the protein. Acid hydrolysis and column chromatography allows the isolation of $N^2$-(1-carboxyethyl)-[$^{14}$C] histidine and an even larger amount of $N^1$-(1-carboxyethyl)-[$^{14}$C]histamine (15). Repetition of the experiment in the presence of labeled histamine yielded only the latter of the two hydrolytic products. Such results make clear that the pyruvyl group of the enzyme participates in Schiff base formation with both the substrate and product of the enzyme reaction, and support the mechanism depicted in Fig. 5.

(14) W. D. Riley and E. E. Snell, *B* 7, 3520–3528 (1968).
(15) P. A. Recsei and E. E. Snell, *B* 9, 1492–1497 (1970).

# Phosphoenolpyruvate Carboxykinase
# [EC 4.1.1.32]

The last of the carboxy lyases we shall consider here catalyzes the reversible decarboxylation of oxaloacetate as follows:

$$\text{Oxaloacetate} + \text{GTP} \underset{}{\overset{Mn^{2+}}{\rightleftharpoons}} \text{phosphoenolpyruvate} + CO_2 + \text{GDP} \quad (1)$$

In kidney and liver, this reaction provides P-enolpyruvate for gluconeogenesis. As phosphoryl donor, ITP serves as well as GTP. An absolute requirement which the enzyme has for a divalent ion is best filled by $Mn^{2+}$. Magnetic resonance measurements on the enzymes from pig liver and sheep kidney make clear that the $Mn^{2+}$ is liganded directly to the enzyme (16), acting the part of a prosthetic group in the catalytic process. These measurements indicate further that, except for carbon dioxide, all of the substrates coordinate with the $Mn^{2+}$, in the manner shown in Fig. 6, to form complexes at intermediate stages of the reversible decarboxylation. It is plain from the figure that, during its transit between pyruvate and GDP, the transferring phosphoryl group remains linked through one of its oxygens to the $Mn^{2+}$; that is, to the holoenzyme.

Though Fig. 6 conveys the sense of a "concerted" reaction, there is reason to believe that the chemical reaction may actually go forward in stages. Unable to catalyze a pyruvate–oxaloacetate exchange, the enzyme (from sheep kidney) does catalyze a $Mn^{2+}$-dependent $CO_2$–oxaloacetate exchange reaction in the absence of added nucleotide (17). GDP (or IDP) causes a large enhancement of the exchange rate, *but with no phosphoryl transfer*. Thus, the behavior of the nucleotide diphosphates in the $CO_2$–oxaloacetate exchange is that of a substrate synergist. In its independence of phosphoryl transfer the exchange reaction hints strongly at the existence of an enzyme–pyruvyl intermediate, which undergoes phosphorylation to P-enolpyruvate or carboxylation to oxaloacetate. The same intermediate may be the source of the pyruvate formed in a secondary reaction catalyzed by the enzyme from chicken liver; namely, the *irreversible* decarboxylation of oxaloacetate in the presence of *catalytic* amounts of GDP or IDP (18).

$$\text{OOCCH}_2\text{COCOO} + H^+ \xrightarrow{\text{GDP or IDP}} CH_3\text{COCOO} + CO_2$$

Also consonant with a stepwise chemical mechanism for phosphoenolpyruvate carboxykinase is the finding that the enzyme (from rat liver) catalyzes a rapid GDP–GTP exchange in the absence of P-enolpyruvate, oxaloacetate, and $CO_2$ (19). The rate of the exchange is 45% of the $V_{max}$ of P-enolpyruvate

---

(16) R. S. Miller, A. S. Mildvan, H.-C. Chang, R. L. Easterday, H. Maruyama, and M. D. Lane, *JBC* 243, 6030–6040 (1968); R. J. Barns, D. B. Keech, and W. J. O'Sullivan, *BBA* 289, 212–224 (1972).
(17) R. J. Barns and D. B. Keech, *BBA* 276, 284–296 (1972).
(18) P. S. Noce and M. F. Utter, *JBC* 250, 9099–9105 (1975).
(19) M. Jomain-Baum and V. L. Schramm, *JBC* 253, 3648–3659 (1978).

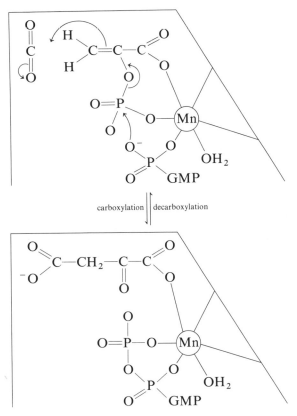

**Fig. 6.** A proposed mechanism of action of phosphoenolpyruvate carboxykinase showing the coordination of substrates, including the transferring phosphoryl group, to the holoenzyme through the prosthetic metal ion (16, 17).

formation in the forward direction of reaction 1. A low concentration of bicarbonate (2 m$M$) raises the rate to 140% of $V_{max}$, but a higher concentration (20 m$M$) tends to become inhibitory (80% of $V_{max}$). On the other hand, P-enolpyruvate (at 100 $\mu M$) severely inhibits the GDP–GTP exchange, reducing it to nearly zero. In reaction 1, therefore, we may suppose that a phosphoryl group of GTP transfers to the enzyme prior to its ultimate capture as P-enolpyruvate.


## Fructose Diphosphate Aldolase [EC 4.1.2.13]

Of the several aldehyde-lyases that have been studied as to their chemical mechanism, the enzyme which participates in glycolysis (long known simply as aldolase) is the one about which we know the most. Aldolases from all species catalyze the reversible aldol cleavage of fructose 1,6-diphosphate into glyceraldehyde-3-P and dihydroxyacetone-P.

$$
\begin{array}{ccc}
\text{CH}_2\text{—OP} & & \text{CH}_2\text{—OP} \\
\text{O}=\text{C} & & \text{O}=\text{C} \\
\text{HO}\text{—C—H} & \longleftarrow & \text{HO—CH} \\
\text{H—C—OH} & & \text{H} \\
\text{H—C—OH} & + & \text{HC}=\text{O} \\
\text{CH}_2\text{OP} & & \text{H—C—OH} \\
& & \text{CH}_2\text{OP}
\end{array}
$$

There are, however, two very different chemical pathways by which this catalysis proceeds. Aldolases from animals and higher plants (the Class I aldolases) form a Schiff base with their substrate. They are inactivated upon reduction with sodium borohydride in the presence of substrate (20). Class II aldolases (found in bacteria and molds) are unaffected by such treatment. They are, however, inhibited by EDTA, and require a metal ion ($Zn^{2+}$) for activity (21).

As a Class I aldolase, the enzyme from rabbit muscle catalyzes the dealdolization of fructose diphosphate by a sequence of chemical reactions shown in Fig. 7. The cycle begins with the formation of a Schiff base between fructose diphosphate and the ε-amino group of a lysine in the active center of the enzyme. The positively charged azomethine group induces the cleavage of the bond between C-3 and C-4 of the sugar. Glyceraldehyde-3-P, formed in the cleavage, diffuses out of the active center leaving behind the other three-carbon fragment as a ketimine carbanion. When the latter acquires a proton (ultimately from the medium), a second Schiff base intermediate is formed. Hydrolysis of the latter results in dihydroxyacetone-P, as the catalytic cycle is completed.

The Schiff base with fructose diphosphate can be trapped in reduced form by the addition of sodium borohydride to the enzyme in the presence of this substrate (22). Some of the second Schiff base is similarly trapped at the same time, or is accessible from dihydroxyacetone-P alone (20, 23). In keeping with these observations is the rapid exchange of the carbonyl oxygen of substrate with oxygens of water (24). The carbanion intermediate is detected by the

---

(20) E. Grazi, T. Cheng, and B. L. Horecker, *Biochem. Biophys. Res. Commun.* 7, 250–253 (1962); B. L. Horecker, P. T. Rowley, E. Grazi, T. Cheng, and O. Tchola, *Biochem. Z.* 338, 36–51 (1963).

(21) W. J. Rutter, *Fed. Proc.* 23, 1248–1257 (1964).

(22) G. Avigad and S. England, *Arch. Biochem. Biophys.* 153, 337–346 (1972); G. Trombetta, G. Balboni, A. di Iasio, and E. Grazi, *Biochem. Biophys. Res. Commun.* 74, 1297–1301 (1977).

(23) E. Grazi, P. T. Rowley, T. Cheng, O. Tchola, and B. L. Horecker, *Biochem. Biophys. Res. Commun.* 9, 38–43 (1962); C. Y. Lai, O. Tchola, T. Cheng, and B. L. Horecker, *JBC* 240, 1347–1350 (1965).

(24) P. Model, L. Ponticorvo and D. Rittenberg, *B* 7, 1339–1347 (1968); E. J. Heron and R. M. Caprioli, *B* 13, 4371–4375 (1974).

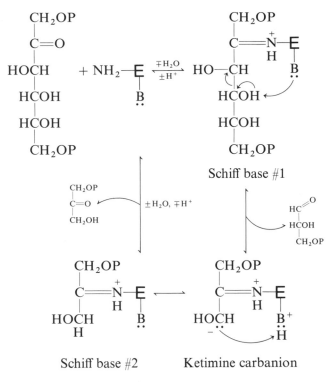

**Fig. 7.** Reaction mechanism proposed for the action of fructose diphosphate aldolase, showing the carbanion and the Schiff base intermediates.

use of tetranitromethane or ferricyanide (25). Its existence can account for the stereospecific exchange of a proton of dihydroxyacetone-P with the protons of water (26). That glyceraldehyde-3-P exchanges more rapidly into fructose diphosphate than does dihydroxyacetone-P is also compatible with a role for the carbanion (27). Protonation of the carbanion is thought to occur through the mediation of a histidine residue in the active center (28).

Typical of the Class II aldolases is the enzyme from yeast, which contains tightly bound zinc. Our knowledge of how Class II aldolases achieve catalysis is but poorly defined. Still uncontested, however, is the original assertion that the metal ion, as an electrophile, induces carbon–carbon cleavage through

(25) P. Christen and J. F. Riordan, *B* 7, 1531–1538 (1968); M. J. Healy and P. Christen, *B* 12, 35–41 (1973).

(26) B. Bloom and Y. J. Topper, *Science* 124, 982–983 (1956); I. A. Rose and S. V. Rieder, *JBC* 231, 315–329 (1958); S. V. Rieder and I. A. Rose, *JBC* 234, 1007–1010 (1959).

(27) I. A. Rose, *PNAS* 44, 10–15 (1958).

(28) P. Hoffee, C. Y. Lai, E. L. Pugh, and B. L. Horecker, *PNAS* 57, 107–113 (1967); L. C. Davis, L. W. Brox, R. W. Gracey, G. Ribereau-Gayon, and B. L. Horecker, *Arch. Biochem. Biophys.* 140, 215–222 (1970); F. C. Hartman and J. P. Brown, *JBC* 251, 3057–3062 (1976).

polarization of the carbonyl group of substrate, in a manner analogous to the Schiff base function in Class I enzymes (21, 29). The carbanion intermediate, as formed in the reaction catalyzed by yeast aldolase, is detectable by its reactivity with tetranitromethane (30).

## 3-Hydroxy-3-Methylglutaryl-Coenzyme A Synthase [EC 4.1.3.5]

The biosynthesis of isoprenes and steroids begins with the synthesis of 3-hydroxy-3-methylglutaryl-CoA in an aldol reaction which unites acetyl-CoA and acetoacetyl-CoA.

$$CH_3CO\!-\!CoA + CH_3COCH_2CO\!-\!CoA + H_2O \longrightarrow$$

$$
\begin{array}{c}
CH_2CO\!-\!CoA \\
| \\
CH_3\!-\!C\!-\!OH \qquad + CoA \quad (2)\\
| \\
CH_2COO
\end{array}
$$

The enzyme catalyzes this reaction by a kinetic mechanism which is partially ping-pong (31), implying a role for an acetyl-enzyme intermediate. On incubation of the enzyme with [$^{14}$C]acetyl-CoA, followed by gel filtration, a radioactive acetyl enzyme can be isolated (32). Acetylation of the enzyme is stoichiometric, and occurs with a "burst" release of CoA proportional to the amount of enzyme present (33). Radioactivity is removed from the acetyl enzyme on treatment with CoA or with acetoacetyl-CoA, yielding as radioactive products [$^{14}$C]acetyl-CoA and [$^{14}$C]3-hydroxy-3-methylglutaryl-CoA, respectively (33). Hydroxylamine and alkali, but not acid, also remove the acetyl group from the protein. Catalysis by the enzyme of an exchange of acetyl between CoA and 3'-dephospho-CoA, in the absence of other reaction components, is also consistent with an acetyl-enzyme intermediate; and the more so, as the exchange rate is comparable to the net rate of reaction 2, affirming thus that the acetyl enzyme is kinetically competent (33, 34). The lability of the acetyl enzyme to performic acid points to a sulfur atom as the connecting link between enzyme and acetyl. This supposition is confirmed

(29) H. A. O. Hill, R. R. Lobb, S. L. Sharp, A. M. Stokes, J. I. Harris, and R. S. Jack, *Biochem. J.* 153, 551–560 (1976).

(30) J. F. Riordan and P. Christen, *B* 8, 2381–2386 (1969).

(31) B. Middleton, *Biochem. J.* 126, 35–47 (1972); M. A. Page and P. K. Tubbs, *Biochem. J.* 173, 925–928 (1978).

(32) P. R. Stewart and H. Rudney, *JBC* 241, 1222–1225 (1966); B. Middleton, *Biochem. J.* 103, 6P (1967).

(33) B. Middleton and P. K. Tubbs, *Biochem. J.* 137, 15–23 (1974).

(34) H. M. Miziorko, K. D. Clinkenbeard, W. D. Reed, and M. D. Lane, *JBC* 250, 5768–5773 (1975).

in the isolation of $N$-acetylcysteic acid upon proteolysis of the acetyl enzyme followed by performic acid oxidation (34).

Reaction 2 actually proceeds in three stages, the first of which is the acetylation of the enzyme by acetyl-CoA.

$$CH_3CO—CoA + HS—E \leftrightharpoons CH_3CO—S—E + CoA$$

There follows the enolization of the acetyl enzyme and its addition, stereoselectively, to the carbonyl carbon of acetoacetyl-CoA to form a second acyl-enzyme intermediate.

$$
CH_3COCH_2CO—CoA + CH_3CO—S—E \leftrightharpoons CH_3—\underset{\underset{CH_2CO—S—E}{\displaystyle |}}{\overset{\overset{CH_2CO—CoA}{\displaystyle |}}{C}}—OH
$$

This, too, can be isolated, by operating in alcoholic solution at $-25°C$ (35). Under these conditions, hydrolysis of the second acyl enzyme

$$
CH_3—\underset{\underset{CH_2CO—S—E}{\displaystyle |}}{\overset{\overset{CH_2CO—CoA}{\displaystyle |}}{C}}—OH \quad + H_2O \longrightarrow CH_3—\underset{\underset{CH_2COOH}{\displaystyle |}}{\overset{\overset{CH_2CO—CoA}{\displaystyle |}}{C}}—OH \quad + HS—E
$$

is sufficiently slowed to permit isolation. Both acyl enzymes are kinetically competent to participate in reaction 2 (35).


# Citrate Lyase [EC 4.1.3.6]

This is a substrate-induced bacterial enzyme which catalyzes the reversible aldol cleavage of citrate to acetate and oxaloacetate.

$$\text{Citrate} \xrightarrow{\text{Mg}^{2+}} \text{acetate} + \text{oxaloacetate} \tag{3}$$

In its native form citrate lyase occurs as an acetyl enzyme (36), being numbered among those enzymes that are isolated already bound covalently to a fragment of substrate. The enzyme is composed of three kinds of subunits (37), one of which—the smallest—carries the acetyl group. A covalently fixed prosthetic group (a modified dephospho coenzyme A) is the functional part of the acyl-carrier protein (38). The sulfhydryl terminus of the prosthetic

(35) H. M. Miziorko, D. Shortle, and M. D. Lane, Biochem. Biophys. Res. Commun. 69, 92–98 (1976); H. M. Miziorko and M. D. Lane, JBC 252, 1414–1420 (1977).
(36) W. Buckel, V. Buschmeier, and H. Eggerer, Z. Physiol. Chem. 352, 1195–1205 (1971).
(37) P. Dimroth and H. Eggerer, EJB 53, 227–235 (1975); D. E. Carpenter, M. Singh, E. G. Richards, and P. A. Srere, JBC 250, 3254–3260 (1975).
(38) N. J. Oppenheimer, M. Singh, C. C. Sweeley, S.-J. Sung, and P. A. Srere, JBC 254, 1000–1002 (1979).

group carries the acetyl in thiolester linkage. Cleavage of citrate, or its reversal, proceeds in two stages: the acyl exchange

$$\text{Acetyl-E} + \text{citrate} \longleftarrow \text{citryl-E} + \text{acetate}$$

and the acyl lyase reaction

$$\text{Citryl-E} \xrightarrow{\text{Mg}^{2+}} \text{acetyl-E} + \text{oxaloacetate}$$

the sum of which yields reaction 3 (36, 39). Inactivation of the enzyme follows upon removal of the acetyl group (with hydroxylamine, for instance), but reactivation is readily achieved by acetylation with acetic anhydride (39). The three subunits of the enzyme can be separately isolated in pure condition, and it is clear that the largest of these catalyzes the acyl exchange, while the second largest catalyzes the lyase reaction (40). How the acyl exchange reaction is brought off is not known; but a mixed anhydride of citric and acetic acids suggests itself as an enzyme-bound intermediate. In the magnesium-requiring lyase phase of the reaction, the citryl enzyme is cleaved at the *pro*-S carboxymethyl group, which is released as acetyl enzyme. The release occurs with inversion of configuration at the methylene carbon as it acquires a proton to become methyl (41).

## ATP Citrate Lyase [EC 4.1.3.8]

Though it catalyzes a seemingly simple dealdolization of citrate—as does citrate lyase (see above)—ATP citrate lyase may, nonetheless, be unequaled among enzymes in the complexity of the total reaction which it catalyzes (42).

$$
\begin{array}{c}
\text{COO} \\
| \\
\text{OOCCH}_2\text{—C—CH}_2\text{COO} + \text{ATP} + \text{CoA} \xrightarrow{\text{Mg}^{2+}} \\
| \\
\text{OH}
\end{array}
$$

$$\text{OOCCH}_2\text{COCOO} + \text{CH}_3\text{COSCoA} + \text{ADP} + \text{P}_i$$

Through this reaction the enzyme furnishes acetyl-CoA for lipogenesis and oxaloacetate for gluconeogenesis in the cytoplasm of many tissues. Like citrate lyase, ATP citrate lyase detaches (with stereochemical inversion) the *pro*-S carboxymethyl group of citrate as a two-carbon fragment (43). But now

(39) W. Buckel, K. Ziegert, and H. Eggerer, *EJB* 37, 295–304 (1973).

(40) P. Dimroth and H. Eggerer, *PNAS* 72, 3458–3462 (1975).

(41) W. Buckel, H. Lenz, P. Wunderwald, V. Buschmeier, H. Eggerer, and G. Gottschalk, *EJB* 24, 201–206 (1971).

(42) P. A. Srere, *JBC* 234, 2544–2547 (1959).

(43) H. Eggerer, W. Buckel, H. Lenz, P. Wunderwald, G. Gottschalk, J. W. Cornforth, C. Donninger, R. Mallaby, and J. W. Redmond, *Nature (London)* 226, 517–519 (1970).

the fragment appears as acetyl-CoA, not as free acetate. To provide for the energy-rich character of acetyl-CoA ATP is consumed in activating a carboxyl group. Activation of the *pro*-S carboxyl of citrate takes place—as it does for acetate kinase (p. 92)—through the prior phosphorylation of the enzyme. Thus, when ATP citrate lyase is incubated briefly with ATP and $Mg^{2+}$, followed by gel filtration, the phosphorylated enzyme, E ~ P, can be isolated (44, 45). The existence of E ~ P is also evident in the rapid ADP–ATP exchange which is catalyzed by the enzyme in the absence of citrate and CoA. When E ~ P is allowed to react with citrate and CoA, orthophosphate is released, accompanied by oxaloacetate and acetyl-CoA in amounts stoichiometric with the orthophosphate. Neither ATP nor $Mg^{2+}$ is required for this conversion. In reacting with citrate, E ~ P is believed to form enzyme-bound citryl phosphate. Citryl phosphate has not been isolated from incubation mixtures, but it can be synthesized chemically as a racemic mixture. With only CoA present, the enzyme catalyzes the stereospecific cleavage of citryl phosphate into oxaloacetate and acetyl-CoA at a rate equal to the rate of the overall reaction starting with ATP, citrate, CoA, and $Mg^{2+}$ (46, 47).

Enzyme-bound citryl phosphate carries the reaction forward by acting upon a catalytic group of the enzyme to form a second covalent enzyme–substrate intermediate—citryl ~ enzyme—which is easily isolated. Incubating the enzyme with synthetic citryl phosphate (47), or with citrate, ATP, and $Mg^{2+}$ in the absence of CoA (44), followed by gel filtration, yields E ~ citryl. The addition of CoA to E ~ citryl elicits the lyase transformation of the latter into oxaloacetate and acetyl-CoA. In view of the above-described work on citrate lyase (36), it seems reasonable to suppose that E ~ citryl undergoes the lyase reaction directly—in the presence of CoA—yielding E ~ acetyl and oxaloacetate as immediate products. E ~ acetyl thereafter acts upon CoA to give acetyl-CoA. E ~ acetyl can in fact be isolated by gel filtration of a mixture of enzyme with acetyl-CoA (48).

Thus the sequence of partial reactions catalyzed by ATP citrate lyase is:

$$E + ATP \xrightarrow{Mg^{2+}} E \sim P + ADP$$

$$E \sim P + citrate \longrightarrow E\cdots citryl \sim P$$

$$E\cdots citryl \sim P \longrightarrow E \sim citryl + P_i$$

$$E \sim citryl \longrightarrow E \sim acetyl + oxaloacetate$$

$$E \sim acetyl + CoA \longrightarrow E + acetyl \sim CoA$$

(44) H. Inoue, F. Suzuki, H. Tanioka, and Y. Takeda, *Biochem. Biophys. Res. Commun.* 26, 602–608 (1967).
(45) H. Inoue, F. Suzuki, H. Tanioka, and Y. Takeda, *J. Biochem.* (*Tokyo*) 63, 89–100 (1968).
(46) C. T. Walsh and L. B. Spector, *JBC* 243, 446–448 (1968).
(47) C. T. Walsh and L. B. Spector, *JBC* 244, 4366–4374 (1969).
(48) F. Suzuki, in *Molecular Mechanisms of Enzyme Action*, Y. Ogura, Y. Tonomura, and T. Nakamura, eds., University Park Press, Baltimore, 1972, pp. 265–280.

These five partial reactions involve the participation of three covalent enzyme–substrate intermediates, all of which are isolable. Citryl phosphate, which can be synthesized chemically, enters into the reaction as a non-covalent, enzyme-bound intermediate. Since E ~ acetyl stems directly from E ~ citryl through dealdolization, it is probable that the acyl bond to the enzyme is the same in both intermediates; namely, to the $\gamma$-carboxyl of a glutamyl residue (48). In E ~ P, the phosphoryl has been found fixed covalently to a histidyl residue (49) and to the $\gamma$-carboxyl of a glutamyl residue of the active center (50), depending on the chemical method used for the detection of these linkages. Possibly the phosphoryl is obliged to traverse both sites during catalysis as part of a "surface walk" (cf. Chapter 1, p. 14). In this remarkable multistage reaction, the enzyme manages to conserve the energy of the terminal pyrophosphate bond of ATP through a succession of four diverse intermediates until it reappears in the thiolester bond of acetyl-CoA.

## Tryptophanase [EC 4.1.99.1]

The lyase reaction catalyzed by this enzyme has the net effect of a hydrolysis.

$$\text{L-Tryptophan} + H_2O \longleftrightarrow \text{pyruvate} + NH_3 + \text{indole}$$

A key step in the reaction is the $\beta$-elimination of indole from tryptophan held in Schiff base bonding to the prosthetic group (pyridoxal-5'-P) of the enzyme (Fig. 8) (51). As the carbon–carbon bond to indole is broken, the carbon–carbon double bond of aminoacrylate appears, still bound to the coenzyme. Setting the stage for the $\beta$-elimination is the prior ionization of the $\alpha$-hydrogen of tryptophan, yielding the quinoid structure **1-1a** (52). The quinoid intermediate manifests itself in the prompt appearance of an absorption band near 500 nm upon the addition of tryptophan to the enzyme. As the degradation of tryptophan proceeds, the absorption at 500 nm disappears. When L-alanine, an inhibitor of tryptophanase, is added, the same absorption band appears; but it remains unchanged with time, since alanine is not further acted upon by the enzyme. Elimination of indole leads to the enzyme-

(49) G. L. Cottam and P. A. Srere, *Biochem. Biophys. Res. Commun.* 35, 895–900 (1969); S. Mårdh, O. Ljungström, S. Högstedt, and Ö. Zetterqvist, *BBA* 251, 419–426 (1971); S. Ramakrishna and W. B. Benjamin, *JBC* 254, 9232–9236 (1979).

(50) F. Suzuki, K. Fukunishi, and Y. Takeda, *J. Biochem.* (*Tokyo*) 66, 767–774 (1969).

(51) T. Watanabe and E. E. Snell, *PNAS* 69, 1086–1090 (1972); E. E. Snell, *Adv. Enzymol.* 42, 287–333 (1975).

(52) Y. Morino and E. E. Snell, *JBC* 242, 2800–2809 (1967); *idem.*, *J. Biochem.* (*Tokyo*) 82, 733–746 (1977).

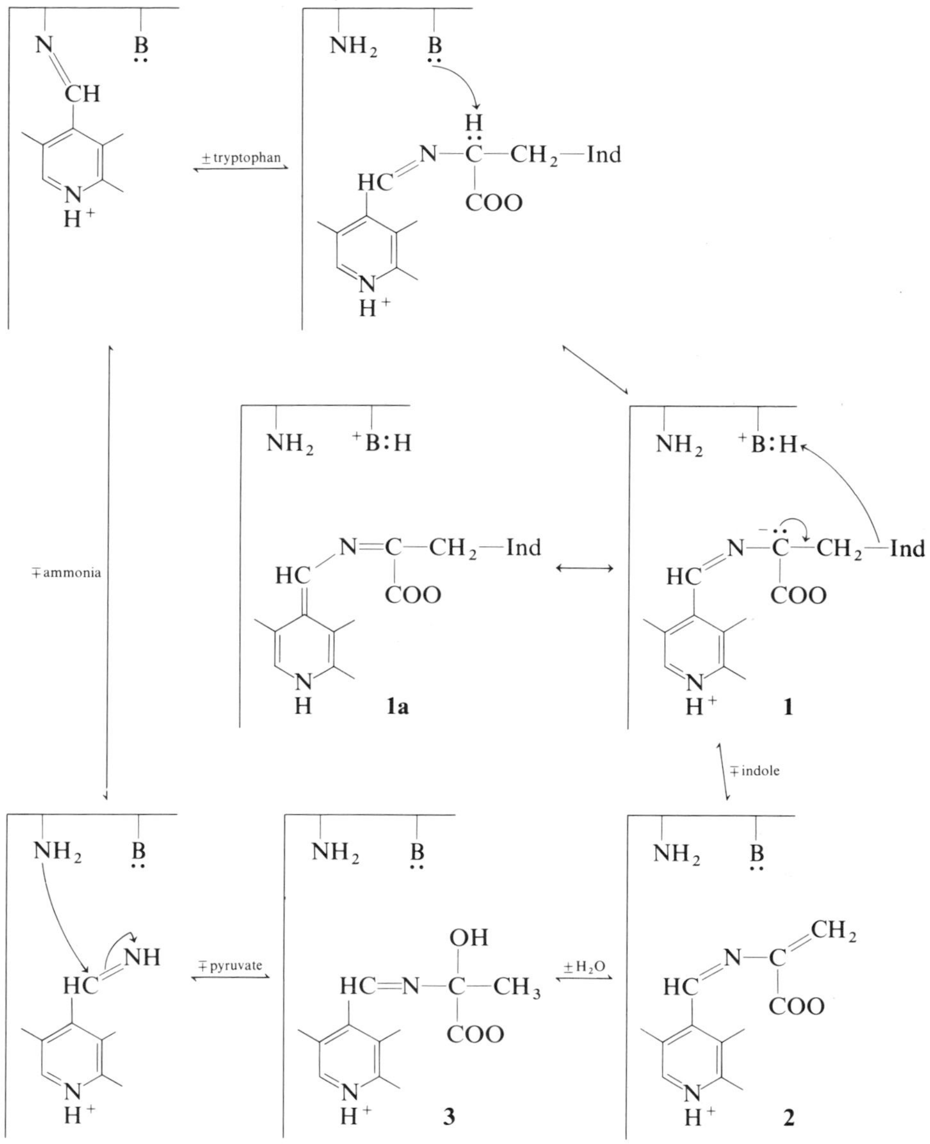

**Fig. 8.** A proposed mechanism for the tryptophanase reaction (51).

linked aminoacrylate (structure **2**), which is thought to add water, breaking down ultimately into pyruvate and ammonia. The particular suite of intermediates (**2** ⇋ **3** ⇋ **4**) is suggested by the observation that the tryptophanase reaction is readily reversible, and that the kinetics of tryptophan synthesis from ammonia, pyruvate and indole show that ammonia adds first to the enzyme, followed by pyruvate and indole in that order (51). The degradation of tryptophan by tryptophanase must of course proceed in the reverse order. The proton which appears on C-3 of indole, after cleavage, originates on C-2 of the amino acid side chain (Fig. 8). Such intramolecular proton transfer implies the mediation of a unique base in the active center of the enzyme (53). The same base may be active in the retention of configuration when a proton replaces indole on C-3 of the amino acid side chain, as this carbon is transforming itself into the methyl group of pyruvate.

## Carbonic Anhydrase [EC 4.2.1.1]

The reversible hydration of carbon dioxide

$$CO_2 + H_2O \longleftrightarrow HOCO_2^- + H^+ \tag{4}$$

is accomplished by carbonic anhydrase in one of the speediest enzymic reactions known. Other reactions are also catalyzed by the enzyme, but less efficiently; for instance, the hydration of acetaldehyde (54).

the hydrolysis of esters (55),

the hydrolysis of certain sultones (56),

(53) E. Schleicher, K. Mascaro, R. Potts, D. R. Mann, and H. G. Floss, *JACS* 98, 1043–1044 (1976); J. C. Vederas, E. Schleicher, M.-D. Tsai, and H. G. Floss, *JBC* 253, 5350–5354 (1978).
(54) Y. Pocker and F. E. Meany, *JACS* 87, 1809–1811 (1965); D. Cheshnovsky and G. Navon, *B* 19, 1866–1873 (1980).
(55) J. A. Verpoorte, S. Mehta, and J. T. Edsall, *JBC* 242, 4221–4229 (1967); Y. Pocker and J. T. Stone, *B* 7, 4139–4145 (1968).
(56) E. T. Kaiser and K. Lo, *JACS* 91, 4912–4918 (1969).

the hydrolysis of 2,4-dinitrofluorobenzene (57),

$$NO_2 \langle \bigcirc \rangle{-}F + H_2O \longrightarrow NO_2 \langle \bigcirc \rangle{-}OH + HF$$

(with NO₂ substituent at top position on each ring)

and the hydrolysis of dimethyl 2,4-dinitrophenyl phosphate (58).

$$NO_2 \langle \bigcirc \rangle{-}O{-}\overset{\overset{\displaystyle O}{\|}}{\underset{\underset{\displaystyle OCH_3}{|}}{P}}{-}OCH_3 + H_2O \longrightarrow$$

$$NO_2 \langle \bigcirc \rangle{-}OH + HO{-}\overset{\overset{\displaystyle O}{\|}}{\underset{\underset{\displaystyle OCH_3}{|}}{P}}{-}OCH_3$$

Carbonic anhydrase is a metalloenzyme bearing a single zinc ion per molecule as prosthetic group. X-ray crystallography at high resolution of the enzyme from human erythrocytes reveals that the zinc ion is coordinated tetrahedrally to three imidazole rings at the bottom of a deep conical cleft (59). The fourth coordination site points out into the cleft and is occupied by a water molecule or a hydroxide ion. It is generally agreed that of the components of reaction 4, the water molecule (or derived hydroxide ion) is *always* coordinated to the zinc ion during catalysis (60). This being so, the zinc-bound water (or hydroxide ion) of carbonic anhydrase may be regarded as a species of "activated water." On this view, the plausible union of zinc-bound hydroxide with carbon dioxide would give zinc-bound bicarbonate (61). Exchange of water for bicarbonate in the coordination site of zinc completes the catalytic cycle (Fig. 9). Recent research on nonenzymic systems makes plain that metal-bound hydroxide is superior to the free hydroxide ion—at neutral pH—in its capacity to add to carbonyl substrates (62).

(57) P. Henkart, G. Guidotti, and J. T. Edsall, *JBC* 243, 2447–2449 (1968).

(58) Y. Pocker and S. Sarkanen, *B* 17, 1110–1118 (1978).

(59) A. Liljas, K. K. Kannan, P. C. Bergstén, I. Waara, K. Fridborg, B. Strandberg, U. Carlbom, L. Järup, S. Lövgren, and M. Petef, *Nature (London)*, *New Biol.* 235, 131–137 (1972); K. K. Kannan, B. Notrand, K. Fridborg, S. Lövgren, A. Ohlsson, and M. Petef, *PNAS* 72, 51–55 (1975).

(60) Y. Pocker and S. Sarkanen, *Adv. Enzymol.* 47, 252–265 (1978).

(61) J. E. Coleman, *JBC* 242, 5212–5219 (1967); M. E. Riepe and J. H. Wang, *JBC* 243, 2779–2787 (1968); R. H. Prince and P. R. Woolley, *Angew. Chem. Int. Ed.* 11, 408–417 (1972); C. K. Tu and D. N. Silverman, *JACS* 97, 5935–5936 (1975); K. K. Kannan, M. Petef, K. Fridborg, H. Cid-Dresdner, and S. Lövgren, *FEBS lett.* 73, 115–119 (1977); A. J. M. S. Uiterkamp, I. M. Armitage, and J. E. Coleman, *JBC* 255, 3911–3917 (1980).

(62) D. A. Buckingham and L. M. Engelhardt, *JACS* 97, 5915–5917 (1975); P. Woolley, *Nature (London)* 258, 677–682 (1975).

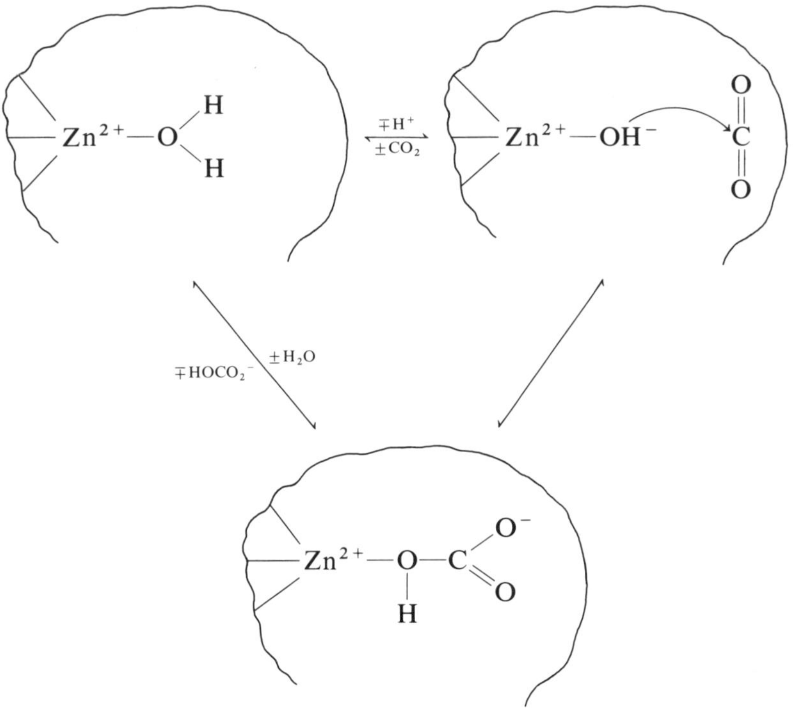

**Fig. 9.** A proposed mechanism for the reversible hydration of carbon dioxide by carbonic anhydrase (61).

## Dehydroquinase [EC 4.2.1.10]

An unusual generation of a double bond through water elimination is seen in the enzymic dehydration of 5-dehydroquinate to 5-dehydroshikimate.

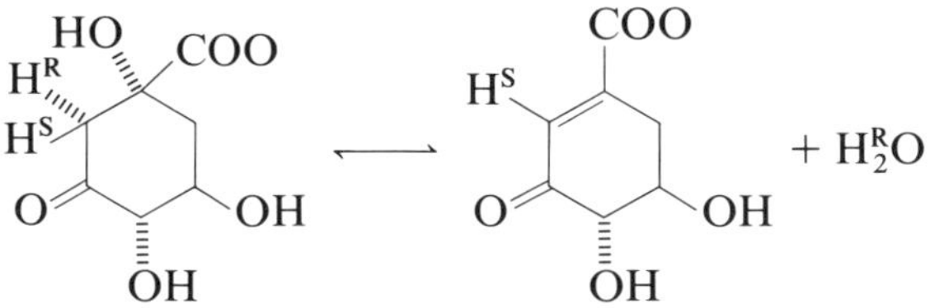

Dehydroquinase has no prosthetic group. It makes use instead of the lysyl residue in its active center to act upon the carbonyl group of the substrate, forming a Schiff base intermediate (Fig. 10) (63). The positively charged imine function of the intermediate exerts a potent labilizing action on one of the

(63) J. R. Butler, W. L. Alworth, and M. J. Nugent, *JACS* 96, 1617–1618 (1974).

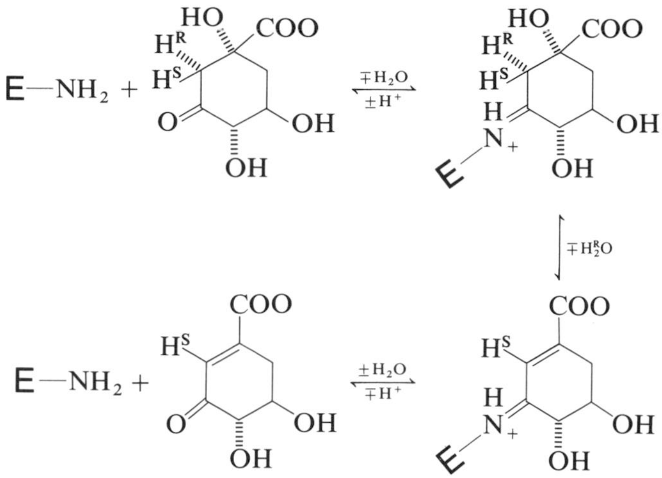

**Fig. 10.** A Schiff base mechanism for the action of dehydroquinase (63, 65).

C-6 protons—the *pro*-R proton—to initiate the *cis* elimination of water (64). Since most elimination reactions occur in the *trans* fashion, the present lyase reaction clearly follows an unusual steric, as well as mechanistic, course. The *cis* dehydration of 5-dehydroquinate is thought to result when the substrate's cyclohexane ring changes to a particular skew-boat conformation when it engages the active site of the enzyme (65).

## Enolase [EC 4.2.1.11]

A hydrolyase reaction of another sort is the one catalyzed by enolase. With a magnesium ion as prosthetic group, enolase catalyzes the reversible, *trans* dehydration of 2-phospho-D-glycerate to phosphoenolpyruvate (66).

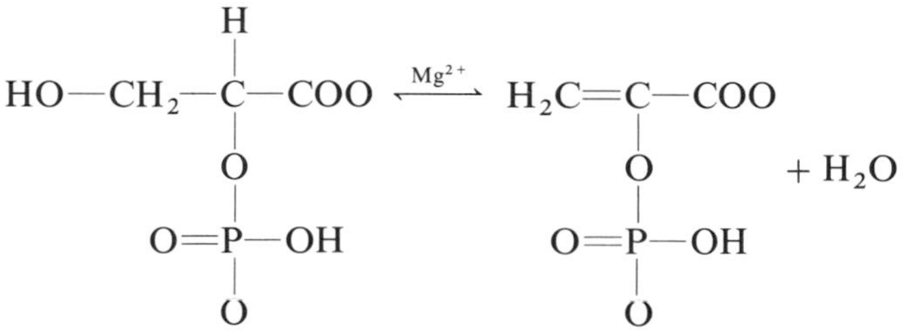

(64) K. R. Hanson and I. A. Rose, *PNAS* 50, 981–988 (1963).
(65) A. D. N. Vaz, J. R. Butler, and M. J. Nugent, *JACS* 97, 5914–5915 (1975).
(66) M. Cohn, J. E. Pearson, E. L. O'Connell, and I. A. Rose, *JACS* 92, 4095–4098 (1970).

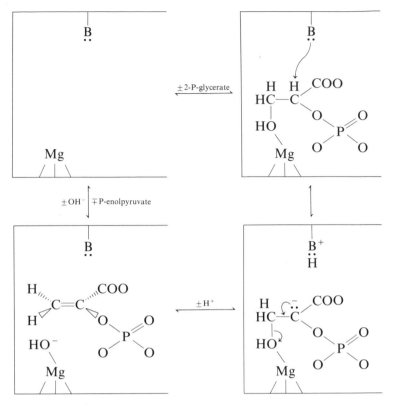

**Fig. 11.** A mechanism proposed for the enolase reaction, showing coordination of substrate to the prosthetic metal ion. In the reverse direction, hydration of P-enolpyruvate requires the activation of water by the metal ion (69).

The enzyme also catalyzes, at a much slower rate, the $\beta,\gamma-\alpha,\beta$ isomerization of 2-phospho-3-butenoic acid to $(Z)$-phosphoenol-$\alpha$-ketobutyrate (67).

$$
\begin{array}{c}
\underset{H}{\overset{\displaystyle HO-\overset{\displaystyle O}{\overset{\|}{P}}-O}{\underset{\displaystyle |}{\underset{\displaystyle O}{\underset{\displaystyle |}{H_2C=C-C-COO}}}}}
\quad\longrightarrow\quad
\underset{H}{\overset{CH_3}{\phantom{x}}}\!\!\diagdown\!\!C\!\!=\!\!C\!\!\diagup\!\!\overset{HO-\overset{O}{\overset{\|}{P}}-O}{\underset{COO}{O}}
\end{array}
\qquad (6)
$$

Since $Mn^{2+}$ can serve in place of $Mg^{2+}$, a magnetic resonance study of enolase action is possible. It shows that the enzyme acts upon normal

(67) J. Applebaum and J. Stubbe, *B* 14, 3908–3912 (1975).

substrate via an enzyme–metal–substrate bridge complex (68, 69). The dehydration cycle begins with the coordination of the C-3 hydroxyl of 2-phosphoglycerate directly to the metal in the active center (Fig. 11). This has the effect of making the hydroxyl a good leaving group. It may also have an activating influence on the C-2 proton of the substrate (70). Removal of the proton by a basic group of the enzyme leaves a transient carbanion, which in the end gives up its C-3 hydroxyl entirely to the metal. The same basic catalytic group probably initiates the $\beta,\gamma-\alpha,\beta$ isomerization of reaction 6 by removing the C-2 proton of the substrate analogue. Consistent with the role of a carbanion in enolase catalysis is the observation that the enzyme catalyzes the incorporation of a water proton into the C-2 position of unreacted 2-phosphoglycerate during the net formation of phosphoenolpyruvate (70, 71). The rate of proton exchange is substantially greater than the rate of oxygen exchange between medium and the C-3 hydroxyl. According to the mechanism of Fig. 11, the reversal of reaction 5 by enolase requires an "activated water" (69).

## Propanediol Dehydrase [EC 4.2.1.28]

One of the best studied of the adenosylcobalamin-dependent enzymes is the one which catalyzes the dehydration of propanediol to propionaldehyde

$$CH_3-\underset{\underset{OH}{|}}{\overset{\overset{H}{|}}{C}}-\underset{\underset{OH}{|}}{\overset{\overset{H}{|}}{CH}} \longrightarrow CH_3-CH_2-\overset{\overset{O}{||}}{CH} + H_2O$$

and the dehydration of ethylene glycol to acetaldehyde.

$$\underset{\underset{OH}{|}}{\overset{\overset{H}{|}}{HC}}-\underset{\underset{OH}{|}}{\overset{\overset{H}{|}}{CH}} \longrightarrow CH_3-\overset{\overset{O}{||}}{CH} + H_2O$$

Though the net reaction is a dehydration, the process is in detail an isomerization reaction followed by a hydro-lyase reaction. In the isomerization, one of the C-1 hydrogens of the substrate is stereoselected (72) to exchange places

(68) M. Cohn and J. S. Leigh, *Nature (London)* 193, 1037–1040 (1962); M. Cohn, *B* 2, 623–629 (1963); T. Nowak and A. S. Mildvan, *JBC* 245, 6057–6064 (1970).

(69) T. Nowak, A. S. Mildvan, and G. L. Kenyon, *B* 12, 1690–1701 (1973).

(70) J. Stubbe and R. H. Abeles, *B* 19, 5505–5512 (1980).

(71) E. C. Dinovo and P. D. Boyer, *JBC* 246, 4586–4593 (1971); T. Y. S. Shen and E. W. Westhead, *B* 12, 3333–3337 (1973).

(72) K. W. Moore and J. H. Richards, *Biochem. Biophys. Res. Commun.* 87, 1052–1057 (1979).

with the C-2 hydroxyl to form an aldehyde hydrate, which loses a molecule
of water (73).

$$
\underset{\substack{|\ \ \ | \\ OH\ OH}}{CH_3\!-\!\overset{\substack{H\ \ \ H \\ |\ \ \ |}}{C\!-\!CH}} \longrightarrow \underset{\substack{|\ \ \ | \\ H\ \ OH}}{CH_3\!-\!\overset{\substack{H\ \ \ H \\ |\ \ \ |}}{C\!-\!C}\!-\!OH} \longrightarrow CH_3\!-\!CH_2\!-\!\overset{\substack{O \\ ||}}{CH} + H_2O
$$

The loss of water from propionaldehyde hydrate—in the hydro-lyase phase
of the reaction—is enzyme-catalyzed and stereospecific (74).

Of the several stages which make up the net reaction, the best understood
is the transfer of hydrogen. It is now established beyond doubt that the
prosthetic group of the enzyme—adenosylcobalamin—is the mediating
hydrogen carrier (75, 76). Though the transferring hydrogen is not exchangeable with solvent protons, the reaction is not strictly *intra*molecular. For when
the enzyme acts simultaneously on unlabeled ethylene glycol and L-$[1,1$-$^2$H$]$
propanediol, deuterium appears not only in propionaldehyde—as predicted
for an *intra*molecular hydrogen transfer—but in acetaldehyde too (77).
Evidently, the hydrogen transfer is at least in part *inter*molecular, the holoenzyme accepting a deuterium atom from a molecule of propanediol and
returning it into either acetaldehyde or propionaldehyde.

Transfer of hydrogen to and from adenosylcobalamin is as rapid as the
overall reaction, and is therefore kinetically competent (78). The transfer is
to the 5'-carbon of the coenzyme, after the enzyme somehow effects the
homolytic rupture of the cobalt–5'-carbon bond (Fig. 12). This latter event
is revealed in the electron spin resonance and other measurements made on
the dehydrase reaction, and suggests the participation of radical species in
the catalysis (79). When substrate hydrogen joins with the 5'-carbon of
adenosylcobalamin a necessary consequence is the formation of enzyme-
bound 5'-deoxyadenosine with three equivalent hydrogens on the 5'-carbon
(75, 78, 80, 81).

(73) B. Zagalak, P. A. Frey, G. L. Karabatsos, and R. H. Abeles, *JBC* 241, 3028–3035 (1966);
P. A. Frey, G. L. Karabatsos, and R. H. Abeles, *Biochem. Biophys. Res. Commun.* 18, 551–556
(1965); J. Rétey, A. Umani-Ronchi, and D. Arigoni, *Experientia* 22, 72–73 (1966).

(74) J. Rétey, A. Umani-Ronchi, J. Seibl, and D. Arigoni, *Experientia* 22, 502–503 (1966);
J. E. Valinsky and R. H. Abeles, *Arch. Biochem. Biophys.* 166, 608–609 (1975).

(75) P. A. Frey, M. K. Essenberg, and R. H. Abeles, *JBC* 242, 5369–5377 (1967).

(76) P. A. Frey and R. H. Abeles, *JBC* 241, 2732–2733 (1966); R. H. Abeles and P. A. Frey,
*Fed. Proc.* 25, 1639–1641 (1966).

(77) R. H. Abeles and B. Zagalak, *JBC* 241, 1245–1246 (1966).

(78) M. K. Essenberg, P. A. Frey, and R. H. Abeles, *JACS* 93, 1242–1251 (1971).

(79) T. H. Findlay, J. Valinsky, A. S. Mildvan, and R. H. Abeles, *JBC* 248, 1285–1290 (1973);
R. G. Eagar, Jr., W. W. Bachovchin, and J. H. Richards, *B* 14, 5523–5528 (1975); T. Toraya,
E. Krodel, A. S. Mildvan, and R. H. Abeles, *B* 18, 417–426 (1979).

(80) O. H. Wagner, H. A. Lee, P. A. Frey, and R. H. Abeles, *JBC* 241, 1751–1762 (1966);
R. H. Abeles, *Adv. Chem. Ser.* 100, 346–364 (1971); K. W. Moore, W. W. Bachovchin, J. B.
Gunter, and J. H. Richards, *B* 18, 2776–2782 (1979).

(81) T. H. Findlay, J. Valinsky, K. Sato, and R. H. Abeles, *JBC* 247, 4197–4207 (1972).

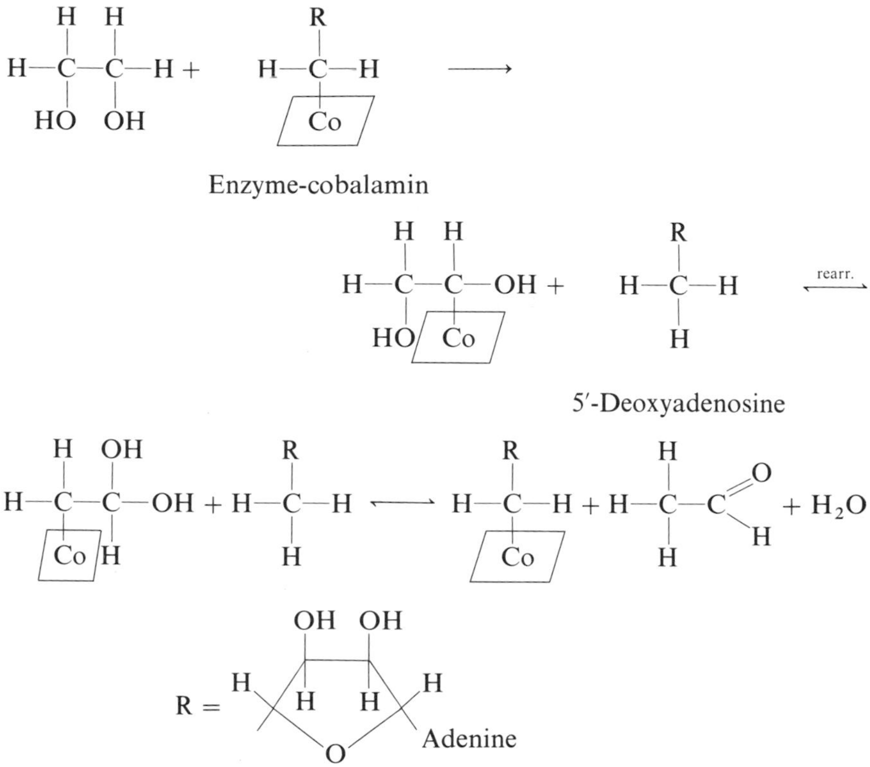

$$\text{H}_3\text{C(H)(OH)}-\text{C(H)(OH)}-\text{H} + \text{H}-\text{C(R)}-\text{H (Co)} \longrightarrow$$

Enzyme-cobalamin

$$\text{H}-\text{C(H)(HO)}-\text{C(H)(Co)}-\text{OH} + \text{H}-\text{C(R)(H)}-\text{H} \xleftarrow{\text{rearr.}}$$

5′-Deoxyadenosine

$$\text{H}-\text{C(H)(Co)}-\text{C(OH)(H)}-\text{OH} + \text{H}-\text{C(R)(H)}-\text{H} \longrightarrow \text{H}-\text{C(R)(Co)}-\text{H} + \text{H}-\text{C(H)(H)}-\text{C} \backslash^{O}_{H} + \text{H}_2\text{O}$$

R = (ribose ring: OH OH, H H H, O, Adenine)

**Fig. 12.** A mechanism proposed for the action of propanediol dehydrase, shown here acting upon ethylene glycol (81).

The substrate fragment remaining after hydrogen transfer to the coenzyme is thought to bind to the cobalt atom, and then to undergo rearrangement prior to regaining the hydrogen atom from the coenzyme (Fig. 12). How the rearrangement takes place is unresolved.

## Phenylalanine Ammonia-Lyase [EC 4.3.1.5]

This carbon–nitrogen lyase of long-standing interest catalyzes the elimination of ammonia from L-phenylalanine to form *trans*-cinnamate.

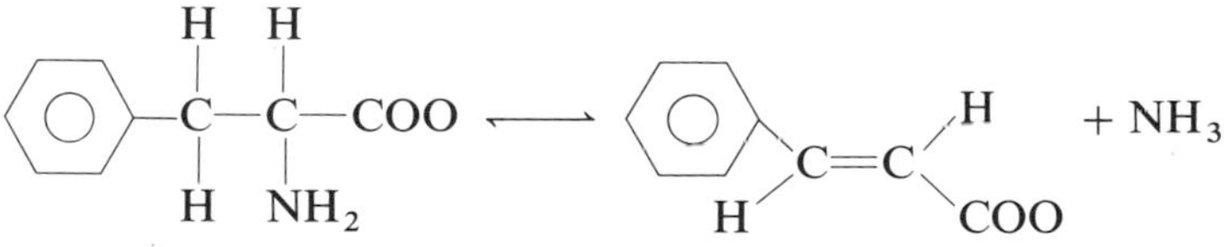

$$\text{Ph}-\text{C(H)(H)}-\text{C(H)(NH}_2)-\text{COO} \longrightarrow \text{Ph}-\underset{H}{C}=\underset{COO}{C}\backslash^{H} + \text{NH}_3$$

The enzyme from some plant sources can also deaminate L-tyrosine to *trans*-p-hydroxycinnamate.

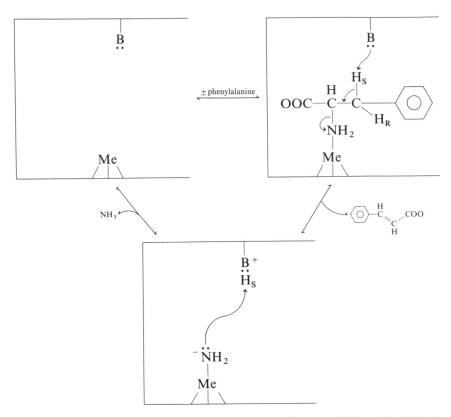

**Fig. 13.** A proposed mechanism for the elimination of ammonia from phenylalanine by phenylalanine ammonia-lyase, showing coordination of substrate through its amino group to prosthetic metal ion.

Eliminated with the amino group of each substrate is the *pro*-S proton on C-3 (82). Within its active center the enzyme has an electrophile which combines with the amino function of the substrate, forming what must be a better leaving group than ammonia. The electrophile is thought to be a dehydro-alanyl residue of the protein (83, 84). Also suggested as an electrophilic pros-

(82) R. H. Wightman, J. Staunton, A. R. Battersby, and K. R. Hanson, *J. Chem. Soc. Perkin Trans. 1* 2355–2364 (1972); P. G. Strange, J. Staunton, H. R. Wiltshire, A. R. Battersby, K. R. Hanson, and E. A. Havir, *J. Chem. Soc. Perkin Trans. 1* 2364–2372 (1972).
(83) K. K. Hanson and E. A. Havir, *Arch. Biochem. Biophys.* 141, 1–17 (1970).
(84) D. S. Hodgins, *JBC* 246, 2977–2985 (1971); J. R. Parkhurst and D. S. Hodgins, *Arch. Biochem. Biophys.* 152, 597–605 (1972); E. A. Havir and K. R. Hanson, *B* 14, 1620–1626 (1975).

thetic group is a tightly held transition metal ion to which the α-amino group of phenylalanine coordinates during catalysis (Fig. 13) (85). Such coordination could activate the *pro*-S hydrogen on C-3 for removal as a proton, as well as aid the amino group in leaving.

Phenylalanine ammonia-lyase follows "Ordered Uni Bi" type of kinetics with the olefinic acid released before ammonia. On following the deamination of unlabeled phenylalanine with the potato enzyme in the presence of [$^{14}$C]cinnamate it is found that the rate of formation of [$^{14}$C]phenylalanine from cinnamate is very much greater than can be explained by the reversal of the overall reaction (86). The same is seen with the enzymes from wheat (87) and a fungus (88). Such findings suggest the existence of an amino enzyme which participates in catalysis as follows:

$$E + phenylalanine \longrightarrow \textit{trans}\text{-cinnamate} + H^+ + amino\text{-}E \qquad (fast)$$

$$Amino\text{-}E + H^+ \longleftarrow E + ammonia \qquad (slow)$$

Also observed are a pair of pertinent "exchange" reactions. With the enzyme from maize, L-phenylalanine reacts with [$^{14}$C]p-hydroxycinnamate to generate [$^{14}$C]L-tyrosine; and L-tyrosine, reacting with [$^{14}$C]cinnamate, affords [$^{14}$C]L-phenylalanine (82). Clearly, the same amino-enzyme intermediate is formed from both tyrosine and phenylalanine.

Histidine ammonia-lyase [EC 4.3.1.3], which catalyzes the deamination of L-histidine to *trans*-urocanate,

parallels phenylalanine ammonia-lyase in many of its catalytic properties. Again, an amino-enzyme is thought to figure in the reaction (89), and a similar, if not the same, prosthetic group is believed to mediate the catalysis (90). Through magnetic resonance studies using $Mn^{2+}$, an enzyme–metal–histidine bridge complex is detectable (91). Also supportive of a role for a metal ion is the inhibition of ammonia-lyase activity by EDTA, and its restoration by the addition of $Mn^{2+}$, $Co^{2+}$, $Zn^{2+}$, $Ni^{2+}$, $Mg^{2+}$, or $Ca^{2+}$ (92).

(85) N. E. Dixon, C. Gazzola, R. L. Blakeley, and B. Zerner, *Science* 191, 1144–1150 (1976); K. R. Hanson and E. A. Havir, *Arch. Biochem. Biophys.* 180, 102–113 (1977).
(86) E. A. Havir and K. R. Hanson, *B* 7, 1904–1914 (1968).
(87) M. R. Young and A. C. Neish, *Phytochemistry* 5, 1121–1132 (1966).
(88) P. V. SubbaRao, K. Moore, and G. H. N. Towers, *Can. J. Biochem.* 45, 1863–1872 (1967).
(89) A. Peterkofsky, *JBC* 237, 787–795 (1962).
(90) T. L. Givot, T. A. Smith, and R. H. Abeles, *JBC* 244, 6341–6453 (1969); R. B. Wickner, *JBC* 244, 6550–6552 (1969).
(91) I. L. Givot, A. S. Mildvan, and R. H. Abeles, *Fed. Proc.* 29, 531 (1970).
(92) T. Shibatani, T. Kakimoto, and I. Chibata, *EJB* 55, 263–269 (1975).

**Table 5.1.** Lyases Known to Act by Covalent Catalysis

| EC no. | Familiar name of enzyme[a] | Criteria[b] | References |
|--------|----------------------------|-------------|------------|
| 4.1.1.1 | Pyruvate decarboxylase [TPP] | I | (93) |
| 4.1.1.4 | Acetoacetate decarboxylase | S | (94) |
| 4.1.1.12 | Aspartate $\beta$-decarboxylase [PLP] | I[c], M | (95) |
| 4.1.1.15 | Glutamate decarboxylase [PLP] | M | (96) |
| 4.1.1.17 | Ornithine decarboxylase [PLP] | M | (97) |
| 4.1.1.19 | Arginine decarboxylase [PLP] | M | (98) |
| 4.1.1.22 | Histidine decarboxylase [pyruvyl] | S | (99) |
| 4.1.1.25 | Tyrosine decarboxylase [PLP] | S | (100) |
| 4.1.1.28 | Aromatic-L-amino-acid decarboxylase [PLP] | I[c], M | (101) |
| 4.1.1.32 | Phosphoenolpyruvate carboxykinase [$Mn^{2+}$] | E, Ma | (102) |
| 4.1.1.35 | UDPglucuronate decarboxylase [NAD] | M | (103) |
| 4.1.1.47 | Glyoxylate carbo-ligase [TPP] | I | (104) |
| 4.1.1.64 | Dialkylamino-acid decarboxylase (pyruvate) [PLP] | I | (105) |
| 4.1.1.65 | Phosphatidylserine decarboxylase [pyruvate] | M | (106) |
| 4.1.1. | UDP-D-apiose synthase [NAD] | M | (107) |
| 4.1.2.4 | Deoxyribose-phosphate aldolase | S | (108) |
| 4.1.2.9 | Phosphoketolase [TPP] | I | (109) |
| 4.1.2.13 | Fructose-diphosphate aldolase—class I | S | (110) |
| 4.1.2.13 | Fructose-diphosphate aldolase—class II [$Zn^{2+}$ or $Mn^{2+}$] | Ma | (111) |
| 4.1.2.14 | Phospho-2-keto-3-deoxygluconate aldolase | S | (112) |
| 4.1.3.3 | N-Acetylneuraminate lyase | S | (113) |
| 4.1.3.5 | Hydroxymethylglutaryl-CoA lyase | I, E, K, B | (114) |
| 4.1.3.6 | Citrate lyase [a derivative of dephospho-CoA] | I[d] | (115) |
| 4.1.3.8 | ATP citrate lyase | I[e], E, K | (116) |
| 4.1.3.16 | 4-Hydroxy-2-ketoglutarate aldolase | S | (117) |
| 4.1.3.17 | 4-Hydroxy-4-methyl-2-oxoglutarate aldolase [$Mg^{2+}$] | E, M | (118) |
| 4.1.3.18 | Acetolactate synthase [TPP] | M | (119) |
| 4.1.3.22 | Citramalate lyase [a derivative of dephospho-CoA] | I[d] | (120) |
| 4.1.3.27 | Anthranilate synthase | M, G | (121) |
| 4.1.3. | p-Aminobenzoate synthase | M, G | (122) |
| 4.1.99.1 | Tryptophanase [PLP] | I[c], K | (123) |
| 4.1.99.2 | Tyrosine phenol-lyase [PLP] | I[c] | (124) |
| 4.2.1.1 | Carbonic anhydrase [$Zn^{2+}$] | M | (125) |
| 4.2.1.3 | Aconitase [$Fe^{2+}$] | M, Ma | (126) |
| 4.2.1.10 | 3-Dehydroquinate dehydratase | S | (127) |
| 4.2.1.11 | Enolase [$Mg^{2+}$] | M, Ma | (128) |
| 4.2.1.14 | D-Serine dehydratase [PLP] | M, S | (129) |
| 4.2.1.16 | Threonine dehydratase [PLP or $\alpha$-ketobutyryl] | I[c], M | (130) |
| 4.2.1.20 | Tryptophan synthase [PLP] | I[c], M | (131) |
| 4.2.1.24 | Aminolevulinate dehydratase | S | (132) |
| 4.2.1.28 | Propanediol dehydratase [cobalamin] | I, M | (133) |
| 4.2.1.30 | Glycerol dehydratase [cobalamin] | M | (134) |
| 4.2.1.41 | 5-Keto-4-deoxy-D-glucarate dehydratase | S | (135) |
| 4.2.1.43 | 2-Keto-3-deoxy-L-arabonate dehydratase | S | (136) |
| 4.2.1.46 | dTDPglucose 4,6-dehydratase [NAD] | I, M | (137) |

**Table 5.1.** Lyases Known to Act by Covalent Catalysis *(Continued)*

| EC no. | Familiar name of enzyme[a] | Criteria[b] | References |
|---|---|---|---|
| 4.2.1.52 | Dihydropicolinate synthase | S | (138) |
| 4.2.99.8 | Cysteine synthase [PLP] | I[c], E, K, M | (139) |
| 4.2.99.9 | Cystathionine γ-synthase [PLP] | I[c], E, M | (140) |
| 4.3.1.3 | Histidine ammonia-lyase [dehydroalanine or a metal] | E, M, Ma | (141) |
| 4.3.1.5 | Phenylalanine ammonia-lyase [dehydroalanine or a metal] | E, K | (142) |
| 4.3.1.7 | Ethanolamine ammonia-lyase [cobalamin] | I | (143) |
| 4.3.1.12 | Ornithine cyclase (deaminating) [NAD] | I[c] | (144) |
| 4.4.1.1 | Cystathionine γ-lyase [PLP] | M | (145) |
| 4.4.1.5 | Glyoxalase I [$Zn^{2+}$, $Mg^{2+}$] | M | (146) |
| 4.4.1.9 | β-Cyanoalanine synthase [PLP] | E, M | (147) |
| 4.4.1.10 | Cysteine lyase [PLP] | S | (148) |

[a] Any parenthetical expression is part of the official name of the enzyme. Prosthetic groups are indicated in brackets. Abbreviations: TPP, thiamine diphosphate; PLP, pyridoxal-5′-P.

[b] The symbols mean the following:

I, the holoenzyme links covalently with the substrate or a fragment of it to form a chemically competent intermediate;

E, the enzyme catalyzes one or more exchange reactions consistent with the participation of a covalent enzyme–substrate intermediate;

K, the enzyme exhibits kinetic properties (exclusive of "burst" kinetics) consistent with the participation of a covalent enzyme–substrate intermediate;

B, the enzyme displays "burst" kinetics;

M, miscellaneous data and derivative arguments which are peculiar to the enzyme in question;

Ma, magnetic resonance studies show that this enzyme binds a substrate molecule by coordinate linkage to its prosthetic group—a metal ion;

S, the holoenzyme forms a Schiff base with substrate;

G, the enzyme is irreversibly inactivated for glutamine utilization by stoichiometric alkylation with a glutamine analogue; inferred from this is the participation of a glutamyl-enzyme intermediate.

[c] Identified spectroscopically.

[d] This enzyme is isolated from tissue as the covalent enzyme–substrate intermediate; that is, an acetyl group is already fixed to a sulfhydryl of the holoenzyme.

[e] ATP citrate lyase, a multisubstrate enzyme, catalyzes a reaction mediated by three distinct covalent enzyme–substrate complexes: a phosphoenzyme, a citryl enzyme, and an acetyl enzyme (see p. 168 of the text).

(93) Refs. (1–3); L. O. Krampitz, *Ann. Rev. Biochem.* 38, 213–340 (1969).

(94) Refs. (4–8).

(95) A. Novogrodsky, J. S. Nishimura, and A. Meister, *JBC* 238, PC1903–1905 (1963); S. S. Tate and A. Meister, *Adv. Enzymol.* 35, 503–543 (1971); N. M. Relyea, S. S. Tate, and A. Meister, *JBC* 249, 1519–1524 (1974).

(96) Refs. (9–13).

(97) M. H. O'Leary and R. M. Herreid, *B* 17, 1010–1014 (1978).

(98) S. L. Blethen, E. A. Boeker, and E. E. Snell, *JBC* 243, 1671–1677 (1968).

(99) Refs. (14, 15).

(100) J. C. Vederas, I. D. Reingold, and H. W. Sellers, *JBC* 254, 5053–5057 (1979).

(101) A. Fiori, C. Turano, C. Borri Voltattorni, A. Minelli, and M. Codini, *FEBS lett.* 54, 122–125 (1975); M. H. O'Leary and R. L. Baugn, *JBC* 252, 7168–7173 (1977); A. L. Maycock, S. D. Aster, and A. A. Patchett, *B* 19, 709–718 (1980).

(102) Refs. (16–19).

(103) J. S. Schutzbach and D. S. Feingold, *JBC* 245, 2476–2482 (1970).

(104) L. Jaenicke and J. Koch, *Biochem. Z.* 336, 432–443 (1962); G. Kohlhaw, B. Deus, and H. Holzer, *JBC* 240, 2135–2141 (1965).

(105) G. B. Bailey, T. Kusamrarn, and K. Vuttivej, *Fed. Proc.* 29, 857 (1970).

(106) M. Satre and E. P. Kennedy, *JBC* 253, 479–483 (1978).

(107) W. J. Kelleher and H. Grisebach, *EJB* 23, 136–142 (1971); O. Gabriel, in *Carbohydrates in Solution*, H. S. Isbell, ed., Amer. Chem. Soc., 1973, p. 387; D. Baron and H. Grisebach, *EJB* 38, 153–159 (1973); C. Gebb, D. Baron, and H. Grisebach, *EJB* 54, 493–498 (1975).

(108) E. Grazi, H. Meloche, G. Martinez, W. A. Wood, and B. L. Horecker, *Biochem. Biophys. Res. Commun.* 10, 4–10 (1963); O. M. Rosen, P. Hoffee, and B. L. Horecker, *JBC* 240, 1517–1524 (1965); P. A. Hoffee, *Arch. Biochem. Biophys.* 126, 795–802 (1968).

(109) M. L. Goldberg and E. Racker, *JBC* 237, 3841–3842 (1962); W. Schröter and H. Holzer, *BBA* 77, 474–481 (1963); R. Votaw, W. T. Williamson, L. O. Krampitz, and W. A. Wood, *Biochem. Z.* 338, 756–762 (1963).

(110) Refs. (20, 22–28).

(111) Refs. (21, 29, 30); A. S. Mildvan, R. D. Kobes, and W. J. Rutter, *B* 10, 1191–1204 (1971).

(112) First citation of Ref. (108); H. P. Meloche, *JBC* 248, 6945–6951 (1973).

(113) G. H. DeVries and S. B. Binkley, *Arch. Biochem. Biophys.* 151, 234–242 (1972); J. E. G. Barnett, D. L. Corina, and G. Rasool, *Biochem. J.* 125, 275–285 (1971).

(114) Refs. (31–35).

(115) Refs. (36–41).

(116) Refs. (44–50).

(117) R. G. Rosso and E. Adams, *JBC* 242, 5524–5534 (1967); R. D. Kobes and E. E. Decker, *B* 10, 388–395 (1971); B. A. Hansen, R. S. Lane, and E. E. Decker, *JBC* 249, 4891–4896 (1974).

(118) A. Marcus and L. M. Shannon, *JBC* 237, 3348–3353 (1962).

(119) H. Holzer and G. Kohlhaw, *Biochem. Biophys. Res. Commun.* 5, 452–456 (1961).

(120) W. Buckel, *Z. Physiol. Chem.* 356, 223–224 (1975); W. Buckel and A. Bobi, *EJB* 64, 255–262 (1976).

(121) H. Nagano, H. Zalkin, and E. J. Henderson, *JBC* 245, 3810–3820 (1970); H. Tamir and P. R. Srinivasan, *JBC* 247, 1153–1155 (1972); Y. Goto, H. Zalkin, P. S. keim, and R. L. Heinrikson, *JBC* 251, 941–949 (1976); M. Kawamura, P. S. Keim, Y. Goto, H. Zalkin, and R. L. Heinrikson, *JBC* 253, 4659–4668 (1978).

(122) P. R. Srinivasan, in *The Enzymes of Glutamine Metabolism*, S. Prusiner and E. R. Stadtman, eds., Academic Press, New York, 1973, pp. 545–568.

(123) Refs. (51–53).

(124) H. Kumagai, H. Yamada, H. Matsui, H. Ohkishi, and K. Ogata, *JBC* 245, 1773–1777 (1970); H. Yamada, H. Kumagai, N. Kashima, and H. Torii, *Biochem. Biophys. Res. Commun.* 46, 370–374 (1972); T. Muro, H. Nakatani, K. Hiromi, H. Kumagai, and H. Yamada, *J. Biochem. (Tokyo)* 84, 633–640 (1978).

(125) Refs. (59–61).

(126) I. A. Rose and E. L. O'Connell, *JBC* 242, 1870–1879 (1967); J. P. Glusker, *JMB* 38, 149–162 (1968); J. J. Villafranca and A. S. Mildvan, *JBC* 247, 3454–3463 (1972).

(127) Ref. (63).

(128) Refs. (68–71).

(129) W. Dowhan, Jr., and E. E. Snell, *JBC* 245, 4629–4635 (1970); I. Y. Yang, Y. Z. Huang, and E. E. Snell, *Fed. Proc.* 34, 496 (1975); Y. F. Cheung and C. T. Walsh, *JACS* 98, 3397–3398 (1976); K. D. Schnackerz, J. H. Ehrlich, W. Giesemann, and T. A. Reed, *B* 18, 3557–3563 (1979).

(130) M. Tokushiga, A. Nakazawa, Y. Shizuta, Y. Okada, and O. Hayaishi, in *Symposium on Pyridoxal Enzymes*, K. Yamada, N. Katunuma, and H. Wada, eds., Maruzen, Tokyo, 1968, p. 105; R. A. Niederman, K. W. Rabinowitz, and W. A. Wood, *Biochem. Biophys. Res. Commun.* 36, 951–956 (1969); Y. Shizuta, A. Kurosawa, T. Tanabe, K. Inoue, and O. Hayaishi, *JBC* 248, 4213–4219 (1973); K. W. Rabinowitz, R. A. Niederman, and W. A. Wood, *JBC* 248, 8207–8215 (1973); G. Kapke and L. Davis, *B* 15, 3745–3749 (1976).

(131) M. E. Goldberg and R. L. Baldwin, *B* 6, 2113–2119 (1967); E. W. Miles, M. Hatanaka, and I. P. Crawford, *B* 7, 2742–2753 (1968); E. W. Miles, *Biochem. Biophys. Res. Commun.* 66, 94–102 (1975); M.-D. Tsai, E. Schleicher, R. Potts, G. E. Skye, and H. G. Floss, *JBC* 253, 5344–5349 (1978).

(132) D. L. Nandi and D. Shemin, *JBC* 243, 1236–1242 (1968).

(133) Refs. (75–81).

(134) B. Zagalak, *Bull. Acad. Pol. Sci., Ser. Sci., Biol.* 16, 67 (1968); R. H. Abeles, *The Enzymes*, 3rd ed., P. D. Boyer, ed., Academic Press, New York, 1971, Vol. 5, p. 485.

(135) R. Jeffcoat, H. Hassall, and S. Dagley, *Biochem. J.* 115, 977–983 (1969).

(136) D. Portsmouth, A. C. Stoolmiller, and R. H. Abeles, *JBC* 242, 2751–2759 (1967).

(137) S. F. Wang and O. Gabriel, *JBC* 245, 8–14 (1970); C. E. Snipes, G.-U. Brillinger, L. Sellers, L. Mascaro, and H. G. Floss, *JBC* 252, 8113–8117 (1977).

(138) J. G. Shedlarski and C. Gilvarg, *JBC* 245, 1362–1373 (1970).

(139) M. A. Becker, N. M. Kredich, and G. M. Tompkins, *JBC* 244, 2418–2427 (1969); P. F. Cook and R. T. Wedding, *JBC* 251, 2023–2029 (1976); H. G. Floss, E. Schleicher, and R. Potts, *JBC* 251, 5478–5482 (1976); P. F. Cook and R. T. Wedding, *JBC* 252, 3459 (1977).

(140) M. M. Kaplan and M. Flavin, *JBC* 241, 5781–5789 (1966); S. Guggenheim and M. Flavin, *BBA* 151, 664–669 (1968); S. Nagai and M. Flavin, *JBC* 242, 3884–3895 (1967); M. Flavin, in *Metabolism of Sulfur Compounds*, D. M. Greenberg, ed., Academic Press, New York, 1975, pp. 469–473.

(141) Refs. (90, 91); K. R. Hanson and E. A. Havir, in *The Enzymes*, 3rd ed., P. D. Boyer, ed., Academic Press, New York, 1972, Vol. 7, pp. 137–166.

(142) Refs. (82–88).

(143) B. Babior, *BBA* 167, 456–458 (1968); B. M. Babior, *JBC* 245, 6125–6133 (1970); T. J. Carty, B. M. Babior, and R. H. Abeles, *JBC* 249, 1683–1688 (1974); K. Sato, J. C. Orr, B. M. Babior, and R. H. Abeles, *JBC* 251, 3734–3737 (1976); S. W. Graves, J. S. Krouwer, and B. M. Babior, *JBC* 255, 7444–7448 (1980).

(144) W. L. Muth and R. N. Costilow, *JBC* 249, 7463–7467 (1974).

(145) M. Krongelb, T. A. Smith, and R. H. Abeles, *BBA* 167, 474–475 (1968); M. Flavin and C. Slaughter, *JBC* 244, 1434–1444 (1969); F. C. Brown, W. R. Hudgins, and J. A. Roszell, *JBC* 244, 2809–2815 (1969); R. H. Abeles and C. T. Walsh, *JACS* 95, 6124–6125 (1973); R. B. Silverman and R. H. Abeles, *B* 16, 5515–5520 (1977).

(146) S. S. Hall, A. M. Doweyko, and F. Jordan, *JACS* 98, 7460–7461 (1976).

(147) H. R. Hendrickson and E. E. Conn, *JBC* 244, 2632–2640 (1969).

(148) E. A. Tolosa, E. V. Goryachenkova, L. N. Stesina, and A. E. Braunshtein, *Mol. Biol.* 1, 610–616 (1967) (Engl. transl.).

# Chapter 6
# Isomerases

Isomerases, as the name implies, catalyze chemical changes within one and the same molecule. These changes are of different sorts, the (seemingly) simplest being the transfer of a proton from one position to another in the same molecule, as, for instance, in the reaction catalyzed by steroid $\Delta$-isomerase. In such cases, the migrating proton is believed to be covalently linked to the enzyme at one or more stages of its transit; which is not to exclude the possibility that some other part(s) of the substrate molecule may be similarly linked during catalysis. The change in the position of a hydrogen may be a steric one, as with the racemases and epimerases, or part of an intramolecular redox reaction, as with the sugar isomerases. Other isomerases catalyze the rearrangement of larger atomic groupings. A sampling of the different kinds of enzymic isomerizations is given in this chapter. Next to the ligases, the isomerases form the least numerous of the six major classes of enzymes (cf. Table 8.1, p. 226). Listed in Table 6.1 at the end of the present chapter are 35 isomerases that almost certainly have a covalent enzyme–substrate intermediate as a component of the catalytic cycle. They comprise about 35% of the isomerases recognized by the Enzyme Commission. By chemical analogy, this extrapolates to nearly 90% of all the known isomerases (cf. Chapter 8).

## Alanine Racemase [EC 5.1.1.1]

This pyridoxal-phosphate-requiring enzyme catalyzes the racemization of D- and L-alanine.

$$\text{D-Alanine} \longleftrightarrow \text{L-alanine}$$

Like other reactions that make coenzymic use of pyridoxal phosphate, the racemization process is thought to begin with the union of substrate and

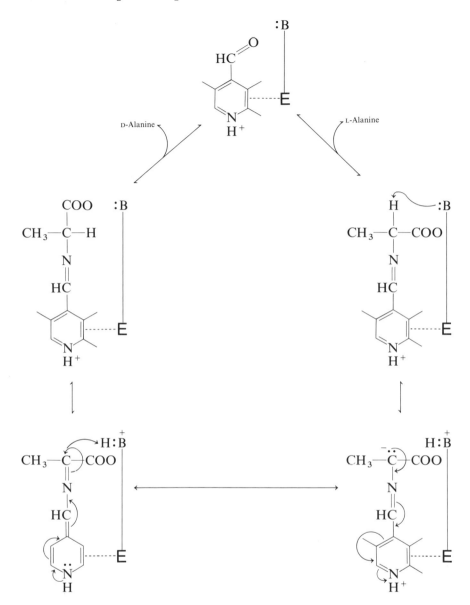

**Fig. 1.** A mechanism for the racemization of D- and L-alanine by alanine racemase. Destruction of the chiral center follows loss of the α-proton of alanine (as a Schiff base with the holoenzyme) to a basic group in the active site. The removed proton is exchangeable with the medium. Regain of the proton on to either side of the planar intermediate restores chirality with racemization.

prosthetic group as a Schiff base (Fig. 1) (1). Though the reasonableness and high probability of this mechanism is generally accepted, direct evidence for it has not been forthcoming through the use of alanine racemase itself. It happens, however, that tyrosine phenol-lyase [EC 4.1.99.2], which catalyzes a number of α, β-elimination and β-replacement reactions of tyrosine, also catalyzes the racemization of D- or L-alanine, acting in this way as an alanine racemase (2). The addition of D-alanine or L-alanine to the holoenzyme of tyrosine phenol-lyase (*E. intermedia*) results in the appearance of a prominent spectral band at about 500 nm. A similar band is familiar in the action of other pyridoxal-phosphate-dependent enzymes (cf., for instance, trypto-phanase, p. 170), and is ascribed to the deprotonated Schiff base formed between substrate and the coenzyme (3). Consonant with such deprotonation is the exchange of the α (but not the β) hydrogen of L- or D-alanine in $D_2O$ in the presence of the alanine racemase of *B. subtilis* (4). A basic group on the enzyme is believed to abstract the α-proton and exchange it with the medium. Alanine racemase (*E. coli*) is irreversibly inhibited by the "suicide" substrates, D- and L-β-fluoroalanine and D- and L-β-chloroalanine. The facts sur-rounding the inhibition by these four inhibitors are such as to suggest a common intermediate in their action; namely, the eneamino-acid–pyridoxal-phosphate–enzyme complex, formed in the β-elimination of halide from the Schiff base between holoenzyme and inhibitor (5).

## Proline Racemase [EC 5.1.1.4]

The interconversion of the D- and L-isomers of proline

is catalyzed enzymically without the aid of pyridoxal phosphate. The enzyme is powerfully inhibited by the planar molecule pyrrol-2-carboxylate,

(1) J. Olivard, D. E. Metzler, and E. E. Snell, *JBC* 199, 669–674 (1952); D. E. Metzler, M. Ikawa, and E. E. Snell, *JACS* 76, 648–652 (1954).
(2) H. Kumagai, N. Kashima, and H. Yamada, *Biochem. Biophys. Res. Commun.* 39, 796–801 (1970).
(3) Y. Morino and E. E. Snell, *JBC* 242, 2800–2809 (1967).
(4) U. M. Babu and R. B. Johnston, *Biochem. Biophys. Res. Commun.* 58, 460–466 (1974).
(5) E. Wang and C. Walsh, *B* 17, 1313–1321 (1978).

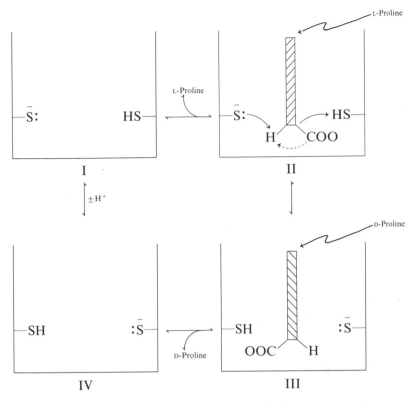

**Fig. 2.** Mechanism of action of proline racemase, showing the two species of protonated enzyme, I and IV. Species I, which received its proton from D-proline (via III → II → I), can react with L-proline but not with D-proline. Species IV, which received its proton from L-proline (via II → III → IV), can react with D-proline but not with L-proline. Adapted from (7).

suggesting a planar transition state for proline during racemization (6). It hints, too, at the possibility that the steric inversion is effected by a pair of bases in the active center operating on opposite sides of the proline ring. At any one time only one of the bases is protonated, and the proton is exchangeable with solvent. As one base removes the α-hydrogen as a proton, the second (protonated) base donates its proton to the opposite side of the α-carbon (Fig. 2).

(6) G. J. Cardinale and R. H. Abeles, *B* 7, 3970–3978 (1968).

Consonant with this picture of the inversion are several experimental facts. Comparison of the kinetics of inversion with the kinetics of deuterium incorporation from $D_2O$ in the first few percent (that is, the initial phase) of the reaction reveals that the rates are the same, regardless of which isomer is used as substrate. It is also clear that when L-proline is the substrate the first product formed is deutero-D-proline. Deutero-L-proline does not appear in the earliest phase of the reaction. The converse holds when D-proline is the substrate; only deutero-L-proline is the initial product (6).

The two-base inversion mechanism implies the existence of two species of the protonated enzyme (I and IV, Fig. 2). The species which received its proton from L-proline can react only with D-proline, and the one which received its proton from D-proline can react only with L-proline. Equilibration of the two protonated species occurs only by the rapid exchange of enzyme-bound protons with solvent, and such exchange seems to be possible only when the active site is empty of proline. Thus, the enzymic release of tritium to the solvent from L-$[\alpha\text{-}^3H]$proline is unaffected as the concentration of L-proline is increased, but the release from DL-$[\alpha\text{-}^3H]$proline is depressed under the same conditions. The product isomer is evidently responsible for the inhibition of tritium release; or stated otherwise, "the enzyme-bound proton derived from one isomer can only be captured by the other isomer" (7).

The bases in the active center which alternate as proton donor and acceptor are believed to be a pair of sulfhydryl groups, present as thiol and thiolate in the active enzyme (7). Iodoacetate alkylates two sulfhydryl groups per binding site, but the alkylation of only one suffices to inactivate the enzyme. Proline and pyrrol-2-carboxylate protect the enzyme from inactivation by iodoacetate. How the proton on the $\alpha$-carbon of proline is activated for transfer to the enzyme is still unclear. It is surmised, however, that the loss of asymmetry at the $\alpha$-carbon requires the participation of the adjacent carboxyl group in some way as yet undetermined. Possibly a metal ion provides a coordination site for the carboxyl group (6).


# Phenylalanine Racemase [EC 5.1.1.11]

Yet another kind of amino acid racemization is evident in the conversion of L-phenylalanine into D-phenylalanine at an early stage in the biosynthesis of the antibiotics gramicidin S and tyrocidine (8, 9). The isomerization

(7) G. Rudnick and R. H. Abeles, *B* 14, 4515–4522 (1975).
(8) W. Gevers, H. Kleinkauf, and F. Lipmann, *PNAS* 63, 1335–1342 (1969).
(9) K. Kurahashi, M. Yamada, K. Mori, K. Fujikawa, M. Kambe, Y. Imae, E. Sato, H. Takahashi, and Y. Sakamoto, *Cold Spring Harbor Symp. Quant. Biol.* 34, 815–826 (1969).

requires no prosthetic group, but it does require the prior activation of the amino acid at the expense of ATP.

$$E + \text{L-phenylalanine} + ATP \longleftrightarrow E \cdot \text{L-phenylalanyl} \sim AMP + PP_i$$

$$E \cdot \text{L-phenylalanyl} \sim AMP \longleftrightarrow E \sim \text{L-phenylalanyl} + AMP \qquad (1)$$

$$E \sim \text{L-phenylalanyl} \longleftrightarrow E \sim \text{D-phenylalanyl} \qquad (2)$$

Sum:  $E + \text{L-phenylalanine} + ATP \longleftrightarrow E \sim \text{D-phenylalanyl} + AMP + PP_i$

Once formed, the isomeric D-phenylalanyl residue—covalently linked to the enzyme—is poised for incorporation into the polypeptide chain of the antibiotics.

In common with other amino-acid–activating enzymes phenylalanine racemase catalyzes an ATP–$PP_i$ exchange in the presence of L- or D-phenylalanine, pointing to the noncovalently enzyme-bound phenylalanyl $\sim$ AMP as an intermediate of the reaction. In addition, the enzyme catalyzes a phenylalanine-dependent ATP–AMP exchange, implying the release of the phenylalanyl group from linkage to AMP and its transfer to some site on the enzyme in which the energy-rich bonding is maintained (Eq. 1) (10). The site is believed to be a sulfhydryl group, and the bond a thiolester (8). When phenylalanine racemase is incubated with L- or D-phenylalanine along with ATP and $Mg^{2+}$, gel filtration of the mixture yields an enzyme–substrate complex in which half of the phenylalanine is present as phenylalanyl $\sim$ AMP and the rest as phenylalanyl $\sim$ enzyme (Eq. 1). It is the phenylalanyl $\sim$ enzyme which undergoes racemization (Eq. 2) (11).

In acting upon L-[$\alpha$-$^3$H]phenylalanine, the racemase transfers some of the tritium into the $\alpha$-position of the product D-phenylalanyl $\sim$ enzyme (11). Continued action of the enzyme after optical equilibrium is reached results ultimately in the total loss of tritium to the solvent. The exchange of the $\alpha$-proton with solvent is therefore much slower than the isomerization. Also notable is the more rapid loss of $\alpha$-$^3$H from D- than from L-phenylalanine. Thus, while proline racemase uses a two-base catalytic cycle (Fig. 2), the action of phenylalanine racemase is more consistent with the operation of a single-base mechanism (Fig. 3). Activation of the amino acid by thio-esterification to the enzyme opens the way to loss of asymmetry at the $\alpha$-carbon atom through formation of a planar intermediate. Flipping over in place enables the intermediate to present opposite sides of its plane to the protonated base for retrieval of the proton, accounting thus for the isomerization.

(10) H. Kleinkauf, W. Gevers, and F. Lipmann, *PNAS* 62, 226–233 (1969).

(11) S. G. Lee and F. Lipmann, personal communication.

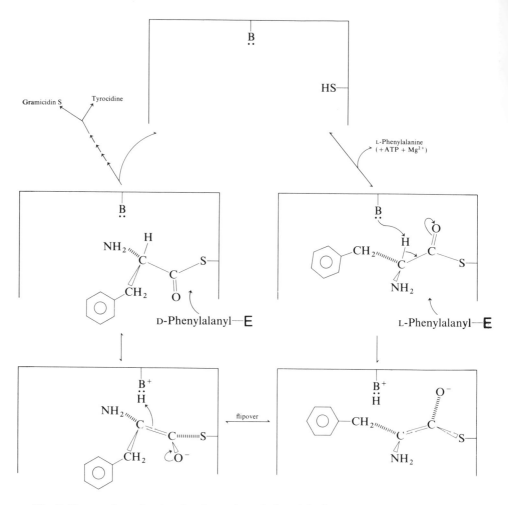

**Fig. 3.** Proposed mechanism for the action of phenylalanine racemase. Racemization occurs while phenylalanine is thioesterified to the enzyme.

## Mandelate Racemase [EC 5.1.2.2]

The reversible transformation of D-mandelate into L-mandelate entails steric inversion at a carbon atom bearing a hydroxyl group.

Mandelate racemase has absolute need of a divalent cation—$Mg^{2+}$ is best—as a prosthetic group in the reaction (12). Concurrent with racemization is the exchange of the α-proton of mandelate with protons of the medium (13). But, as is true of phenylalanine inversion (see above), the rate of isomerization is greater than the rate of exchange, suggesting an "intramolecular" transfer of hydrogen; which is to say that the proton is substantially shielded from solvent during isomerization. A one-base catalytic mechanism suggests itself, in which the α-proton is removed and returned at opposite sides of the α-carbon of mandelate (14).

A clue as to what the proton-removing group of the enzyme may be comes from the irreversible inhibition of the enzyme by α-phenylglycidate, a close structural analogue of mandelate (15).

α-Phenylglycidate

The inhibition, like the racemization process itself, requires $Mg^{2+}$, and one molecule of inhibitor binds covalently to each of the four subunits of the enzyme. D- or L-Mandelate protects the enzyme from inactivation. The inhibitor can be stripped from the protein rapidly with hydroxylamine and more slowly by hydrolysis at pH 9. In the hydrolysis, some of the inhibitor is recovered as α-phenylglyceric acid. These results tend to the conclusion that in the active site a carboxylate ion (the β-carboxyl of aspartate or the γ-carboxyl of glutamate) acts as the proton-removing base during the inversion of mandelic acid.

As a prosthetic group, the divalent cation acts to form a bridge between the enzyme and D- or L-mandelate (16). Within the complex so formed the mandelate is in the second coordination sphere of the metal, with an intervening water molecule liganded directly to the metal (Fig. 4). Such is the picture of the active ternary complex suggested by magnetic resonance measurements, in which, for technical necessity, $Mn^{2+}$ replaces $Mg^{2+}$. By polarizing the carboxylate group through hydrogen bonding or protonation, the inner sphere water could exert a loosening effect on the α-proton of mandelate. Transfer of that proton to the basic group of the enzyme coincides with adoption of a planar structure by the substrate. And if the basic group is a carboxylate ion swinging at the end of an aspartyl or a glutamyl chain, then return of the α-proton is conceivable to either side of the planar structure.

(12) J. A. Fee, G. D. Hegeman, and G. L. Kenyon, B 13, 2528–2532 (1974).
(13) G. L. Kenyon and G. D. Hegeman, B 9, 4036–4043 (1970).
(14) T. R. Sharp, G. D. Hegeman, and G. L. Kenyon, B 16, 1123–1128 (1977).
(15) J. A. Fee, G. D. Hegeman, and G. L. Kenyon, B 13, 2533–2538 (1974).
(16) E. T. Maggio, G. L. Kenyon, A. S. Mildvan, and G. D. Hegeman, B 14, 1131–1139 (1975).

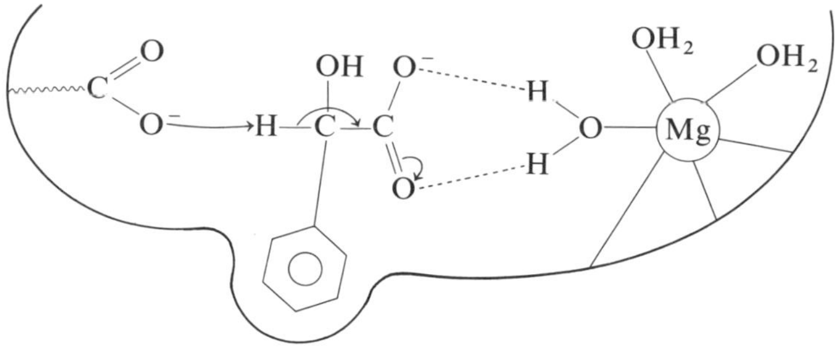

**Fig. 4.** Proposed role of the metal ion in activating the deprotonation of mandelate during racemase action. Adapted from (16).

## Uridine Diphosphoglucose Epimerase [EC 5.1.3.2]

Indispensable to the metabolism of D-galactose is its isomeric relationship to D-glucose, the two differing solely in their configuration at C-4. As the respective UDPglycosides, the sugars are reversibly interconverted through inversion at this carbon atom.

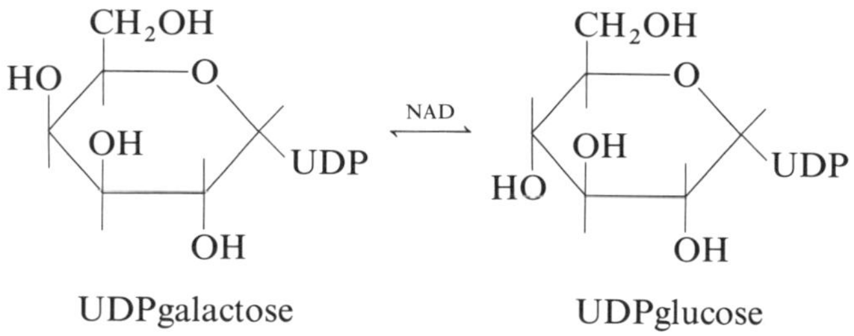

UDPgalactose                                       UDPglucose

Though it gives the appearance of a straightforward Walden inversion, the isomerization is, in its intimate details, the consequence of a cycle of redox reactions, with NAD mediating the electron transfers (17). While the epimerase from liver needs catalytic amounts of exogenous NAD for activity, the epimerases of yeast and *E. coli* have no such need, since they are isolated with NAD (1 mole per mole of enzyme) already firmly fixed in the active center.

Inversion at C-4 requires no oxygen or hydrogen exchange with the medium. Rather, the swap of steric spaces by the oxygen and by the carbon-bound hydrogen at C-4 is strictly intramolecular (18). It was early observed that the addition of substrate to the epimerase (yeast) causes the appearance of a spectral band at 345 nm, suggesting the formation of enzyme-bound NADH (19). This was later borne out by experiments in which enzyme

(17) E. S. Maxwell, *JBC* 229, 139–151 (1957).
(18) L. Glaser and L. Ward, *BBA* 198, 613–615 (1970).
(19) D. B. Wilson and D. S. Hogness, *JBC* 239, 2469–2481 (1964).

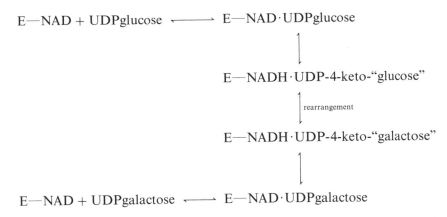

**Fig. 5.** The reversible isomerization of UDPglucose into UDPgalactose mediated by NAD bound to UDPglucose epimerase. Reduction of the holoenzyme by substrate yields a 4-keto sugar as an enzyme-bound intermediate, which manages somehow to regain its lost hydrogen on to the side of the pyranose ring opposite to the side from which it was lost.

(*E. coli*), incubated with TDP-[4-³H]glucose (an active analogue of UDP-glucose), becomes labeled with tritium in its attached NADH (20). The same experiment repeated with TDP-[3-³H]glucose causes no such labeling. Thus, proton and a pair of electrons transfer from C-4 of the substrate to the holoenzyme, leaving as intermediate the 4-keto form of the substrate adhering firmly to the enzyme (21) (Fig. 5). In some way which is as yet unclear, the intermediary 4-keto sugar rearranges on the enzyme in such wise as to yield TDPgalactose upon regain of the proton and electrons from the NADH.

## Triosephosphate Isomerase [EC 5.3.1.1]

Prominent among the reactions of the glycolytic pathway is the conversion of dihydroxyacetone phosphate into D-glyceraldehyde-3-P.

$$
\begin{array}{ccc}
\overset{H}{\underset{H}{}}C\text{—OH} & & HC\overset{\nearrow O}{} \\
| & & | \\
C{=}O & \longleftrightarrow & HCOH \\
| & & | \\
CH_2OP & & CH_2OP
\end{array}
$$

As an aldo-keto isomerization, the reaction is an intramolecular oxidoreduction, with the *pro*-R proton of C-1 and a pair of electrons moving reversibly

(20) H. M. Kalckar, in *Biochemistry of the Glycosidic Linkage*, R. Piras and H. G. Pontis, eds., Academic Press, New York, 1972, pp. 171–175; W. L. Adair, Jr., O. Gabriel, D. Ullrey, and H. M. Kalckar, *JBC* 248, 4635–4639 (1973).

(21) G. L. Nelsestuen and S. Kirkwood, *JBC* 246, 7533–7543 (1971); U. S. Maitra and H. Ankel, *PNAS* 68, 2660–2663 (1971); U. S. Maitra and H. Ankel, *JBC* 248, 1477–1479 (1973).

to C-2. Proton transfer occurs on the *re-re* face of the proposed *cis*-enediol intermediate (22, 23). A single basic group of the isomerase, positioned above the plane of the *cis*-enediol, is believed to shuttle the proton between the two carbon atoms (23, 24). While attached to the basic group of the enzyme, the proton is exchangeable with the solvent. Nonetheless, when the *pro*-R hydrogen on C-1 is replaced by tritium, some 3–6% of the tritium appears on C-2 of the product during isomerase action (24). Indications are that the basic group of the enzyme is the γ-carboxylate ion of a glutamyl residue; this because the active-site-directed inhibitors—halohydroxyacetone phosphates (25) and D- and L-glycidol phosphate (26)—all form an ester with the same glutamyl residue of the enzyme.

Enolization of dihydroxyacetone phosphate and D-glyceraldehyde-3-P is $10^9$ times faster in the presence of triosephosphate isomerase than in its absence (27). An electrophilic group in the active center is thought to have a part in the enolization by polarizing the carbonyl groups of the substrates and thus labilizing the α-protons. Consonant with this notion is the observation that dihydroxyacetone phosphate—in the presence of isomerase—is reduced by sodium borohydride in a manner completely stereoselective, and at a rate eightfold faster than in free solution (28).

## Steroid Δ-Isomerase [EC 5.3.3.1]

The enzymic transposition of a carbon–carbon double bond is a key step in the biosynthesis of steroid hormones. Rearrangement of the double bond is concurrent with a 1,3-proton shift, as exemplified by the transformation of $Δ^5$-androstene-3,17-dione into $Δ^4$-androstene-3,17-dione.

(22) S. V. Rieder and I. A. Rose, *JBC* 234, 1007–1010 (1959).

(23) I. A. Rose, *Brookhaven Symp. Biol.* 15, 293–309 (1962).

(24) J. M. Herlihy, S. G. Maister, W. J. Albery, and J. R. Knowles, *B* 15, 5601–5607 (1976).

(25) A. F. W. Coulson, J. R. Knowles, and R. E. Offerd, *J. Chem. Soc., Chem. Commun.* 7–8 (1970); F. C. Hartman, *B* 10, 146–154 (1971).

(26) I. A. Rose and E. L. O'Connell, *JBC* 244, 6548–6550 (1969); J. C. Miller and S. G. Waley, *Biochem. J.* 123, 163–170 (1971).

(27) A. Hall and J. R. Knowles, *B* 14, 4348–4352 (1975).

(28) M. R. Webb and J. R. Knowles, *Biochem. J.* 141, 589–592 (1974); *idem.*, *B* 14, 4692–4698 (1975); J. G. Belasco and J. R. Knowles, *B* 19, 472–477 (1980).

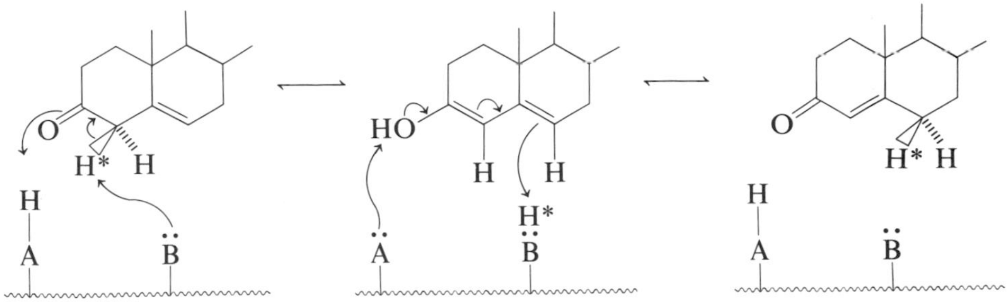

Fig. 6. Mechanism proposed for the action of steroid $\Delta$-isomerase. Adapted from (30).

This isomerization, as catalyzed by steroid $\Delta$-isomerase, is the fastest enzymic reaction known (29). The hydrogen transfer is stereospecific and intramolecular (30). An axial–axial transfer of the proton from C-4$\beta$ to C-6$\beta$ proceeds with but partial exchange of the proton with the medium (31). The stereochemistry of the reaction speaks for the presence in the active center of a single basic group which effects the suprafacial transfer of the proton (29, 30, 32). It is proposed, moreover, that an electrophilic or proton-donating function in the active center polarizes the C-3 carbonyl group, thereby labilizing the proton for removal to the enzyme and stabilizing the enolic intermediate (Fig. 6).

# Phosphoglycerate Mutase [EC 5.4.2.1]

The enzyme from wheat germ catalyzes the *intra*molecular transfer of a phosphoryl group in a reaction which is independent of 2,3-bisphospho-D-glycerate as a cofactor.

$$\text{2-P-D-glycerate} \longleftrightarrow \text{3-P-D-glycerate} \qquad (3)$$

This is in notable contrast to the phosphoglycerate mutase [EC 2.7.5.3] (from animal sources and yeast) which *requires* the cofactor and which catalyzes the *inter*molecular transfer of a phosphoryl group.

2,3-Bisphospho-D-glycerate + 2-P-D-glycerate $\longleftrightarrow$
3-P-D-glycerate + 2,3-bisphospho-D-glycerate

In this latter reaction the cofactor keeps the enzyme phosphorylated, in which condition it can phosphorylate one or the other of the monophospho-substrates (33).

(29) S.-F. Wang, F. S. Kawahara, and P. Talalay, *JBC* 238, 576–585 (1963).
(30) S. K. Malhotra and H. J. Ringold, *JACS* 87, 3228–3236 (1965); A Viger, S. Coustal, and A. Marquet, *JACS* 103, 451–458 (1981).
(31) P. Talalay and V. S. Wang, *BBA* 18, 300–301 (1955); S. B. Smith, J. W. Richards, and W. F. Benisek, *JBC* 255, 2685–2689 (1980).
(32) K. R. Hanson and I. A. Rose, *Acc. Chem. Res.* 8, 1–10 (1975).
(33) L. I. Pizer, *JACS* 80, 4431–4432 (1958); L. I. Pizer, *JBC* 235, 895–901 (1960); Z. B. Rose, *Arch. Biochem. Biophys.* 140, 508–513 (1970).

The cofactor-independent phosphoglycerate mutase catalyzes a reaction (Eq. 3) which is clearly *intra*molecular (34, 35), and which is therefore a true isomerase. But though *intra*molecular, reaction 3, like the cofactor-dependent reaction, proceeds over a pathway which includes a phosphoenzyme intermediate. Cyclic glycerate 2,3-phosphate, though an *a priori* reasonable intermediate in the phosphoryl transfer, plays no part whatever in the transfer (34, 36). This and the results of a variety of isotope and substrate analogue experiments converge on a phosphoenzyme as mediator of reaction 3 (36). Confirming this conclusion is the steric course of the *intra*molecular phosphoryl transfer, which is a net retention of configuration at the phosphorus atom (37), in accordance with double-displacement catalysis by this enzyme.

## DNA Topoisomerase [EC 5.4.2.-]

This enzyme catalyzes a topological isomerization when it removes superhelical turns from a closed circular DNA molecule. It does this by introducing a transient single-strand break into the DNA. The break (or "nick") allows the strands to rotate (or unwind) relative to the helical axis before the break is resealed. The nicking process requires that the enzyme join covalently to the phosphorus atom at the break. The enzyme (from rat liver) links to the 3'-phosphate end of the broken strand in the nicked intermediate (38). The intermediate (with covalently attached enzyme) can be trapped when the reaction is conducted at a high ratio of enzyme to DNA and then quenched by adjusting the pH to 12.5 or to 4.5 (39). The break is resealed upon reversal of the reaction, when the 5'-hydroxyl at the nick displaces the enzyme from the phosphorus as the phosphodiester bond is resynthesized. Resynthesis of the bond requires no expenditure of energy, since the latter is evidently conserved in the enzyme–phosphorus bond of the intermediate. The link to the enzyme is through the phenolic oxygen of a tyrosyl residue (40). A very similar nicking–closing activity is shown by the DNA topoisomerase of *E. coli* (41) and by the $\phi$X174 cistron A protein (42). But these enzymes, unlike the enzyme from rat liver, are reported to bond covalently at the 5'-phosphate side of the break (42, 43). Despite this difference the basic chemical mechanism of all nicking–closing enzymes is probably the same.

(34) H. G. Britton, J. Carreras, and S. Grisolia, *B* 10, 4522–4532 (1971).

(35) J. A. Gatehouse and J. R. Knowles, *B* 16, 3045–3053 (1977).

(36) R. Breathnach and J. R. Knowles, *B* 16, 3054–3060 (1977).

(37) W. A. Blättler and J. R. Knowles, *B* 19, 738–743 (1980).

(38) J. J. Champoux, *JMB* 118, 441–446 (1978).

(39) J. J. Champoux, *PNAS* 74, 3800–3804 (1977).

(40) Y. C. Tse, K. Kirkegaard, and J. C. Wang, *JBC* 255, 5560–5565 (1980).

(41) J. C. Wang *JMB* 55, 523–533 (1971).

(42) J.-E. Ikeda, A. Yudelevich, and J. Hurwitz, *PNAS* 73, 2669–2673 (1976); S. Eisenberg, J. Griffith, and A. Kornberg, *PNAS* 74, 3198–3202 (1977).

(43) R. E. Depew, L. F. Liu, and J. C. Wang, *Fed. Proc.* 35, 1493 (1976).

# D-α-Lysine Mutase [EC 5.4.3.4]

This mutase catalyzes the following isomerization:

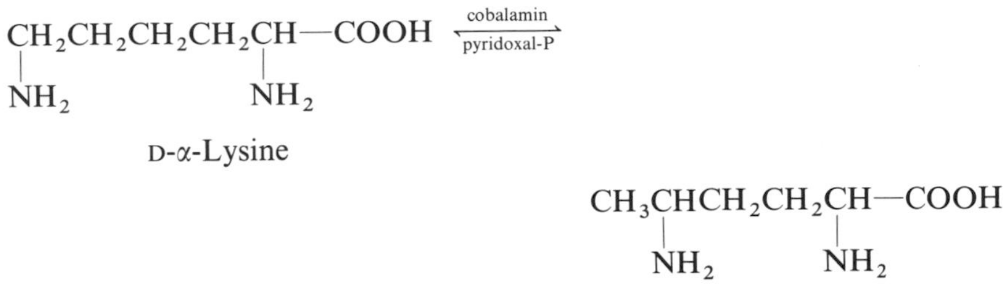

$$\underset{\underset{NH_2}{|}}{CH_2}CH_2CH_2CH_2\underset{\underset{NH_2}{|}}{CH}-COOH \quad \xrightarrow[\text{pyridoxal-P}]{\text{cobalamin}}$$

D-α-Lysine

$$CH_3\underset{\underset{NH_2}{|}}{CH}CH_2CH_2\underset{\underset{NH_2}{|}}{CH}-COOH$$

2,5-Diaminohexanoate

The enzyme has cobalamin as one of its coenzymes. In effect, a hydrogen on C-5 of lysine exchanges places stereospecifically with the amino group on C-6 to yield the isomeric product. As in other cobalamin-dependent reactions (see, for instance, propanediol dehydrase, p. 177), the cobalamin has the role of hydrogen carrier. Reversible transfer of hydrogen occurs between C-5 of lysine, C-5' of the coenzyme, and C-6 of 2,5-diaminohexanoate (44).

Lysine mutase, like the other aminomutases, also has a requirement for pyridoxal-P. It is suggested that the amino group migrates while bound as a Schiff base to the coenzyme (45).

**Table 6.1.** Isomerases Known to Act by Covalent Catalysis

| EC no. | Familiar name of enzyme[a] | Criteria[b] | References |
|--------|----------------------------|-------------|------------|
| 5.1.1.1 | Alanine racemase [PLP] | I[c], M | (46) |
| 5.1.1.4 | Proline racemase | M | (47) |
| 5.1.1.8 | Hydroxyproline epimerase | M | (48) |
| 5.1.1.9 | Arginine racemase [PLP] | M | (49) |
| 5.1.1.11 | Phenylalanine racemase (ATP-hydrolysing) | I | (50) |
| 5.1.2.1 | Lactate racemase | M | (51) |
| 5.1.2.2 | Mandelate racemase [$Mg^{2+}$] | M | (52) |
| 5.1.3.1 | Ribulosephosphate 3-epimerase | M | (53) |
| 5.1.3.2 | UDPglucose 4-epimerase [NAD] | I | (54) |
| 5.1.3.4 | L-Ribulosephosphate 4-epimerase [$Mn^{2+}$] | M | (55) |
| 5.2.1. | Maleylacetone cis-trans-isomerase [glutathione] | M | (56) |
| 5.3.1.1 | Triosephosphate isomerase | M | (57) |
| 5.3.1.3 | Arabinose isomerase [$Mn^{2+}$] | M | (58) |
| 5.3.1.4 | L-Arabinose isomerase [$Mn^{2+}$] | M | (59) |
| 5.3.1.5 | Xylose isomerase [$Mn^{2+}$] | M | (60) |
| 5.3.1.6 | Ribosephosphate isomerase | M | (61) |
| 5.3.1.8 | Mannosephosphate isomerase [$Zn^{2+}$] | M | (62) |
| 5.3.1.9 | Glucosephosphate isomerase | M | (63) |
| 5.3.1.10 | Glucosamine-6-phosphate isomerase | M | (64) |

**Table 6.1.** Isomerases Known to Act by Covalent Catalysis  *(Continued)*

| EC no. | Familiar name of enzyme[a] | Criteria[b] | References |
|---|---|---|---|
| 5.3.1. | Fucose isomerase | M | (65) |
| 5.3.3.1 | Steroid Δ-isomerase | M | (66) |
| 5.3.3.2 | Isopentenylpyrophosphate isomerase | M | (67) |
| 5.3.3.3 | Vinylacetyl-CoA Δ-isomerase | M | (68) |
| 5.3.3.7 | Aconitate Δ-isomerase | M | (69) |
| 5.3.3. | β-Carboxymuconolactone Δ-isomerase | M | (70) |
| 5.4.2.1 | Phosphoglycerate phosphomutase | E, M | (71) |
| 5.4.2. | DNA topoisomerase | I[d] | (72) |
| 5.4.2. | DNA gyrase | I[d] | (73) |
| 5.4.3.3 | β-Lysine 5,6-aminomutase [PLP, cobamide] | I, M | (74) |
| 5.4.3.4 | D-Lysine 5,6-aminomutase [PLP, cobamide] | I, M | (75) |
| 5.4.3.5 | D-Ornithine 4,5-aminomutase [PLP, cobamide] | M | (76) |
| 5.4.99.1 | Methylaspartate mutase [cobamide] | M | (77) |
| 5.4.99.2 | Methylmalonyl-CoA mutase [cobamide] | I, M | (78) |
| 5.4.99.4 | 2-Methylene-glutarate mutase [cobamide] | I, M | (79) |
| 5.5.1.4 | *myo*-Inositol-1-phosphate synthase [NAD] | M | (80) |

[a] Any parenthetical expression is part of the official name of the enzyme. Prosthetic groups are indicated in brackets. Abbreviation: PLP, pyridoxal-5'-P.

[b] The symbols mean the following:

   I, the holoenzyme links covalently with the substrate or some fragment of it to form a chemically competent intermediate;

   M, miscellaneous data and derivative arguments which are peculiar to the enzyme in question;

   E, the enzyme catalyzes an exchange reaction consistent with the participation of a covalent enzyme–substrate intermediate.

[c] Identified spectroscopically.

[d] The covalent enzyme–substrate intermediate was isolated in denatured condition.

---

(44) C. G. D. Morley and T. C. Stadtman, *B* 10, 2325–2329 (1971).

(45) C. G. D. Morley and T. C. Stadtman, *B* 11, 600–605 (1972).

(46) Refs. (1–5).

(47) Refs. (6, 7).

(48) T. H. Finlay and E. Adams, *JBC* 245, 5248–5260 (1970); E. Adams, *Adv. Enzymol.* 44, 117–124 (1976).

(49) T. Yorifugi, H. Misono, and K. Soda, *JBC* 246, 5093–5101 (1971).

(50) Refs. (8–11).

(51) S. S. Shapiro and D. Dennis, *B* 4, 2283–2288 (1965); I. A. Rose, *The Enzymes*, 3rd ed., P. D. Boyer, ed., Academic Press, New York, 1970, Vol. 2, pp. 296–297; A. Cantwell and D. Dennis, *B* 13, 287–291 (1974).

(52) Refs. (12–16).

(53) L. Davis, N. Lee, and L. Glaser, *JBC* 247, 5862–5866 (1972).

(54) Refs. (17–21).

(55) M. W. McDonough and W. A. Wood, *JBC* 236, 1220–1224 (1960); J. D. Deupree and W. A. Wood, *JBC* 247, 3093–3097 (1972); Ref. (53); second citation of Ref. (51).

(56) S. Seltzer and M. Lin, *JACS* 101, 3091–3097 (1979).

(57) Refs. (23–28, 32).

(58) Ref. (32).

(59) I. A. Rose, E. L. O'Connell, and R. P. Mortlock, *BBA* 178, 376–379 (1969); Ref. (32).

(60) Ref. (59).

(61) Ref. (32).

(62) H. S. Simon and R. Medina, *Z. Naturforsch.* 21B, 496–497 (1966); I. A. Rose and E. L. O'Connell, *JBC* 248, 2232–2234 (1973); Ref. (32).

(63) I. A. Rose and E. L. O'Connell, *JBC* 236, 3086–3092 (1961); I. A. Rose, *Brookhaven Symp. Biol.* 15, 293–309 (1962); P. J. Shaw and H. Muirhead, *FEBS lett.* 65, 50–55 (1976); Ref. (32).

(64) C. F. Midelfort and I. A. Rose, *B* 16, 1590–1596 (1977); Ref. (32).

(65) First citation on Ref. (59).

(66) Ref. (29–32).

(67) D. H. Shah, W. W. Cleland, and J. W. Porter, *JBC* 240, 1946–1956 (1965); I. A. Rose, in *The Enzymes*, 3rd ed., P. D. Boyer, ed., Academic Press, New York, 1970, Vol. 2, p. 301; T. Koyama, K. Ogura, and S. Sato, *JBC* 248, 8043–8051 (1973).

(68) H. C. Rilling and M. J. Coon, *JBC* 235, 3087–3092 (1960); H. Hashimoto, H. Günter, and H. Simon, *FEBS lett.* 33, 81–83 (1973); E. Schleicher and H. Simon, *Z. Physiol. Chem.* 357, 535–542 (1976); Refs. (32).

(69) J. P. Klinman and I. A. Rose, *B* 10, 2259–2266 (1971); Ref. (32).

(70) R. A. Hill, G. W. Kirby, and D. J. Robins, *J. Chem. Soc., Chem. Commun*, 459–460 (1977).

(71) Refs. (33–37).

(72) Refs. (38–43).

(73) A. Sugino, C. L. Peebles, K. N. Kreuzer, and N. R. Cozzarelli, *PNAS* 74, 4767–4771 (1977); M. Gellert, K. Mizuchi, M. H. O'Dea, T. Itoh, and J. Tomizawa, *PNAS* 74, 4772–4776 (1977); A. Morrison and N. R. Cozzarelli, *Cell* 17, 175–184 (1979); Ref. (40).

(74) J. Rétey, F. Kunz, T. C. Stadtman, and D. Arigoni, *Experientia* 25, 801–802 (1969); J. J. Baker, C. van der Drift, and T. C. Stadtman, *B* 12, 1054–1063 (1973).

(75) Refs. (44, 45).

(76) R. Somack and R. N. Costilow, *B* 12, 2597–2604 (1973).

(77) R. G. Eager, Jr., B. G. Baltimore, M. M. Herbst, H. A. Barker, and J. H. Richards, *B* 11, 253–264 (1972).

(78) J. Rétey and D. Arigoni, *Experienta* 22, 783–784 (1966); G. J. Cardinale and R. H. Abeles, *BBA* 132, 517–518 (1967); W. W. Miller and J. H. Richards, *JACS* 91, 1498–1507 (1969); B. M. Babior, A. D. Woodams, and J. D. Brodie, *JBC* 248, 1445–1450 (1973).

(79) H. F. Kung and L. Tsai, *JBC* 246, 6436–6443 (1971).

(80) J. E. G. Barnett, A. Rasheed, and D. L. Corina, *Biochem. J.* 131, 21–30 (1973).

# Chapter 7
# Ligases

The ligases (or synthetases) form the least numerous of the six major classes of enzymes (cf. Table 8.1, p. 226). Accordingly, Table 7.1 at the end of the present chapter numbers only thirty enzymes; yet these amount to 33% of all the ligases recognized by the Enzyme Commission.

Ligases catalyze the joining together of two molecules at the expense of the chemical energy in ATP (or other nucleoside triphosphate). The great majority of ligase reactions are therefore three-substrate reactions. As such they are intrinsically complex. They proceed nevertheless in discernible stages, each of which is familiar as a simple transferase reaction of the kind exemplified in Chapter 3 of this book. Of the two molecules being joined in the ligase reaction, one is most commonly an acid needing activation by phosphorylation before the joining can occur. In this first (phosphorylative) phase of ligase activity the enzyme acts as a phosphotransferase, transferring a phosphoryl (or adenylyl) group from ATP to the carboxylic acid. We saw earlier that the phosphorylation of acetate—the prototypic carboxylic acid—to acetyl phosphate proceeds by a reaction pathway which includes a phosphoenzyme intermediate (p. 92). The same appears to hold true in the three-substrate reaction catalyzed by a ligase. But in this case the product of phosphotransferase activity—an acyl phosphate or acyl adenylate—is not set free in solution as it is in acetyl phosphate synthesis. Instead, it is held firmly and noncovalently on the enzyme preparatory to the second phase of ligase activity—the joining to cosubstrate.

Such joining is the outcome of transacylase activity on the part of the ligase, the acyl group being transferred from the intermediary acyl phosphate to the ultimate acceptor. And transacylases, we have seen, are apt to catalyze their reactions by way of an acyl-enzyme intermediate (Chapter 3). It follows from all this that, as a multisubstrate enzyme, a ligase wears the aspect of two enzymes packed into one; in effect a double transferase (1). The double transferase character of a ligase is doubtless imposed upon it by the innate instability of the acyl phosphate formed as the product of the first (phosphorylative) transferase reaction. Whereas acetyl phosphate and other simple acyl phosphates have a rather wide range of pH stability (2, 3), the more complex succinyl phosphate (3), $\gamma$-glutamyl phosphate (4), acetyl adenylate (5), tryptophanyl adenylate (6), and the like have each a narrower range of pH stability, and a generally more marked tendency toward chemical instability. Holding such compounds bound (noncovalently) to the enzyme must preserve them from destruction. But this, we see, forces upon the ligase the double duty of catalyzing two distinct transferase reactions.

From the standpoint of chemical mechanism, it is only the rare ligase which can boast of demonstrated covalent catalysis in both of its transferase activities. For the most part, covalent catalysis by any individual ligase has been demonstrated for one or the other of its transferase activities, but not for both. This state of affairs is owing to the inherent complexity of these enzymes and to the relative disinterest in their study. No ligase has ever enjoyed the attention lavished on such popular enzymes as chymotrypsin, hexokinase, or carbonic anhydrase. Despite these troubles it is possible to discern the basic mechanistic theme—with a few variations here and there—which pervades the chemical activities of this class of enzymes (1). With the enzymes discussed in this chapter we attempt to delineate this theme.

## Tryptophanyl-tRNA Synthetase [EC 6.1.1.2]

This enzyme catalyzes the activation of tryptophan

$$\text{ATP} + \text{L-tryptophan} + \text{tRNA}^{\text{Trp}} \xrightarrow{\text{Mg}^{2+}}$$
$$\text{AMP} + \text{PP}_i + \text{L-tryptophanyl-tRNA}^{\text{Trp}} \tag{1}$$

preparatory to its incorporation into protein. As in the activation of valine (7), isoleucine (8), and probably other amino acids, an acyl adenylate—tryptophanyl adenylate—is an intermediate of reaction 1. Tryptophanyl adenylate, noncovalently bound to the enzyme, is isolable from reaction

(1) L. Spector, in *Energy, Biosynthesis, and Regulation in Molecular Biology*, D. Richter, ed., Walter de Gruyter, Berlin, 1974, pp. 564–574.
(2) D. E. Koshland, Jr., *JACS* 74, 2286–2292 (1952).
(3) C. T. Walsh, J. G. Hildebrand, and L. B. Spector, *JBC* 245, 5699–5708 (1970).
(4) L. Levintow and A. Meister, *Fed. Proc.* 15, 299 (1956).
(5) P. Berg, *JBC* 222, 1015–1023 (1956).
(6) H. S. Kingdon, L. T. Webster, Jr., and E. W. Davie, *PNAS* 44, 757–765 (1958).
(7) C. F. Midelfort, K. Chakraburtty, A. Steinschneider, and A. H. Mehler, *JBC* 250, 3866–3873 (1975).
(8) D. V. Santi and R. W. Webster, Jr., *JBC* 250, 3874–3877 (1975).

$$ATP + E \xrightarrow{\text{Mg}^{2+}} E \sim AMP + PP_i$$
$$E \sim AMP + AA \longleftrightarrow E \cdots AA \sim AMP$$

$\left.\right\}$ Adenylyltransferase

$$E \cdots AA \sim AMP \longleftrightarrow E \sim AA + AMP$$
$$E \sim AA + tRNA^{AA} \longleftrightarrow E + AA \sim tRNA^{AA}$$

$\left.\right\}$ Aminoacyltransferase

Sum: $ATP + AA + tRNA^{AA} \longleftrightarrow AMP + PP_i + AA \sim tRNA^{AA}$

$$[AA = \text{amino acid}]$$

**Fig. 1.** The proposed suite of reactions leading to the activation of amino acids for protein synthesis. The scheme expresses solely the chemical mechanism of activation, irrespective of kinetic considerations. Apropos of the latter, it is a possibility, if not a probability, that the amino acid is already adsorbed to the enzyme at the moment when $E \sim AMP$ is synthesized, and that tRNA is also adsorbed when $E \sim AA$ is synthesized.

mixtures containing the enzyme, ATP, $Mg^{2+}$, and tryptophan, but omitting $tRNA^{Trp}$ (6, 9, 10). Chemically competent, the enzyme-bound tryptophanyl adenylate acts on pyrophosphate to give ATP, and on $tRNA^{Trp}$ to give tryptophanyl-$tRNA^{Trp}$ (10).

The enzymic synthesis of tryptophanyl adenylate is of course the consequence of adenylyl transfer from ATP to tryptophan (Fig. 1). As a transferase reaction it ought to proceed over a pathway which includes an adenylyl enzyme ($E \sim AMP$) as intermediate (11, 12). That it seems indeed to do so is hinted at in the isolation by gel filtration of just such an adenylyl enzyme after incubation of ATP and $Mg^{2+}$ with tryptophanyl-tRNA synthetase in the absence of added tryptophan (13). The $E \sim AMP$ so prepared exhibits its chemical competence in making ATP from inorganic pyrophosphate; but its other properties are as yet unreported.

As aminoacyl donor, the enzyme-bound tryptophanyl adenylate is the reactant in the aminoacyltransferase phase of the ligase reaction (Fig. 1). Expectation, therefore, is that an aminoacyl enzyme ($E \sim AA$) mediates the reaction. It happens that tryptophanyl-tRNA synthetase can be isolated from beef pancreas with tryptophan (1 mole per mole of enzyme) already covalently joined to it (14). The tryptophanyl group is in the activated state

(9) M. Karasek, P. Castelfranco, P. P. Krishnaswamy, and A. Meister, *JACS* 80, 2335–2336 (1958); P. V. Graves, J. de Bony, J. P. Mazat, and B. Labouesse, *Biochimie* 62, 33–41 (1980).

(10) M. Dorizzi, B. Labouesse, and J. Labouesse, *EJB* 19, 563–572 (1971).

(11) A. A. Kraeveskii, L. L. Kisselev, and B. P. Gottikh, *Mol. Biol.* 7, 634–639 (1973) (Engl. transl.).

(12) L. B. Spector, *Bioorg. Chem.* 2, 311–321 (1973).

(13) L. L. Kisselev and L. L. Kochkina, *Dokl. Biochem.* 214, 7–9 (1974) (Engl. transl.); L. L. Kisselev and O. O. Favorova, *Adv. Enzymol.* 40, 141–238 (1974).

(14) L. L. Kisselev, O. O. Favorova, and G. K. Kovaleva, *Methods Enzymol.* 59, 234–257 (1979).

since it reacts specifically with tRNA$^{Trp}$ to aminoacylate it (15). A carboxyl group of the active center holds the tryptophan as an acid anhydride. In conformity with such linkage, the fixed tryptophan exchanges with exogenous tryptophan, but not with any other amino acid. It reacts, too, with hydroxylamine to yield tryptophanyl hydroxamate. With $[^{14}C]CH_3ONH_2$, most of the radioactivity appears as the corresponding tryptophanyl methylhydroxamate. But some of the radioactivity remains fixed to the enzyme, presumably in the carboxyl of the active center as the methylhydroxamate. The enzyme, thus modified, can no longer aminoacylate tRNA, but it can still catalyze the usual ATP–PP$_i$ exchange. Upon denaturation of the tryptophanyl enzyme with urea, the tryptophanyl group migrates from the carboxyl of the enzyme to a sulfhydryl—that is, from acid anhydride to thiolester linkage, maintaining thus the activated condition of the tryptophanyl group (16).

Figure 1 shows in simple outline the basic theme of all ligase activity. There is first the activation of the acidic group of the substrate, which converts the substrate into an activated acyl donor—noncovalently bound to the enzyme. Thereupon the transfer of the acyl group to the acceptor completes the cycle of ligase catalysis. With appropriate changes of detail, the partial reactions depicted in Fig. 1 recur over and over again in the chemistry of ligase catalysis.

# Phenylalanyl-tRNA Synthetase [EC 6.1.1.20]

To activate phenylalanine for incorporation into polypeptides, phenylalanine-tRNA synthetase catalyzes the following reaction:

$$ATP + \text{L-phenylalanine} + tRNA^{Phe} \xrightarrow{\quad Mg^{2+} \quad}$$

$$AMP + PP_i + \text{L-phenylalanyl-tRNA}^{Phe} \tag{2}$$

It is obvious that reaction 2 has the same chemical form as reaction 1. A reasonable supposition therefore is that the two reactions have the same chemical mechanism. Accordingly, Fig. 1 predicts that the activation of phenylalanine, like the activation of tryptophan, requires the participation of three intermediates—E ~ AMP, phenylalanyl adenylate (AA ~ AMP), and phenylalanyl enzyme (E ~ AA). No evidence is available at this time for E ~ AMP. But evidence for the latter two intermediates does exist. Thus,

(15) G. K. Kovaleva, S. G. Moroz, O. O. Favorova, and L. L. Kisselev, *FEBS lett.* 95, 81–84 (1978); L. L. Kisselev, O. O. Favorova, and G. K. Kovaleva, in *Transfer RNA: Structure, Properties, and Recognition*, P. R. Schimmel, D. Söll, and J. N. Abelson, eds., Cold Spring Harbor Laboratory, 1979, pp. 235–246.
(16) Does this herald a "surface walk" in the normal activity of this enzyme?

phenylalanyl adenylate, complexed noncovalently with phenylalanyl-tRNA synthetase (*E. coli*), can be isolated from incubation mixtures containing ATP, $Mg^{2+}$, and phenylalanine, but with tRNA omitted. The complex from *E. coli* is chemically competent (17), and the one from yeast is also kinetically competent (18).

According to Fig. 1, a phenylalanyl enzyme ($E \sim AA$) ought to mediate the transfer of the phenylalanyl group from phenylalanyl adenylate to $tRNA^{Phe}$. Evidence to this effect is available in the case of the phenylalanyl-tRNA synthetase from yeast. The enzyme can of course catalyze an AMP–ATP exchange in the presence of all components of the total reaction (Eq. 2), including the $tRNA^{Phe}$. The exchange so observed is usually regarded as being due to reversal of the total reaction, since, in the absence of $tRNA^{Phe}$, there is no exchange. Yet if the normal $tRNA^{Phe}$ is replaced by 3'-deoxy $tRNA^{Phe}$—*which cannot accept the phenylalanyl residue*—a lively AMP–ATP exchange nonetheless occurs, though at a somewhat reduced rate relative to the exchange observed in the presence of normal $tRNA^{Phe}$ (19). From Fig. 1 it is clear that reversal of the four partial reactions, in the presence of normal $tRNA^{Phe}$, can indeed account for the AMP–ATP exchange. But if the fourth reaction of the suite is impossible (with 3'-deoxy $tRNA^{Phe}$), then reversal of the *overall* reaction is also impossible. It follows that the AMP–ATP exchange in the presence of the 3'-deoxy analogue of $tRNA^{Phe}$ is best ascribed to catalysis of the first three reactions of Fig. 1, which includes a role for $E \sim AA$—the phenylalanyl enzyme.

Further to the same conclusion is the actual isolation by gel filtration of the $E \sim AA$ intermediate of phenylalanine activation. To fulfill the space-filling function of normal $tRNA^{Phe}$—without its aminoacyl-accepting capability—periodate-oxidized $tRNA^{Phe}$ is used. After incubation of phenyl-alanyl-tRNA synthetase (from yeast) with ATP, $Mg^{2+}$, phenylalanine, and periodate-oxidized $tRNA^{Phe}$, there is isolated an enzyme complex containing AMP and amino acid in the proportion of 1 to 2 (20). One phenylalanine is fixed to AMP, the other to the enzyme. Both phenylalanyl residues in the complex are transferable to normal $tRNA^{Phe}$. Repetition of the experiment in the absence of all tRNA leads to the familiar phenylalanyl-adenylate–enzyme complex, with AMP and amino acid present in the ratio 1 to 1. Clearly, the presence of a nonaccepting $tRNA^{Phe}$ calls forth the activation of a second molecule of phenylalanine, in addition to the expected phenylalanyl adeny-late. Binding of the second phenylalanine to the enzyme requires ATP, and the energy-rich character of the phenylalanyl group within the complex is evident in its transferability to normal $tRNA^{Phe}$. The phenylalanyl bond to the enzyme can hardly be other than covalent.

(17) P. Bartmann, T. Hanke, and E. Holler, *JBC* 250, 7668–7674 (1975).

(18) P. Bartmann, T. Hanke, and E. Holler, *B* 14, 4777–4786 (1975); F. Fasolio and A. R. Fersht, *EJB* 85, 85–88 (1978).

(19) P. Remy and J. P. Ebel, *FEBS lett.* 61, 28–31 (1976).

(20) R. Thiebe, *FEBS lett.* 60, 342–345 (1975).

The passage of phenylalanine through two stages of activation—phenyl-lalanyl adenylate and phenylalanyl enzyme—before ultimately joining to tRNA$^{Phe}$ accords generally with the chemical mechanism of Fig. 1. Activation of amino acids in this fashion is also found in the biosynthesis of polypeptide antibiotics. Thus, aminoacyl-enzyme intermediates as well as aminoacyl adenylates figure prominently in the assembly of the polypeptide chains of gramicidin S (21) and tyrocidine (22). There is even the same 1 to 2 proportion of covalently fixed AMP and amino acid. The aminoacyl groups are joined to the enzymes as thiolesters. The same may be true of the phenylalanine activating enzyme, since modification of two of its sulfhydryl groups inhibits aminoacylation of tRNA$^{Phe}$, but has no effect on the aminoacylation of AMP (23).

# Carnosine Synthetase [EC 6.3.2.11]

Carnosine synthetase is a ligase capable of synthesizing the carbon–nitrogen bond of a dipeptide.

$$\text{ATP} + \beta\text{-alanine} + \text{L-histidine} \xrightarrow{\text{Mg}^{2+}} \text{AMP} + \text{PP}_i + \text{carnosine} \quad (3)$$

From the chemical form of reaction 3 we infer that an E ~ AMP mediates the first phase of the reaction, that $\beta$-alanyl adenylate is a noncovalently enzyme-bound intermediate, and that $\beta$-alanyl enzyme intervenes in the second—acyl transferring—phase of the catalytic cycle (Fig. 2). Of these three anticipated intermediates, evidence is available for only the first two. Carnosine synthetase catalyzes a $\text{PP}_i$–ATP exchange in the absence of any detectable $\beta$-alanine (24). $\beta$-Alanine, in moderate concentration, has little effect on the exchange—but

$$\text{ATP} + \text{E} \xrightarrow{\text{Mg}^{2+}} \text{E} \sim \text{AMP} + \text{PP}_i \qquad \left.\right\} \text{Adenylyltransferase}$$

$$\text{E} \sim \text{AMP} + \beta\text{-alanine} \longleftarrow\!\!\!\longrightarrow \text{E} \cdots \beta\text{-alanyl} \sim \text{AMP}$$

$$\text{E} \cdots \beta\text{-alanyl} \sim \text{AMP} \longleftarrow\!\!\!\longrightarrow \text{E} \sim \beta\text{-alanyl} + \text{AMP} \qquad \left.\right\} \beta\text{-Alanyltransferase}$$

$$\text{E} \sim \beta\text{-alanyl} + \text{histidine} \longleftarrow\!\!\!\longrightarrow \text{E} + \text{carnosine}$$

Sum: ATP + $\beta$-alanine + histidine $\longleftarrow\!\!\!\longrightarrow$ AMP + $\text{PP}_i$ + carnosine

**Fig. 2.** The proposed chemical mechanism for the action of carnosine synthetase. There is some evidence for the existence of E ~ AMP and $\beta$-alanyl ~ AMP. E ~ $\beta$-alanyl, however, is conjectural, but is predicted on grounds of analogy.

(21) W. Gevers, H. Kleinkauf, and F. Lipmann, *PNAS* 63, 1335–1342 (1969).
(22) R. Roskoski, Jr., W. Gevers, H. Kleinkauf, and F. Lipmann, *B* 9, 4839–4845 (1970).
(23) A. Murayama, J. P. Raffin, P. Remy, and J. P. Ebel, *FEBS lett.* 53, 15–22 (1975); M. Baltzinger and P. Remy, *FEBS lett.* 79, 117–120 (1977).
(24) J. J. Stenesh and T. Winnick, *Biochem. J.* 77, 575–581 (1960).

at high concentration is strongly inhibitory. $\alpha$-Alanine, by contrast, is altogether without effect on the exchange. Since fluoride ion inhibits both the exchange and the net synthesis of carnosine, the $PP_i$–ATP exchange is believed to be an integral part of the carnosine synthetic apparatus and intimates the existence of an $E \sim AMP$ intermediate.

$\beta$-Alanine acting on $E \sim AMP$ ought to yield $\beta$-alanyl adenylate— noncovalently bound to the enzyme (Fig. 2). This compound has yet to be isolated from an enzymatic mixture. But it can be prepared by chemical synthesis (25, 26). Incubated with carnosine synthetase and $PP_i$, synthetic $\beta$-alanyl adenylate yields ATP. And with histidine, in the absence of ATP, carnosine is synthesized from $\beta$-alanyl adenylate at a rate comparable in magnitude with the rate of its synthesis from $\beta$-alanine, histidine, and ATP (26). $\beta$-alanyl adenylate appears thus to be a genuine intermediate of carnosine synthesis. The anticipated $\beta$-alanyl enzyme has not so far been observed, but on grounds of analogy its participation in reaction 3 is predicted.

## Acetyl-CoA Synthetase [EC 6.2.1.1]

Exemplifying the activation of a carboxylic acid to the level of a thiolester is the transformation of acetate into acetyl-CoA.

$$ATP + acetate + CoA \xrightarrow{Mg^{2+}} AMP + PP_i + acetyl\text{-}CoA \qquad (4)$$

In its chemical form reaction 4 is obviously similar to reactions 1, 2, and 3 for the activation of amino acids. Similarities in chemical mechanism are therefore anticipated (Fig. 3), and indeed are observed. Acetyl adenylate— noncovalently bound to the enzyme—was recognized early as an inter- mediate in acetate activation (27). More recently, an acetyl enzyme was also found to share in the catalysis. Thus, it happens that coenzyme A in Eq. 4 is replaceable by dephospho-coenzyme with the consequent synthesis of acetyl- dephospho-CoA (28). And, as a corollary activity, the enzyme catalyzes the rapid and reversible transfer of an acetyl group between coenzyme A and dephospho-coenzyme A—*in the assured absence of all other components of the total ligase reaction* (29). From general considerations it seems safe to say that dephospho-coenzyme A occupies the same subsite as coenzyme A itself in the active center of the enzyme. The reversible transfer of an acetyl group between coenzyme A and its analogue is thus chemically equivalent to an independent CoA—acetyl-CoA exchange, symbolized by the fourth

(25) G. D. Kalyankar and A. Meister, *JACS* 81, 1515–1516 (1959).

(26) G. D. Kalyankar and A. Meister, *JBS* 234, 3210–3218 (1959).

(27) P. Berg, *JBC* 222, 991–1013 (1956); L. T. Webster and F. Campagnari, *JBC* 237, 1050–1055 (1962); L. T. Webster, *JBC* 238, 4010–4015 (1963).

(28) M. E. Jones and F. Lipmann, *Methods Enzymol.* 1, 585–591 (1955).

(29) H. Anke and L. B. Spector, *Biochem. Biophys. Res. Commun.* 67, 767–773 (1975).

$$ATP + E \xrightarrow{\text{Mg}^{2+}} E \sim AMP + PP_i$$

$$E \sim AMP + acetate \longrightarrow E \cdots acetyl \sim AMP$$

\} Adenylyltransferase

$$E \cdots acetyl \sim AMP \longleftarrow E \sim acetyl + AMP$$

$$E \sim acetyl + CoA \longleftarrow E + acetyl \sim CoA$$

\} Acetyltransferase

Sum: $ATP + acetate + CoA \longleftarrow AMP + PP_i + acetyl \sim CoA$

**Fig. 3.** The proposed chemical mechanism of the acetyl-CoA synthetase reaction. Good evidence for $E \sim AMP$ is still lacking. Adsorption of acetate to the enzyme may be necessary for the formation of $E \sim AMP$, and the adsorption of CoA may be necessary for $E \sim acetyl$ formation.

reaction of Fig. 3. Such exchange, occurring independently of all other components of the total reaction, points to acetyl enzyme as a mediating factor. Since the exchange is an intrinsic activity of acetyl-CoA synthetase, it is inferred that the same acetyl enzyme mediates the transfer of acetyl between acetyl AMP and coenzyme A in the normal course of the reaction (third and fourth reactions of Fig. 3).

Of the three energy-rich intermediates included in Fig. 3 as participants in acetate activation—$E \sim AMP$, $E \cdots acetyl \sim AMP$, and $E \sim acetyl$—only the latter two are supported adequately by experiment. The only evidence relating to the adenylyltransferase phase of acetate activation has to do with the stereochemical course of the reaction. The experimental finding is that there is a net inversion of configuration at the $\alpha$-phosphorus atom of ATP as it is transformed into AMP (30). This, it is claimed, results from a single-displacement catalysis of the reaction between ATP and acetate to yield acetyl adenylate. But experience with the other acetate activating enzyme— acetate kinase (p. 92)—suggests, on the contrary, that the net inversion on phosphorus proceeds from a *triple-displacement* catalysis, in which the adenylyl group forms two successive covalent intermediates with the enzyme on its "surface walk" from ATP to linkage with acetate.

# Succinyl-CoA Synthetase [EC 6.2.1.5]

Like acetyl-CoA synthetase, succinyl-CoA synthetase catalyzes the conversion of a carboxylic acid to a thiolester,

$$ATP + succinate + CoA \xrightarrow{\text{Mg}^{2+}} ADP + P_i + succinyl\text{-}CoA \qquad (5)$$

with the distinction, however, that the ATP is hydrolyzed to ADP plus orthophosphate instead of AMP plus pyrophosphate. Yet the chemical form of reaction 5 intimates that the activation of succinate follows the same

(30) C. F. Midelfort and I. Sarton-Miller, *JBC* 253, 7127–7129 (1978); M.-D. Tsai, *B* 18, 1468–1472 (1979).

$$ATP + E \xrightarrow{Mg^{2+}} E \sim P + ADP$$

$$E \sim P + succinate \longleftrightarrow E \cdots succinyl \sim P$$

$\left.\vphantom{\begin{matrix}a\\b\end{matrix}}\right\}$ Phosphotransferase

$$E \cdots succinyl \sim P \longleftrightarrow E \sim succinyl + P_i$$

$$E \sim succinyl + CoA \longleftrightarrow E + succinyl \sim CoA$$

$\left.\vphantom{\begin{matrix}a\\b\end{matrix}}\right\}$ Succinyltransferase

Sum: $ATP + succinate + CoA \longleftrightarrow ADP + P_i + succinyl \sim CoA$

**Fig. 4.** The proposed chemical mechanism of the succinyl-CoA synthetase reaction. Intermediary roles for $E \sim P$ and succinyl $\sim P$ are established. But only suggestive evidence for $E \sim$ succinyl is available at this time. Since the enzyme follows sequential kinetics, $E \sim P$ and the putative $E \sim$ succinyl do not exist as free entities, but have one or more of the reaction components coadsorbed at all times during catalysis.

general pattern governing that of acetate and the amino acids. Thus, a phosphoenzyme ($E \sim P$ in Fig. 4) replaces the adenylyl enzyme ($E \sim AMP$ of Figs. 1, 2, and 3); and an acyl phosphate replaces the acyl adenylate. As it happens, succinyl-CoA synthetase (from *E. coli*) is actually isolated in the phosphorylated condition (31). The phosphoenzyme is also accessible (a) by incubation of the dephosphoenzyme with ATP and $Mg^{2+}$ in the absence of succinate and coenzyme A, or (b) by incubation with $P_i$ and succinyl-CoA in the absence of ADP. The phosphoenzyme is kinetically competent (32). Its phosphoryl group is joined to a ring nitrogen atom of a histidyl residue. Upon reaction of the phosphoenzyme with succinate and coenzyme A, succinyl-CoA results (Fig. 4). But the more immediate product of the action of succinate (minus coenzyme A) on the phosphoenzyme is succinyl phosphate—noncovalently bound to the enzyme—which is isolable from a mixture of enzyme, ATP, $Mg^{2+}$, and succinate (33). Succinyl phosphate, synthesized chemically (34), has all the properties appropriate to its role as a source of activated phosphorus or activated succinyl (33, 34). It yields ATP with ADP, and succinyl-CoA with coenzyme A. In the absence of these acceptors succinyl phosphate phosphorylates the enzyme (33). Phosphorylation of the enzyme by succinyl phosphate, as well as by ATP, conforms of course with the phosphotransferase activity of the enzyme (first two reactions of Fig. 4).

Succinyl phosphate, in its role as succinyl donor to coenzyme A, probably succinylates the enzyme during succinyl transfer. But the evidence on this point is meager. A succinyl enzyme—prepared by reaction of the *E. coli* enzyme with succinyl-CoA—is indeed known and has the proper chemical composition; but it is devoid of enzymic activity (35, 36). A succinyl enzyme

(31) G. Kreil and P. D. Boyer, *Biochem. Biophys. Res. Commun.* 16, 551–555 (1964).
(32) W. A. Bridger, W. A. Millen, and P. D. Boyer, *B* 7, 3608–3616 (1968).
(33) J. S. Nishimura and A. Meister, *B* 4, 1457–1462 (1965).
(34) J. G. Hildebrand and L. B. Spector, *JBC* 244, 2606–2613 (1969).
(35) R. W. Benson, J. L. Robinson, and P. D. Boyer, *B* 8, 2496–2502 (1969).
(36) P. H. Pearson and W. A. Bridger, *JBC* 250, 8524–8529 (1975).

can also be prepared by reacting the phosphoenzyme (from pigeon breast muscle) with succinate in the absence of coenzyme A and nucleotides. Inorganic phosphate is released, while an amount of succinate stoichiometric with the released $P_i$ remains bound to the enzyme and is stable to gel filtration (37). Unfortunately, details on the enzymic activity—or inactivity—of the new succinyl enzyme are still unreported.

Succinyl-CoA synthetase (from *E. coli*) is composed of two nonidentical subunits assembled in the enzyme according to the formula $\alpha_2\beta_2$ (38). The pure $\alpha$ and $\beta$ subunits can be isolated (39). Of the two subunits only the $\alpha$ can be phosphorylated. Moreover, the pure, isolated $\alpha$ subunit catalyzes its own phosphorylation by ATP in the absence of the $\beta$ subunit, though the presence of the latter considerably enhances the rate (36). Phosphorylation of the $\alpha$ subunit is not possible with succinyl-CoA plus orthophosphate. For such phosphorylation the $\beta$ subunit is necessary. It is concluded that the binding site of succinyl-CoA is probably on the $\beta$ subunit. Consonant with this is the discovery that the succinyl group in the above-cited, inactive succinyl enzyme (*E. coli*) is covalently linked to the $\beta$ subunit only. Thus, an emerging picture of catalysis by succinyl-CoA synthetase assigns chemical functions to each subunit. The $\alpha$ subunit catalyzes the phosphotransferase activity of the enzyme, while the $\beta$ subunit is the succinyltransferase. On this view succinyl phosphate provides the chemical connection between the juxtaposed binding sites on the two subunits. An analogous division of catalytic labor among nonidentical subunits of an enzyme is also manifest in the action of transcarboxylase (p. 65) and citrate lyase (p. 167).

Of singular consequence to our notions of enzymic catalysis was the finding that succinyl-CoA synthetase follows sequential kinetics (40). ATP binds to the enzyme first, followed by random addition of coenzyme A and succinate to form a quaternary complex before the release of any product. Stated otherwise, succinyl-CoA synthetase does *not* have a ping-pong kinetic pattern, despite the proved participation in its catalytic cycle of at least one covalent enzyme–substrate intermediate. This observation is historically significant because it did away with the misconception, which prevailed earlier, that an enzyme *must* exhibit ping-pong kinetics if a covalent enzyme–substrate complex mediates its action.

In the same vein, we may comment on the ADP–ATP exchange catalyzed by succinyl-CoA synthetase in the absence of other substrates. Such an exchange signals, of course, the participation of a phosphoenzyme in the catalysis. But the rate of the exchange happens in this case to be only 2% of the rate of the net reaction, despite abundant proof that the phosphoenzyme is on the reaction pathway (32). A slow exchange—even when proved *bona*

(37) D. G. Mikeladze, L. N. Matveeva, and S. E. Severin, *Biokhimiya* 43, 1144–1152 (1978) (Engl. transl.).
(38) W. A. Bridger, *Biochem. Biophys. Res. Commun.* 42, 948–954 (1971).
(39) P. H. Pearson and W. A. Bridger, *JBC* 250, 4451–4455 (1975).
(40) F. J. Moffet and W. A. Bridger, *JBC* 245, 2758–2762 (1970).

*fide* in all other respects—is often distrusted as a criterion for a covalent enzyme–substrate intermediate, solely on account of its slowness. We have here a clearly distinguishable case of what has come to be called "substrate synergism." Thus, a striking acceleration of the ADP–ATP exchange is induced simply upon addition of succinyl-CoA alone (32) or of coenzyme A alone (41). To elicit the full exchange capacity of the enzyme the presence of all the other substrates in their respective subsites on the enzyme is required, in conformity with the enzyme's sequential kinetics.


# Long-Chain Fatty Acyl-CoA Synthetase [EC 6.2.1.3]

We consider here one more carboxyl-activating mechanism, recently described, and which, at the time of writing, is the only one of its kind known. The enzyme's mode of action is a remarkable departure from the general pattern of carboxyl activation with which we have been familiar till now. On the face of it, the transformation of palmitate into palmityl-CoA

$$\text{ATP} + \text{palmitate} + \text{CoA} \xrightarrow{\text{Mg}^{2+}} \text{AMP} + \text{PP}_i + \text{palmityl-CoA} \qquad (6)$$

has the same chemical form as the transformation of acetate into acetyl-CoA (Eq. 4), the only difference being the length of the hydrocarbon chains of the two carboxylic acid substrates. And, indeed, the first partial reaction of the catalytic cycle of the two enzymes is also the same; namely, the reaction of enzyme with ATP, yielding adenylyl enzyme and $\text{PP}_i$ (cf. Figs. 3 and 5). But while acetate activation has an acyl adenylate as a reaction intermediate, the activation of palmitate has none. Beginning thus with the second reaction of Figs. 3 and 5, the activation process for the long-chain fatty acid diverges sharply from that of acetate (42), and begins to look instead like the coenzyme A transferase reaction (p. 104).

The enzyme catalyzes a PP–ATP exchange, in the absence of cosubstrates, which is comparable in rate to that of the net reaction. In conformity with this finding, an adenylyl enzyme can be prepared and isolated from the reaction of ATP—alone—on the palmitate-activating enzyme. The adenylyl enzyme reacts rapidly with coenzyme A to yield stoichiometric amounts of E-COSCoA and released AMP. E-COSCoA is formed from coenzyme A and enzyme only in the presence of ATP. When palmitate is added to E-COSCoA, coenzyme A is released from the enzyme and appears quantitatively in palmityl-CoA. The transfer of coenzyme A from enzyme to palmitate recalls the chemical mechanism of the coenzyme A transferase reaction, and prompts

(41) F. L. Grinnell and J. S. Nishimura, *B* 8, 562–568 (1969).
(42) S.-Y. Huang and P. Parsons, *Fed. Proc.* 37, 1804 (1978); D. G. Philipp and P. Parsons, *JBC* 254, 10785–10790 (1979).

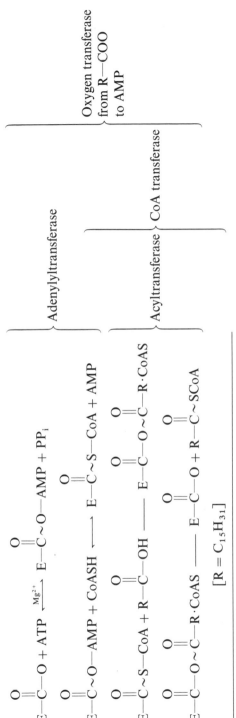

Adenylyltransferase

Oxygen transferase from R—COO to AMP

Acyltransferase } CoA transferase

$$E—C\overset{O}{\|}—O + ATP \xrightarrow{Mg^{2+}} E—C\overset{O}{\|}\sim O—AMP + PP_i$$

$$E—C\overset{O}{\|}\sim O—AMP + CoASH \longrightarrow E—C\overset{O}{\|}\sim S—CoA + AMP$$

$$E—C\overset{O}{\|}\sim S—CoA + R—C\overset{O}{\|}—OH \longrightarrow E—C\overset{O}{\|}—O\sim C\overset{O}{\|}—R\cdot CoAS$$

$$E—C\overset{O}{\|}—O\sim C\overset{O}{\|}—R\cdot CoAS \longrightarrow E—C\overset{O}{\|}—O + R—C\overset{O}{\|}\sim SCoA$$

$$[R = C_{15}H_{31}]$$

Sum: $ATP + palmitate + CoA \longrightarrow AMP + PP_i + palmityl\sim CoA$

**Fig. 5.** A proposed mechanism for the enzymic synthesis of palmityl-CoA. Evidence for E—COOAMP and E—COSCoA has been reported. The existence of the intermediary carboxylic anhydride, E—COOCOR, is conjectural, but, on chemical grounds, seems unavoidable; just as it is in the case of the reaction catalyzed by coenzyme A transferase. In the proposed mechanism the adenylyl group is transferred, in effect, from ATP to a hydroxyl group (of water). The fatty acyl group is transferred from a hydroxyl group to coenzyme A. Coenzyme A is transferred from a proton to the fatty acyl group of the conjectural carboxylic anhydride, E—COOCOR.

inclusion in the catalytic cycle of the chemically plausible anhydride intermediate composed of palmitate and an enzymic carboxyl (Fig. 5). If the
reactions of Fig. 5 truly represent the catalytic pathway of palmityl-CoA
synthetase, then we further anticipate that an oxygen atom of plamitate is
transferred to AMP indirectly via the enzymic carboxyl. Palmityl-CoA
synthetase ought thus to have an "oxygen transferase" activity (like that of
CoA-transferase) in addition to the other transferase activities depicted in
Fig. 5 (43).

Why the synthesis of acetyl-CoA and palmityl-CoA take such divergent
catalytic routes can only be conjectured. It must surely have to do with the
length of the respective hydrocarbon chains of acetate and palmitate. We can
think of acetate as a small, mobile molecule which can move with ease, in
free or combined form, from one locale to another within the active center
of the acetate-activating enzyme. We conceive thus that acetate, reacting with
$E \sim AMP$ to form acetyl adenylate in one subsite, can slide over (as acetyl
adenylate) to another subsite where $E \sim$ acetyl comes into being (Fig. 3). But
such mobility is probably denied to a long-chain fatty acid (44), because the
numerous hydrophobic connections it makes with the enzyme must anchor
it to one place. This immobility of substrate, in turn, forces all of the chemical
events of catalysis to unfold in a relatively confined space. We therefore
hazard the assertion that, as implied in Fig. 5, all of the chemical action goes
forward at one and the same carboxyl group in the active center of the enzyme.
This contrasts notably with the events of succinate activation—which we
think are analogous to those of acetate activation—in which the enzyme is
phosphorylated at one site (the $\alpha$ subunit) and succinylated at another (the
$\beta$ subunit), while succinyl phosphate somehow bridges the distance between
them (Fig. 4). Operating as a "double transferase," the succinate- (or acetate-)
activating enzyme appears thus to form *two* covalent intermediates with
substrates, one each at *two* separate sites on the enzyme. In the activation of
palmitate, on the other hand, all chemical activity takes place at *one* catalytic
group of the enzyme, with covalent unions to *three* substrates. At the end of
a catalytic cycle the enzyme has done duty as a "tetra transferase," if we
include the above-cited oxygen transferase as one of its activities. All this
complexity just to accommodate the immobility of a fatty acid. It provides,
however, a remarkable display of covalent virtuosity by an enzyme.

---

(43) We note in passing that an oxygen transfer from substrate carboxyl to AMP or $P_i$ also
occurs in conventional carboxyl activation during the formation and further reaction of the
enzyme-bound acyl adenylate or acyl phosphate. In these reactions the oxygen atom is simply
transferred *directly* between donor and acceptor, and forms the bridge oxygen in the non-
covalent, enzyme-bound intermediate. Since, in these cases, the transferring oxygen never joins
covalently to the enzyme, there is, strictly speaking, no "oxygen transferase" activity in the
conventional carboxyl-activating enzyme.

(44) H. U. Gally, A. K. Spencer, I. M. Armitage, J. H. Prestegard, and J. E. Cronan, Jr., *B* 17,
5377–5382 (1978).

# Pyruvate Carboxylase [EC 6.4.1.1]

Somewhat typical of the ligases that form carbon–carbon bonds through carbon dioxide fixation is the one which converts pyruvate into oxaloacetate.

$$\text{ATP} + \text{bicarbonate} + \text{pyruvate} \xrightarrow[\text{acetyl-CoA}]{\text{Mg}^{2+}, \text{Mn}^{2+}} \text{ADP} + \text{P}_i + \text{oxaloacetate} \quad (7)$$

Biotin is the covalently bound prosthetic group of such ligases. A firmly held $\text{Mn}^{2+}$ ion is found in the pyruvate carboxylase from avian and vertebrate sources, and is believed to facilitate the enolization of pyruvate prior to its carboxylation. Acetyl-CoA is a regulatory factor and has no chemical role in the catalysis.

It is widely accepted that the carboxylation of pyruvate proceeds in two easily distinguishable stages, the first of which is the carboxylation of the holoenzyme.

$$\text{E—biotin} + \text{ATP} + \text{HCO}_3 \xrightarrow[\text{acetyl-CoA}]{\text{Mg}^{2+}} \text{E—biotin—CO}_2 + \text{ADP} + \text{P}_i \quad (8)$$

Thereafter, pyruvate accepts carbon dioxide from the carboxy enzyme to complete the reaction.

$$\text{E—biotin—CO}_2 + \text{pyruvate} \xrightarrow[\text{acetyl-CoA}]{\text{Mn}^{2+}} \text{E—biotin} + \text{oxaloacetate} \quad (9)$$

The enzyme-biotin-$\text{CO}_2$ is isolable by gel filtration of an incubation mixture containing ATP, bicarbonate, $\text{Mg}^{2+}$, and acetyl-CoA, but omitting pyruvate (45). Transfer of carbon dioxide from E-biotin-$\text{CO}_2$ requires only pyruvate and the firmly held metal ion, but is stimulated by acetyl-CoA. A pyruvate–oxaloacetate exchange, predicted by reaction 9, is catalyzed by the enzyme independently of the other reaction components (46, 47). The properties of the E-biotin-$\text{CO}_2$ complex are consistent with its formulation as 1-$N'$-carboxybiotinyl enzyme.

Division of the net reaction (Eq. 7) into the two partial reactions (Eqs. 8 and 9) is also consistent with the kinetic mechanism of the pyruvate carboxylases from chicken liver (48), rat liver (49), and sheep kidney (47, 50). This mechanism assumes that a spatially separate subsite exists for the reactants of each partial reaction. The biotinyl residue, at the end of a long ("swinging arm") chain of atoms, acts the part of carboxyl carrier, uniting the two

(45) M. C. Scrutton, D. B. Keech, and M. F. Utter, *JBC* 240, 574–581 (1965).
(46) W. R. McClure, H. A. Lardy, and W. W. Cleland, *JBC* 246, 3584–3590 (1971).
(47) L. K. Ashman and D. B. Keech, *JBC* 250, 14–21 (1975).
(48) R. E. Barden, C.-H. Fung, M. F. Utter, and M. C. Scrutton, *JBC* 247, 1323–1333 (1972).
(49) W. R. McClure, H. A. Lardy, M. Wagner, and W. W. Cleland, *JBC* 246, 3579–3583 (1971).
(50) S. B. Easterbrook-Smith, J. C. Wallace, and D. B. Keech, *Biochem. J.* 169, 225–228 (1978).

separated catalytic subsites into one active center of the enzyme, as in the case of transcarboxylase (p. 65).

While the chemistry of reaction 9 is clear enough, the chemistry of reaction 8 is still unresolved. That carboxyl phosphate (presumably enzyme-bound) may be a precursor of E-biotin-$CO_2$ is strongly intimated by the ability of pyruvate carboxylase (sheep kidney) to catalyze the synthesis of ATP from ADP and *carbamyl phosphate* (47). Carbamyl phosphate is here construed as a chemical analogue of carboxyl phosphate.

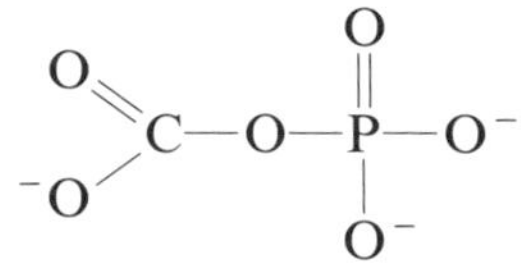
Carboxyl phosphate

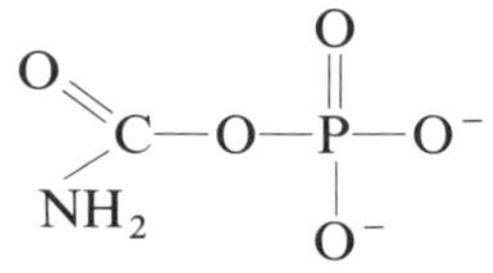
Carbamyl phosphate

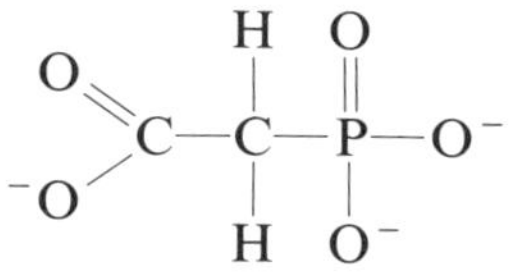
Phosphonacetate

Acetyl-CoA strongly activates the reaction. Moreover, phosphonacetate, another analogue of carboxyl phosphate, is an effective inhibitor of pyruvate carboxylation. Though the rate of ATP synthesis from carbamyl phosphate is only about 0.3% of the rate of pyruvate carboxylation, the reaction is nonetheless regarded as a real activity of the enzyme.

If carboxyl phosphate is indeed an intermediate in pyruvate carboxylase catalysis, then how is it synthesized? Considering bicarbonate as a kind of carboxylate ion and carboxyl phosphate as a kind of acyl phosphate, there is justification for the view, on grounds of precedent, that carboxyl phosphate synthesis follows upon the reaction of bicarbonate with the phosphorylated enzyme (Fig. 6). Thus we have seen that the activation of succinate to enzyme-bound succinyl phosphate proceeds from phosphorylated succinyl-CoA synthetase. And the activation of citrate to enzyme-bound citryl phosphate

$$\left.\begin{array}{l} \text{ATP} + \text{E} \xrightleftharpoons[\text{acetyl-CoA}]{\text{Mg}^{2+}} \text{E}\sim\text{P} + \text{ADP} \\ \text{E}\sim\text{P} + \text{bicarbonate} \xrightleftharpoons[\text{acetyl-CoA}]{\text{Mg}^{2+}} \text{E}\cdots\text{carboxyl}\sim\text{P} \end{array}\right\} \text{Phosphoryltransferase}$$

$$\left.\begin{array}{l} \text{E}\cdots\text{carboxyl}\sim\text{P} \xrightleftharpoons[\text{acetyl-CoA}]{\text{Mg}^{2+}} \text{E}\sim\text{CO}_2 + \text{P}_i \\ \text{E}\sim\text{CO}_2 + \text{pyruvate} \xrightleftharpoons[\text{acetyl-CoA}]{\text{Mn}^{2+}} \text{E} + \text{oxaloacetate} \end{array}\right\} \text{Carboxyltransferase}$$

$$\text{Sum: ATP} + \text{bicarbonate} + \text{pyruvate} \rightleftharpoons \text{ADP} + \text{P}_i + \text{oxaloacetate}$$

**Fig. 6.** Proposed chemical mechanism for pyruvate carboxylase action. Of the three anticipated intermediates—E ~P, carboxyl ~P, and E ~ $CO_2$—only the last is firmly established in experiment. The reality of E ~P and carboxyl ~P is inferred from suggestive data (see text) and from the precedent set in other ligase reactions (1). Pyruvate carboxylase has biotin as a prosthetic group to which the carbon dioxide is covalently joined.

proceeds from a phosphorylated ATP citrate lyase (p. 168). Likewise, the activation of acetate to free acetyl phosphate is mediated by a phosphorylated acetate kinase in a two-substrate reaction (p. 92). The case for a phospho-enzyme intermediate in pyruvate carboxylase action is also implicit in the ADP–ATP exchange which is catalyzed by the chicken liver enzyme in the absence of all other reaction components except $Mg^{2+}$. Proceeding at a rate which is only 0.2–0.4% of pyruvate carboxylation (51), the ADP–ATP exchange is nonetheless as rapid as the above-cited ATP synthesis from carbamyl phosphate, and, like the latter, is regarded as an intrinsic activity of the enzyme. The phosphorylated pyruvate carboxylase is not likely to exist free, but rather as a Michaelis complex with the slowly dissociating ADP, in accord with the sequential kinetics of reaction 8 (49). Figure 6 expresses the basic chemistry of pyruvate carboxylation according to the above conceptions.

# DNA Ligase (NAD) [EC 6.5.1.2]

This enzyme (*E. coli*) catalyzes the restoration of a 2 phosphodiester bond at the site of a single-strand break ("nick") in duplex DNA.

$$+ \text{AMP} + \text{NMN} \tag{10}$$

The acid which is activated here is the 5'-phosphate function of the substrate DNA—not a carboxyl group as in the ligase reactions described above. And the acceptor of the activated phosphate is the 3'-hydroxyl of the same DNA molecule on the opposite side of the break. In the process, the pyrophosphate bond of NAD is ruptured, and AMP and nicotinamide (NMN) appear as products. From phage-infected *E. coli* and from a variety of eukaryotic sources including rat liver, rabbit bone marrow, spleen, thymus, and plants, a DNA ligase [EC 6.5.1.1] can be isolated which uses ATP (with cleavage to AMP and $PP_i$) in place of NAD to catalyze the same joining reaction by the same chemical mechanism. DNA ligase is believed to participate in genetic recombination, repair of damage to DNA, and DNA replication (52).

(51) M. C. Scrutton and M. F. Utter, *JBC* 240, 3714–3723 (1965).
(52) I. R. Lehman, *Science* 186, 790–797 (1974).

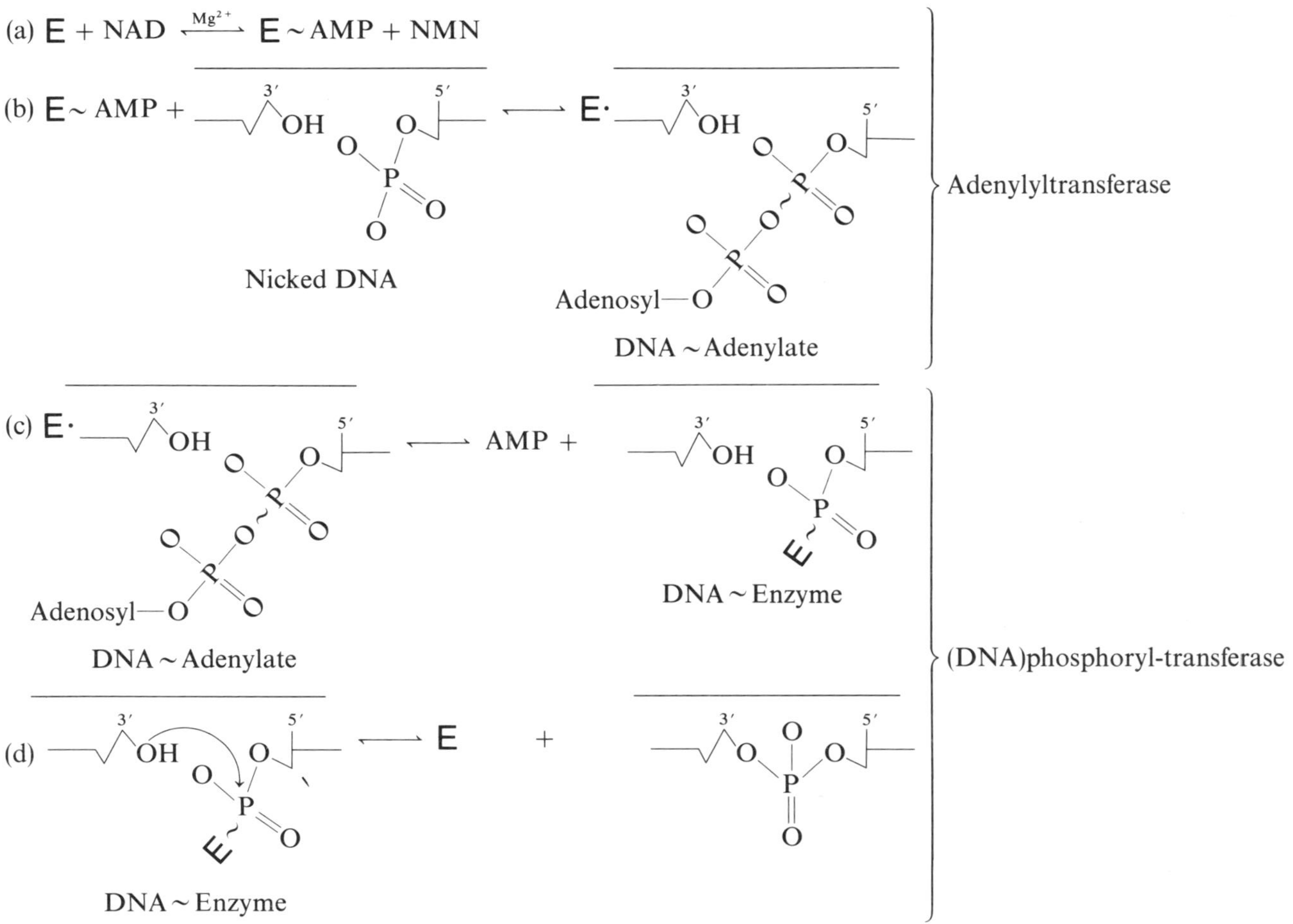

**Fig. 7.** A chemical mechanism proposed for the action of DNA ligase (NAD). Of the three anticipated intermediates—E ~ AMP, DNA ~ adenylate, and DNA ~ enzyme—only the first two have been isolated and characterized. The existence of DNA ~ enzyme is assumed on grounds of precedent.

The catalytic synthesis of a phosphodiester bond between adjacent 5'-phosphoryl and 3'-hydroxyl groups in nicked duplex DNA progresses in a series of discrete steps involving at least two isolable intermediates (Fig. 7). In the first of these steps (reaction a, Fig. 7), the enzyme reacts with NAD to form a ligase adenylate (E ~ AMP) and NMN (53). About one mole of AMP is thus fixed per mole of ligase (54). The isolated ligase adenylate reacts with nicked DNA to repair the nick with a new phosphodiester bond (reactions b–d, Fig. 7). Or, incubated with NMN, the isolated ligase adenylate resynthesizes NAD. The NMN–NAD exchange predicted by reaction a proceeds at a rate manyfold faster than the net joining reaction, attesting the kinetic competence of the ligase-adenylate intermediate (55). The equilibrium constant for the formation of ligase adenylate is 28, which is sufficiently large to suggest that nearly all of the ligase in the *E. coli* cell is in the adenylylated form (55). Within the ligase-adenylate complex the adenylyl group is joined in phosphoamide linkage to the $\varepsilon$-amino group of a lysine residue of the enzyme (56).

When ligase adenylate acts upon nicked DNA (reaction b, Fig. 7) it transfers the adenylyl group to the exposed 5'-phosphate of the substrate. The new intermediate so formed is DNA ~ adenylate. It possesses a pyrophosphate function with the thermodynamic potential of building a phosphodiester bond with the 3'-hydroxyl on the other side of the nick. DNA ~ adenylate does not normally accumulate in the reaction, but small amounts of it can be isolated under special conditions (57). Isolated DNA ~ adenylate is converted enzymically to AMP and reconstituted DNA (reactions c and d) at a rate which is faster than the net joining reaction (55). DNA ~ adenylate can also be made to yield ligase adenylate and nicked DNA, in accordance with reaction b operating in reverse (58). In fact, the entire reaction 10 is readily reversible, the ligase acting as a (AMP- plus NMN-dependent) nuclease.

Closure of the break in the nicked strand of DNA ~ adenylate is depicted in Fig. 7 (reactions c and d) as a two-stage process mediated by a covalent enzyme–substrate intermediate—the DNA–enzyme complex. No direct evidence for the DNA–enzyme complex exists. Yet conformity of the DNA ligase reaction with the general chemical pattern seen in other ligase reactions seems to demand a role for this intermediate. Also to consider is the discovery that the DNA topoisomerase of *E. coli* and of the $\phi$X174 ciston A protein catalyzes the nicking–closing (reaction 11) of a single strand in closed circular DNA via just such a DNA–enzyme intermediate (p. 198).

(53) J. W. Little, S. B. Zimmerman, C. K. Oshinsky, and M. Gellert, *PNAS* 58, 2004–2011 (1967); B. M. Olivera, Z. W. Hall, Y. Anraku, J. R. Chien, and I. R. Lehman, *Cold Spring Harbor Symp. Quant. Biol.* 33, 27–34 (1968).
(54) P. Modrich, Y. Anraku, and I. R. Lehman, *JBC* 248, 7495–7501 (1973).
(55) P. Modrich and I. R. Lehman, *JBC* 248, 7502–7511 (1973).
(56) R. I. Gumport and I. R. Lehman, *PNAS* 68, 2559–2563 (1971).
(57) B. M. Olivera, Z. W. Hall, and I. R. Lehman, *PNAS* 61, 237–244 (1968).
(58) P. Modrich, I. R. Lehman, and J. C. Wang, *JBC* 247, 6370–6372 (1972).

$$\text{(11)}$$

Here the isomerase ruptures the phosphodiester by joining covalently, as shown, to the phosphorus, while displacing the 3'-hydroxyl. Reaction 11 is of course the reverse of reaction d of Fig. 7. With this mechanistic precedent, the reversal of reaction d by DNA ligase, followed by the sequential reversals of reactions c, b, and a, fully accounts for the enzyme's activity as a (AMP-plus NMN-dependent) nuclease. It follows that, in the forward direction, the ligase activity of the enzyme is expressed over the same pathway and the same set of intermediates.

**Table 7.1.** Ligases Known to Act by Covalent Catalysis

| EC no. | Familiar name of enzyme[a] | Criteria[b] | References |
|--------|----------------------------|-------------|------------|
| 6.1.1.2 | Tryptophanyl-tRNA synthetase | I | (59) |
| 6.1.1.20 | Phenylalanyl-tRNA synthetase | I, E | (60) |
| 6.2.1.1 | Acetyl-CoA synthetase | E | (61) |
| 6.2.1.3 | Long-chain fatty acyl-CoA synthetase | I, E | (62) |
| 6.2.1.4 | Succinyl-CoA synthetase (GDP-forming) | I$^c$, E | (63) |
| 6.2.1.5 | Succinyl-CoA synthetase (ADP-forming) | I$^c$, E | (64) |
| 6.2.1.9 | Malyl-CoA synthetase | I, E | (65) |
| 6.3.2.2 | $\gamma$-Glutamyl-cysteine synthetase | E, K, G | (66) |
| 6.3.2.3 | Glutathione synthetase | E, K | (67) |
| 6.3.2.11 | Carnosine synthetase | E | (68) |
| 6.3.2.14 | 2,3-Dihydroxybenzoylserine synthetase | I$^d$ | (69) |
| 6.3.2. | Gramicidin A synthetase | I | (70) |
| 6.3.2. | Gramicidin S synthetase [phosphopantetheine] | I | (71) |
| 6.3.2. | Tyrocidine synthetase [phosphopantetheine] | I | (72) |
| 6.3.2. | Polymyxin synthetase | I$^d$ | (73) |
| 6.3.2. | Bacitracin synthetase [phosphopantetheine] | I | (74) |
| 6.3.4.1 | GMP synthetase | G, M | (75) |
| 6.3.4.2 | CTP synthetase | I$^d$, G$^e$ | (76) |
| 6.3.4.6 | Urea carboxylase (hydrolysing) [biotin] | I | (77) |
| 6.3.5.2 | GMP synthetase (glutamine-hydrolysing) | G | (78) |
| 6.3.5.3 | Phosphoribosylformylglycinamide synthetase | I, E, K, G$^e$ | (79) |
| 6.3.5.4 | Asparagine synthetase (glutamine-hydrolysing) | G$^e$ | (80) |
| 6.3.5.5 | Carbamyl phosphate synthetase (glutamine-hydrolysing) | G | (81) |
| 6.4.1.1 | Pyruvate carboxylase [biotin, Mn$^{2+}$] | I, E, K | (82) |
| 6.4.1.2 | Acetyl-CoA carboxylase [biotin] | I | (83) |
| 6.4.1.3 | Propionyl-CoA carboxylase (ATP-hydrolysing) [biotin] | I | (84) |

**Table 7.1.** Ligases Known to Act by Covalent Catalysis   (*Continued*)

| EC no. | Familiar name of enzyme[a] | Criteria[b] | References |
|--------|----------------------------|-------------|------------|
| 6.4.1.4 | Methylcrotonyl-CoA carboxylase [biotin] | I, E | (85) |
| 6.5.1.1 | DNA ligase (ATP) | I, E | (86) |
| 6.5.1.2 | DNA ligase (NAD) | I, E, K | (87) |
| 6.5.1.3 | RNA ligase | I, E | (88) |

[a] Any parenthetical expression is part of the official name of the enzyme. Prosthetic groups are indicated in brackets.

[b] The symbols mean the following:

I, the holoenzyme links covalently with the substrate or some fragment of it to form a chemically competent intermediate;

E, the enzyme catalyzes an exchange reaction consistent with the participation of a covalent enzyme–substrate intermediate;

K, the enzyme exhibits kinetic properties consistent with the participation of a covalent enzyme–substrate intermediate;

G, the enzyme is irreversibly inactivated by stoichiometric alkylation with a glutamine analogue; inferred from this is the participation of a glutamyl-enzyme intermediate;

M, miscellaneous data and derivative arguments peculiar to the enzyme in question.

[c] This enzyme is actually isolated from tissue as the covalent enzyme–substrate intermediate; that is, the transferring portion of the substrate molecule (a phosphoryl group) is already fixed to the enzyme.

[d] Isolated in denatured condition.

[e] The inactivation of this enzyme by a glutamine analogue applies only to the utilization of glutamine as amino donor. The enzyme remains active with ammonia.

(59) Refs. (13–15).

(60) Refs. (19, 20).

(61) Ref. (29).

(62) Ref. (42).

(63) S. Cha, C. M. Cha, and R. E. Parks, Jr., *JBC* 242, 2582–2592 (1967); D. P. Baccanari and S. Cha, *Fed. Proc.* 31, 500 (1971).

(64) Refs. (31–34, 36, 37, 41).

(65) L. B. Hersh, *JBC* 249, 6264–6271 (1974); M. Elwell and L. B. Hersh, *JBC* 254, 2434–2438 (1979).

(66) G. C. Webster and J. E. Varner, *Arch. Biochem. Biophys.* 52, 22–32 (1954); D. H. Strumeyer and K. Bloch, *JBC* 235, PC27 (1960); J. S. Davis, J. B. Balinsky, J. S. Harington, and J. B. Shepherd, *Biochem. J.* 133, 667–678 (1973); R. Sekura and A. Meister, *JBC* 252, 2606–2610 (1977).

(67) J. E. Snoke and K. Bloch, *JBC* 213, 825–835 (1955); A. Wendel and L. Flohe, *Z. Physiol. Chem.* 353, 523–530 (1972).

(68) Ref. (24).

(69) G. F. Bryce and N. Brot, *B* 11, 1708–1715 (1972).

(70) K. Akashi and K. Kurahashi, *Biochem. Biophys. Res. Commun.* 77, 259–267 (1977).

(71) R. Roskoski, Jr., G. Ryan, H. Kleinkauf, W. Gevers, and F. Lipmann, *Arch. Biochem. Biophys.* 143, 485–492 (1971); H. Kleinkauf, R. Roskoski, Jr., and F. Lipmann, *PNAS* 68, 2069–2072 (1971).

(72) R. Roskoski, Jr., W. Gevers, H. Kleinkauf, and F. Lipmann, *B* 9, 4839–4845 (1970); R. Roskoski, Jr., H. Kleinkauf, W. Gevers, and F. Lipmann, *B* 9, 4846–4851 (1970); last citation in Ref. (71).

(73) S. Komura and K. Kurahashi, *J. Biochem.* (Tokyo) 86, 1013–1021 (1979); *idem.*, *J. Biochem*, (Tokyo) 88, 285–288 (1980).

(74) Ø. Frøyshov, *EJB* 59, 201–206 (1975).

(75) H. Zalkin and C. D. Truitt, *JBC* 252, 5431–5436 (1977).

(76) A Levitski and D. E. Koshland, *B* 10, 3365–3371 (1971).

(77) P. A. Whitney and T. Cooper, *JBC* 248, 325–330 (1973); P. A. Castric and B. Levenberg, *BBA* 438, 574–583 (1976).

(78) B. H. Lee and S. C. Hartman, *Biochem. Biophys. Res. Commun.* 60, 918–925 (1974).

(79) K. Mizobuchi and J. M. Buchanan, *JBC* 243, 4853–4862 (1968); K. Mizobuchi, G. L. Kenyon, and J. M. Buchanan, *JBC* 243, 4863–4877 (1968); D. D. Schroeder, A. J. Allison, and J. M. Buchanan, *JBC* 244, 5856–5865 (1969); S. Y. Chu and J. F. Henderson, *Can. J. Biochem.* 50, 490–500 (1972); S. Ohnoki, B.-H. Hong, and J. M. Buchanan, *B* 16, 1065–1069 (1977).

(80) B. Horowitz and A. Meister, *JBC* 247, 6708–6719 (1972).

(81) V. P. Wellner, P. M. Anderson, and A. Meister, *B* 12, 2061–2066 (1973).

(82) Refs. (45–51).

(83) M. Waite and S. J. Wakil, *JBC* 238, 77–80 (1963); S. Numa, E. Ringelmann, and F. Lynen, *Biochem. Z.* 340, 228–242 (1964).

(84) Y. Kaziro and S. Ochoa, *JBC* 236, 3131–3136 (1961).

(85) F. Lynen, J. Knappe, E. Lorch, G. Jütting, E. Ringelmann, and J. P. Lachance, *Biochem. Z.* 335, 123–167 (1961); R. H. Himes, D. L. Young, E. Ringelmann, and F. Lynen, *Biochem. Z.* 337, 48–61 (1963); J. Knappe, B. Wenger, and U. Wiegand, *Biochem Z.* (1963); J. Knappe, B. Wenger, and U. Wiegand, *Biochem. Z.* 337, 232–246 (1963).

(86) B. Weiss and C. C. Richardson, *JBC* 242, 4270–4272 (1972); B. Weiss, A. Thompson, and C. C. Richardson, *JBC* 243, 4556–4563 (1968); S. Söderhäll and T. Lindahl, *JBC* 248, 672–675 (1973); S. Söderhäll, *EJB* 51, 129–136 (1975).

(87) Refs. (52–58).

(88) J. W. Cranston, R. Silber, V. G. Malathi, and J. Hurwitz, *JBC* 249, 7447–7456 (1974).

# Chapter 8

# Summary

Altogether, the six tables of Chapters 2–7 list a grand total of 465 enzymes, each of which during its catalytic cycle forms a covalent bond with its substrate or some fragment of it. These long lists of enzymes point the contrast between the abundance of positive evidence for covalent catalysis and the total dearth of positive evidence for single-displacement catalysis (1). How these enzymes are apportioned among the six major classes of enzymes recognized by the Enzyme Commission (hereinafter referred to as the EC enzymes) is shown in Table 8.1 at the end of this chapter. It is clear that at least 21% of all the EC enzymes effect their catalysis through covalent enzyme–substrate intermediates. This figure assumes an even greater significance when it is realized that the vast majority of the 2200 EC enzymes have never been investigated from the standpoint of chemical mechanism. Prior to the survey embodied in the tables of Chapters 2–7, the prevalence of so much covalent catalysis by enzymes was hardly suspected.

While the 21% figure is impressive in its own right, it is further of interest to estimate the range of chemical diversity of the reactions catalyzed by the 465 enzymes, as well as the degree to which they are representative of the 2200 EC enzymes. To do this we reproduce at the end of the chapter, and consider briefly, the *Key to Numbering and Classification of Enzymes* as it appears in the official publication of the Enzyme Commission (2). It is seen that each major class of enzymes is divided into a number of subclasses; and each of these in turn is further divided into subsubclasses—or categories—each designated by a code number of three digits. Within each subsubclass are

_____

(1) To document this dearth of evidence I must again direct the reader to the whole body of literature on the chemical mechanism of enzyme action.

(2) *Enzyme Nomenclature*. Recommendations (1978) of the nomenclature committee of The International Union of Biochemistry. Academic Press, New York, 1979, pp. 19–26.

listed all of the enzymes that fall naturally into the same catagory by reason of the chemical resemblance among the reactions they catalyze (3). For, like any system of classification, the Enzyme Commission's system is built upon similarities—in this case, *chemical* similarities. As the hierarchy of classification is descended, the degree of similarity among the chemical reactions sharpens, and may become very sharp indeed at the level of the subsubclass. For instance, all the enzymes catalyzing the phosphorylation of a hydroxyl group by ATP are grouped in one subsubclass (i.e., EC 2.7.1). They differ among themselves mainly in their specificity, especially as it concerns the R-group of the hydroxyl-bearing substrates upon which they act. The 90-odd phosphotransferases of EC 2.7.1, which catalyze the phosphorylation of hydroxyl groups by ATP, catalyze what is in effect the same chemical reaction. And implicit in this sameness is a sameness of chemical mechanism. Thus, if any of the 90-odd phosphotransferases makes use of a phosphoenzyme as intermediate, then it is probable that all the others do the same. This is the venerable argument from analogy. From its earliest days the science of chemistry has depended on the empirical and qualitative rule that like substances react similarly.

In the *Key to Numbering and Classification of Enzymes* reproduced here the subsubclasses exemplified by at least one of the enzymes in the tables of Chapter 2–7 are indicated by a boldface code number. Many of the subsubclasses are, of course, represented by more than one enzyme, and some are not represented at all. These last have their code numbers in regular type. Table 8.2 at the end of this chapter summarizes the findings. Of the total of 171 subsubclasses, 120 are represented by the enzymes of Chapters 2–7, leaving 51 that are not represented. Significantly, about half of the 51 unrepresented subsubclasses contain only three (or less) enzymes each. These tend generally to be of an uncommon reaction type, and have been little studied, if at all. Adding up all of the enzymes in the 51 unrepresented categories yields a total of a mere 174 enzymes, or 8% of all the EC enzymes. It follows, therefore, that 92% of the 2200 EC enzymes are represented in the tables of Chapters 2–7. As to chemical diversity, it is also evident from Table 8.2 that the enzymes of Chapter 2–7 embrace about 70% of all the EC reaction types. Though some uncertainty is unavoidably built into these numbers, I believe that they give, in the main, a credible estimate of the sought-for quantities (4).

---

(3) The individual enzymes are of course omitted from the "Key." They are given in Ref. (2), pp. 28–441, with their respective four-digit EC code numbers.

(4) It is generally true, as stated in the text, that the subsubclasses (categories) in the "Key" differ from each other in the type of chemical reaction catalyzed by their respective enzymes. But there are exceptions to this generalization. One prominent exception is the group of 14 subsubclasses of nuclease (including the restriction enzymes) having code numbers EC 3.1.11 through EC 3.1.31. Within these 14 categories, all of the enzymes catalyze the hydrolysis of the same chemical function—a phosphodiester. The 14 categories differ among themselves, therefore, not in *chemical* type but in *structural* type. Other (less striking) departures from the stated

To recapitulate: the research embodied in this book brings three cogent facts into bold relief:

1. The number of documented cases of covalent catalysis by enzymes is indeed large—amounting to more than 20% of all the officially recognized enzymes (Table 8.1).

2. Since enzymes catalyzing analogous reactions are apt to do so by an analogous mechanism, the documented cases represent about 92% of all the recognized enzymes (Table 8.2).

3. The documented cases embrace nearly three quarters of the entire range of chemical diversity known to enzymic catalysis (Table 8.2).

These findings uphold the one great fact which stands out conspicuous in all mass-law catalysis—the covalent union of catalyst with substrate. Nonenzymic catalysis—in solution and on solid surfaces—always involves a covalent bond between catalyst and substrate, or some fragment of the substrate. The especially strong kinship between enzymic and heterogeneous catalysis was remarked upon earlier (Chapter 1). It is in any case hard to imagine why chemical catalysis by an enzyme should differ *in its fundamentals* from the other two forms of chemical catalysis. And, indeed, the testimony of the tables in Chapters 2–7 argues that there is no real difference.

Covalent catalysis by enzymes may be one of those scientific principles which, first perceived in a limited number of instances, gradually comes to be established through the accumulation of observations consistent with the principle, and through the continued absence of countervailing evidence. In the current debate between covalent catalysis and single-displacement catalysis all the *positive* evidence is on the side of covalent catalysis. Like the hypothetical "phlogiston" in 18th-century chemistry and the illusory "luminiferous aether" in 19th-century physics, single-displacement catalysis has long held sway in 20th-century enzymology. But, like these others, it remains an hypothesis without basis in hard fact (5).

---

generalization are also found in the "Key." In Table 8.2 I attempt to assess the number of reaction types that are subject to catalysis by the enzymes of Chapters 2–7. To this end I think it fair, in the exceptional case of the 14 above-cited subsubclasses, to unite them into a single subsubclass (EC 3.1.4), and to consider them all, from the standpoint of chemistry and reaction type, as a single category with the phosphodiester function as common substrate.

(5) While compiling the list of covalent enzymes in Chapters 2–7 I was obliged from time to time to delete one of them as new evidence threw doubt upon old evidence. And, in the nature of things, it seems a certainty that more such deletions will occasionally be necessary in the future. My own feeling is that these deletions are only temporary, and that the deleted enzymes will in time be restored to the list. Indeed, I believe that all enzymes will one day join the list. Meanwhile, it should be clear that an enzyme deleted from the list of covalent enzymes does not become a single-displacement enzyme by virtue of that deletion. The deleted enzyme merely rejoins that large number of enzymes about which we have no evidence, at this time, as to its single- or double-displacement character. Despite such occasional deletions, the list of covalent enzymes continues to grow, while the list of authentic single-displacement enzymes continues at zero.

Is it possible, despite the argument of this book, that one or more cases of single-displacement catalysis by enzymes will in time be discovered? From the standpoint of pure logic, the possibility cannot of course be denied. But scientific inference by induction—which is the method of this book—does not rest on pure logic. It rests rather on our "animal faith in order and regularity" (6). Scientific inference by induction tells us that the sun will rise tomorrow, even if we cannot prove it in pure logic. It tells us, too, that single-displacement catalysis by enzymes is an unlikely prospect. The reliability of this prediction hinges of course on whether the 465 enzymes of Chapter 2–7 are a sufficiently large number of instances, and of wide enough chemical diversity, to warrant confidence in any inference by induction. Clearly, I think they are sufficient. But we know, too, how hopeless it is to establish a scientific principle as true beyond question. Certainly it is possible that single-displacement catalysis will one day be discovered. If so, I venture the guess that such catalysis will be of an exceptional sort, and that the broad principle of covalent catalysis by enzymes will need but minor qualification.

**Table 8.1.** How the Enzymes of Chapters 2–7 Are Apportioned among the Six Major Classes of EC Enzymes

| Class of enzyme | No. of EC enzymes[a] | No. of enzymes in the tables of Chapters 2–7[b] |
| --- | --- | --- |
| Oxidoreductases | 571 (8) | 139 (24) |
| Transferases | 574 (6) | 105 (18) |
| Hydrolases | 621 (7) | 100 (16) |
| Lyases | 243 (2) | 56 (23) |
| Isomerases | 100 (4) | 35 (35) |
| Ligases | 92 (5) | 30 (33) |
| Total: | 2201 | 465 |

$$\frac{465}{2201} = 21\%$$

[a] Calculated by adding up the enzymes in Ref. (2), pp. 28–441. Each of the large numbers in this column includes a few enzymes not yet listed officially by the Enzyme Commission. How many they are is indicated by the parenthetical number. Their names are given in the corresponding tables of Chapters 2–7, where they are listed with three-digit code numbers.

[b] The parenthetical number in this column gives the percentage of the corresponding large number in the second column.

(6) J. Trusted, *The Logic of Scientific Inference*, The Macmillan Press, London, 1979.

**Table 8.2.** How the Enzymes of Chapters 2–7 Are Apportioned among the Subsubclasses of EC Enzymes

| Class of enzyme | Number of EC subsubclasses[a] | Number of subsubclasses represented by the enzymes of Chapters 2–7[b] | Number of enzymes in the unrepresented subsubclasses[c] |
|---|---|---|---|
| Oxidoreductases | 71 | 52 | 43 |
| Transferases | 25 | 20 | 34 |
| Hydrolases | 37 | 23 | 58 |
| Lyases | 13 | 8 | 20 |
| Isomerases | 16 | 10 | 11 |
| Ligases | 9 | 7 | 8 |
| Total: | 171 | 120 | 174 |

$\dfrac{174}{2201} = \sim 8\%$ of the EC enzymes are *not* represented in the tables of Chapters 2–7.[d]

Or, conversely, 92% of the EC enzymes *are* represented in the tables of Chapters 2–7.

$\dfrac{120}{171} = \sim 70\%$ of the EC reaction types are catalyzed by the enzymes of Chapters 2–7.[e]

[a] Calculated from the *Key to Numbering and Classification of Enzymes* by adding up the code numbers in regular and boldface type.
[b] Calculated by adding up the boldface code numbers in the "Key."
[c] Calculated by adding up the enzymes in the unrepresented (regular type) subsubclasses [Ref. (2), pp. 20–441]. A number in this column, divided by the corresponding number in the second column of Table 8.1, gives the fraction of all the EC enzymes that are unrepresented in the corresponding tables of Chapters 2–7.
[d] The number 2201 is taken from Table 8.1.
[e] On the assumption that each subsubclass corresponds to one type of chemical reaction.

# Key to Numbering and Classification of Enzymes*

Each three-digit code number designates a subsubclass (category) of enzymes. A boldface code number indicates a subsubclass which is represented by one or more enzymes in the tables of Chapters 2–7. The other subsubclasses are unrepresented in the tables of Chapters 2–7.

## 1.    OXIDOREDUCTASES

*1.1    Acting on the CH—OH group of donors*

**1.1.1**    With NAD$^+$ or NADP$^+$ as acceptor
**1.1.2**    With a cytochrome as acceptor

* Ref. (2), reproduced with permission.

**1.1.3**      With oxygen as acceptor
**1.1.99**     With other acceptors

*1.2*      *Acting on the aldehyde or oxo group of donors*

**1.2.1**      With $NAD^+$ or $NADP^+$ as acceptor
**1.2.2**      With a cytochrome as acceptor
**1.2.3**      With oxygen as acceptor
**1.2.4**      With a disulphide compound as acceptor
**1.2.7**      With an iron-sulphur protein as acceptor
1.2.99     With other acceptors

*1.3*      *Acting on the CH—CH group of donors*

**1.3.1**      With $NAD^+$ or $NADP^+$ as acceptor
1.3.2      With a cytochrome as acceptor
**1.3.3**      With oxygen as acceptor
1.3.7      With an iron-sulphur protein as acceptor
**1.3.99**     With other acceptors

*1.4*      *Acting on the CH—NH$_2$ group of donors*

**1.4.1**      With $NAD^+$ or $NADP^+$ as acceptor
1.4.2      With a cytochrome as acceptor
**1.4.3**      With oxygen as acceptor
1.4.4      With a disulphide compound as acceptor
1.4.7      With an iron-sulphur protein as acceptor
**1.4.99**     With other acceptors

*1.5*      *Acting on the CH—NH group of donors*

1.5.1      With $NAD^+$ or $NADP^+$ as acceptor
**1.5.3**      With oxygen as acceptor
**1.5.99**     With other acceptors

*1.6*      *Acting on NADH or NADPH*

**1.6.1**      With $NAD^+$ or $NADP^+$ as acceptor
**1.6.2**      With a cytochrome as acceptor
**1.6.4**      With a disulphide compound as acceptor
1.6.5      With a quinone or related compound as acceptor
**1.6.6**      With a nitrogenous group as acceptor
**1.6.99**     With other acceptors

*1.7*      *Acting on other nitrogenous compounds as donors*

**1.7.2**      With a cytochrome as acceptor
**1.7.3**      With oxygen as acceptor

**1.7.7**     With an iron-sulphur protein as acceptor
**1.7.99**    With other acceptors

*1.8*    *Acting on a sulphur group of donors*

**1.8.1**    With $NAD^+$ or $NADP^+$ as acceptor
1.8.2    With a cytochrome as acceptor
**1.8.3**    With oxygen as acceptor
**1.8.4**    With a disulphide compound as acceptor
1.8.5    With a quinone or related compound as acceptor
1.8.7    With an iron-sulphur protein as acceptor
**1.8.99**   With other acceptors

*1.9*    *Acting on a heme group of donors*

**1.9.3**    With oxygen as acceptor
**1.9.6**    With a nitrogenous group as acceptor
1.9.99   With other acceptors

*1.10*    *Acting on diphenols and related substances as donors*

1.10.1   With $NAD^+$ or $NADP^+$ as acceptor
1.10.2   With a cytochrome as acceptor
**1.10.3**  With oxygen as acceptor

*1.11*    *Acting on hydrogen peroxide as acceptor*

*1.12*    *Acting on hydrogen as donor*

**1.12.1**   With $NAD^+$ or $NADP^+$ as acceptor
**1.12.2**   With a cytochrome as acceptor

*1.13*    *Acting on single donors with incorporation of molecular oxygen (oxygenases)*

**1.13.11**  With incorporation of two atoms of oxygen
**1.13.12**  With incorporation of one atom of oxygen (internal monooxygenases or internal mixed function oxidases)
1.13.99  Miscellaneous (requires further characterization)

*1.14*    *Acting on paired donors with incorporation of molecular oxygen*

1.14.11  With 2-oxoglutarate as one donor, and incorporation of one atom each of oxygen into both donors
**1.14.12**  With NADH or NADPH as one donor, and incorporation of two atoms of oxygen into one donor
**1.14.13**  With NADH or NADPH as one donor, and incorporation of one atom of oxygen

*2.4*     *Glycosyltransferases*

**2.4.1**     Hexosyltransferases
**2.4.2**     Pentosyltransferases
2.4.99    Transferring other glycosyl groups

*2.5*     *Transferring alkyl or aryl groups, other than methyl groups*

*2.6*     *Transferring nitrogenous groups*

**2.6.1**     Aminotransferases
2.6.3     Oximinotransferases

*2.7*     *Transferring phosphorus-containing groups*

**2.7.1**     Phosphotransferases with an alcohol group as acceptor
**2.7.2**     Phosphotransferases with a carboxyl group as acceptor
**2.7.3**     Phosphotransferases with a nitrogenous group as acceptor
**2.7.4**     Phosphotransferases with a phosphate group as acceptor
**2.7.5**     Phosphotransferases with regeneration of donors (apparently catalysing intramolecular transfers)
2.7.6     Diphosphotransferases
**2.7.7**     Nucleotidyltransferases
2.7.8     Transferases for other substituted phosphate groups
**2.7.9**     Phosphotransferases with paired acceptors

*2.8*     *Transferring sulphur-containing groups*

**2.8.1**     Sulphurtransferases
2.8.2     Sulphotransferases
**2.8.3**     CoA-transferases


3.     HYDROLASES

*3.1*     *Acting on ester bonds*

**3.1.1**     Carboxylic ester hydrolases
**3.1.2**     Thiolester hydrolases
**3.1.3**     Phosphoric monoester hydrolases
**3.1.4**     Phosphoric diester hydrolases
3.1.5     Triphosphoric monoester hydrolases
**3.1.6**     Sulphuric ester hydrolases
3.1.7     Diphosphoric monoester hydrolases
3.1.11    Exodeoxyribonucleases producing 5'-phosphomonoesters
3.1.13    Exoribonucleases producing 5'-phosphomonoesters
3.1.14    Exoribonucleases producing other than 5'-phosphomonoesters

*3.5    Acting on carbon-nitrogen bonds, other than peptide bonds*

**3.5.1**    In linear amides
**3.5.2**    In cyclic amides
3.5.3    In linear amidines
**3.5.4**    In cyclic amidines
3.5.5    In nitriles
3.5.99    In other compounds

*3.6    Acting on acid anhydrides*

**3.6.1**    In phosphoryl-containing anhydrides
3.6.2    In sulphonyl-containing anhydrides

*3.7    Acting on carbon-carbon bonds*

**3.7.1**    In ketonic substances

*3.8    Acting on halide bonds*

3.8.1    In C-halide compounds
3.8.2    In P-halide compounds

*3.9    Acting on phosphorus-nitrogen bonds*

*3.10    Acting on sulphur-nitrogen bonds*

*3.11    Acting on carbon-phosphorus bonds*

4.    LYASES

*4.1    Carbon-carbon lyases*

**4.1.1**    Carboxy-lyases
**4.1.2**    Aldehyde-lyases
**4.1.3**    Oxo-acid-lyases
**4.1.99**    Other carbon-carbon lyases

*4.2    Carbon-oxygen lyases*

**4.2.1**    Hydro-lyases
4.2.2    Acting on polysaccharides
**4.2.99**    Other carbon-oxygen lyases

*4.3    Carbon-nitrogen lyases*

**4.3.1**    Ammonia-lyases
4.3.2    Amidine-lyases

6.3    *Forming carbon-nitrogen bonds*

   6.3.1    Acid–ammonia (or amine) ligases (amide synthetases)
   **6.3.2**    Acid–amino-acid ligases (peptide synthetases)
   6.3.3    Cyclo-ligases
   **6.3.4**    Other carbon-nitrogen ligases
   **6.3.5**    Carbon-nitrogen ligases with glutamine as amido-*N*-donor

**6.4**    *Forming carbon-carbon bonds*

**6.5**    *Forming phosphate ester bonds*

# Index

Major discussions and tabular citations are noted in **bold** type.

# Q